Lecture Notes in Computer Science   5174

Commenced Publication in 1973
Founding and Former Series Editors:
Gerhard Goos, Juris Hartmanis, and Jan van

Sergey Balandin   Dmitri Moltchanov
Yevgeni Koucheryavy (Eds.)

Next Generation Teletraffic and Wired/Wireless Advanced Networking

8th International Conference
NEW2AN and ruSMART 2008
St. Petersburg, Russia, September 3-5, 2008
Proceedings

 Springer

Volume Editors

Sergey Balandin
Nokia Research Center
Itamerenkatu 11-13
00180 Helsinki, Finland
E-mail: sergey.balandin@nokia.com

Dmitri Moltchanov
Yevgeni Koucheryavy
Tampere University of Technology
Department of Communications Engineering
Korkeakoulunkatu 10
33720 Tampere, Finland
E-mail: {moltchan, yk}@cs.tut.fi

Library of Congress Control Number: Applied for

CR Subject Classification (1998): C.2, C.4, H.4, D.2, J.1, K.6, K.4

LNCS Sublibrary: SL 5
Computer Communication Networks and Telecommunications

ISSN   0302-9743
ISBN-10   3-540-85499-1 Springer Berlin Heidelberg New York
ISBN-13   978-3-540-85499-9 Springer Berlin Heidelberg New York

Springer is a part of Springer Science+Business Media

springer.com

© Springer-Verlag Berlin Heidelberg 2008
Printed in Germany

Typesetting: Camera-ready by author, data conversion by Scientific Publishing Services, Chennai, India
Printed on acid-free paper   SPIN: 12513628   06/3180   5 4 3 2 1 0

Preface

We welcome you to the joint proceedings of the 8th NEW2AN 2008 (Next Generation Teletraffic and Wired/Wireless Advanced Networking) and the 1st ruSMART 2008 conferences held in St. Petersburg, Russia on September 1–5, 2008.

This year NEW2AN features significant contributions to various aspects of networking. The topics presented encompass several layers of communication networks: from the physical layer to transport protocols. In particular, issues of QoS in wireless and IP-based multi-service networks are dealt with. Cross-layer optimization and traffic characterization are also addressed within the program. It is also worth mentioning the emphasis on wireless networks, including, but not limited to, cellular networks, wireless local area networks, personal area networks, mobile ad hoc networks, and sensor networks.

The NEW2AN 2008 call for papers attracted 60 papers from 19 countries, resulting in an acceptance ratio of 35%. With the help of the excellent Technical Program Committee and a number of associated reviewers, the best 21 high-quality papers were selected for publication. The conference was organized in 7 single-track sessions. We wish to thank the Technical Program Committee members of each of the conferences and the associated reviewers for their hard work and important contribution to the conference.

The first Russian Conference on Smart Spaces, ruSMART 2008, was the first conference in Russia to address this topic. It aimed to attract academic and industrial researchers to an emerging area of Smart Spaces that creates completely new opportunities for making fully-customized applications and services for the users. The conference was a meeting place for leading experts from top affiliations around the world. It attracted active participation and a strong interest from Russian attendees, who have a good reputation for high-quality research and business in innovative service creation and applications development.

The Technical Programs of both of the conferences benefited from three keynote speakers: Ian Oliver, Nokia Research Center, Finland; Cornel Klein and Gerald Kaefer, Siemens, Germany; Arkady Zaslavsky, Luleå University of Technology, Sweden.

This year the conferences were organized in cooperation with ITC (International Teletraffic Congress), IEEE, the Popov Society, COST 290, and with the support of NOKIA (Finland), Siemens (Germany), Ubitel (Russia) and BalticIT Ltd. (Russia). The support of these organizations is gratefully acknowledged.

Finally, we wish to thank the many people who contributed to the organization. In particular, we are grateful to Jakub Jakubiak (TUT), who took charge of the submission and review process and the website maintenance. He also did an excellent job on the compilation of the camera ready papers and the interaction with Springer. Many thanks go to Natalia Avdeenko and Elizaveta Zabavicheva

(Monomax Meetings & Incentives) for their excellent local organization efforts and their preparation of the conference's social program.

We believe that the work done for the 8th NEW2AN and the 1st ruSMART conferences provided an interesting and up-to-date scientific program. We hope that participants enjoyed the technical and social conference program, Russian hospitality and the beautiful city of St. Petersburg.

June 2008   Sergey Balandin
                                                          Dmitri Moltchanov
                                                       Yevgeni Koucheryavy

Organization

NEW2AN International Advisory Committee

Ian F. Akyildiz	Georgia Institute of Technology, USA
Nina Bhatti	Hewlett Packard, USA
Igor Faynberg	Alcatel Lucent, USA
Jarmo Harju	Tampere University of Technology, Finland
Andrey Koucheryavy	ZNIIS R&D, Russia
Villy B. Iversen	Technical University of Denmark, Denmark
Paul Kühn	University of Stuttgart, Germany
Kyu Ouk Lee	ETRI, R. Korea
Mohammad S. Obaidat	Monmouth University, USA
Michael Smirnov	Fraunhofer FOKUS, Germany
Manfred Sneps-Sneppe	Ventspils University College, Latvia
Ioannis Stavrakakis	University of Athens, Greece
Sergey Stepanov	Sistema Telecom, Russia
Phuoc Tran-Gia	University of Würzburg, Germany
Gennady Yanovsky	State Univ. of Telecommunications, Russia

NEW2AN Technical Program Committee

Mari Carmen Aguayo-Torres	University of Malaga, Spain
Ozgur B. Akan	METU, Turkey
Khalid Al-Begain	University of Glamorgan, UK
Tricha Anjali	Illinois Institute of Technology, USA
Konstantin Avrachenkov	INRIA, France
Francisco Barcelo	UPC, Spain
Thomas M. Bohnert	University of Coimbra, Portugal
Torsten Braun	University of Bern, Switzerland
Chrysostomos Chrysostomou	University of Cyprus, Cyprus
Georg Carle	University of Tübingen, Germany
Ibrahim Develi	Erciyes University, Turkey
Roman Dunaytsev	Tampere University of Technology, Finland
Eylem Ekici	Ohio State University, USA
Sergey Gorinsky	Washington University in St. Louis, USA
Markus Fidler	NTNU Trondheim, Norway
Giovanni Giambene	University of Siena, Italy
Stefano Giordano	University of Pisa, Italy
Ivan Ganchev	University of Limerick, Ireland
Vitaly Gutin	Popov Society, Russia
Martin Karsten	University of Waterloo, Canada

Andreas Kassler	Karlstad University, Sweden
Maria Kihl	Lund University, Sweden
Tatiana Kozlova Madsen	Aalborg University, Denmark
Yevgeni Koucheryavy	Tampere University of Technology, Finland (Chair)
Jong-Hyouk Lee	Sungkyunkwan University, R. Korea
Vitaly Li	Kangwon National University, R. Korea
Lemin Li	U. of Electronic Science and Techn. of China, China
Leszek T. Lilien	Western Michigan University, USA
Saverio Mascolo	Politecnico di Bari, Italy
Maja Matijaševic	University of Zagreb, Croatia
Paulo Mendes	DoCoMo Euro-Labs, Germany
Ilka Miloucheva	Salzburg Research, Austria
Dmitri Moltchanov	Tampere University of Technology, Finland
Edmundo Monteiro	University of Coimbra, Portugal
Seán Murphy	University College Dublin, Ireland
Marc Necker	University of Stuttgart, Germany
Mairtin O'Droma	University of Limerick, Ireland
Jaudelice Cavalcante de Oliveira	Drexel University, USA
Evgeni Osipov	RWTH Aachen, Germany
George Pavlou	University of Surrey, UK
Simon Pietro Romano	Università degli Studi di Napoli "Federico II", Italy
Stoyan Poryazov	Bulgarian Academy of Sciences, Bulgaria
Alexander Sayenko	University of Jyväskylä, Finland
Dirk Staehle	University of Würzburg, Germany
Sergei Semenov	NOKIA, Finland
Burkhard Stiller	University of Zürich and ETH Zürich, Switzerland
Weilian Su	Naval Postgraduate School, USA
Veselin Rakocevic	City University London, UK
Dmitry Tkachenko	IEEE St. Petersburg BT/CE/COM Chapter, Russia
Vassilis Tsaoussidis	Demokritos University of Thrace, Greece
Christian Tschudin	University of Basel, Switzerland
Kurt Tutschku	University of Würzburg, Germany
Lars Wolf	Technische Universität Braunschweig, Germany

NEW2AN Additional Reviewers

M.C. Aguayo-Torres	A. Amirante	F. Barcelo-Arroyo
K. Al-Begain	K. Avrachenkov	S. Bayhan

ruSMART Executive Technical Program Committee

ruSMART Additional Reviewers

TAMPERE UNIVERSITY OF TECHNOLOGY

NOKIA

Table of Contents

I NEW2AN

Wireless Networks

Multi-hop Wireless Networks

Cross-Layer Design

Teletraffic Theory

Multimedia Communications

Heterogeneous Networks

Network Security

II   ruSMART

Session I - Keynote Talks

Session II

Session III

Decentralized Synchronization and Estimation in Wireless Networks

Nikolai Nefedov

Nokia Research Center, Zurich,
Switzerland
nikolai.nefedov@nokia.com

Abstract. In this paper we address estimation/control methods in complex networks where a global estimate (or decision) is obtained in a distributed fashion without fusion or centralized control centers. The suggested approach is based on local exchange of information among the nearby nodes within a connected (wireless) network that allows, under certain conditions, to reach a global decision based on locally available decisions/measurements. In particular, we consider network nodes as local dynamical systems with impulse-like coupling to establish time synchronization among the transmitted packets together with phase-coupling during packet durations to achieve distributed estimation/control. The suggested method may be used for distributed spectral sensing in cognitive radio and wireless sensor networks.

Key words: distributed estimation, synchronization, cognitive networks.

1 Introduction

In current cellular networks the centralized control and synchronization are widely used to establish and maintain coordination among the nodes, e.g., in joint estimation/detection methods, data fusion schemes, media access protocols and etc. However, centralized methods are known to be sensitive to congestion problems and failures of central (fusion) nodes. It makes these techniques not robust and inefficient in complex network with changing topology and mobility.

On the other hand, future wireless communication systems assume co-existence of various communication systems with dynamical frequency allocation and self-organization. It implies that control functions are to be shifted from dedicated central nodes to end-users nodes. Besides, these systems should meet requirements on mobility and scalability to provide normal operation despite topology and connectivity changes. These trends gave rise to cognitive radio (CR) [1] and CR network architectures.

Currently cognitive processing is mainly applied for dynamic spectrum access and is based on complicated digital signal processing in baseband. On the other hand, there is growing interest in self-organization phenomena observable in natural systems (biology, physics and etc) that mimic cognitive behavior [2]-[15].

S. Balandin et al. (Eds.): NEW2AN 2008, LNCS 5174, pp. 1–12, 2008.

In this paper we apply the concept of self-organization of coupled dynamical systems to establish decentralized synchronization and perform distributed estimation of local parameters in wireless networks. In case of CR systems the local measurements may present interference temperature at certain frequency bands at different locations to facilitate dynamic spectrum access and resolve hidden terminal problem. The proposed method may be implemented in analog domain without a need for power hungry analog-digital converters (ADC) and extensive digital signal processing.

The paper is organized as follows. In Section 2 we overview coupled dynamical systems and then apply this approach for distributed estimation in wireless networks in Section 3. Decentralized synchronization for packed-based transmission is addressed in Section 4, followed by the proposed architecture in Section 5 and conclusions in Section 6.

2   Coupled Dynamical Systems

2.1   System Model

We consider a network where nodes, initialized at different states (local opinions), are interacting trying to reach a global over network stable behavior. The stable system mode may take a form of a consensus (the same state for all nodes) or synchronization mode (synchronous dynamics of states at all nodes).

Dynamics of a local state in the simplest case may be described as dynamics of an oscillator, where a local state is presented by the oscillator phase. At the abstract level the local states may present a decision variable in distributed control or the local estimate of a parameter of interest. This generic model is used to model decentralized operations in the following.

Let's consider a network of N nodes where each n-th node has an access to a common interaction media (e.g, radio or logical channels) and consists of (Fig. 1):

(a) a local decision block which makes a local decision (or measurement) on a (vector-) parameter $y_n(t_k)$, e.g., interference temperature in a certain frequency bands at time instant t_k;
(b) a processing block to calculate a function of the measurement, $g_n(y_n(t_k))$, which presents an initial state of n-th node $x_n(t_0)$; in the simplest case $x_n(t_0)$ $= g_n(y_n(t_k)) = y_n(t_k)$;
(c) a receiver block that senses environment to obtain local decisions from other nodes (a part of front-end at Fig. 1);
(d) a dynamical system, characterized by a state $x_n(t)$ which dynamics depends on a local decision $g_n(y_n(t_k))$ and decisions obtained from other states $x_m(t)$;
(e) an interface block to map a local state $x_n(t)$ on some physical carrier and broadcast it to its neighbors (a part of front-end at Fig. 1).

As a whole this system allows to implement distributed estimation/control without data fusion centers, where each node makes a local decision (or opinion) about some vector value, adjusts its decision based on decisions obtained from other modes and then broadcasts it further its neighbors.

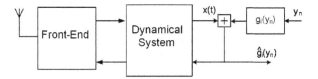

Fig. 1. Node architecture

The system dynamics may be described by motion equations in continuous time

$$\dot{x}_n(t) = g_n(y_n(t)) + \frac{K}{c_n} \sum_{m=1}^{N} a_{nm} h(x_m(t - \tau_{nm}) - x_n(t)) + \eta_n(t), \quad n = 1...N \quad (1)$$

where $h(.)$ is a coupling function; K is a global coupling gain; c_n is local positive coefficient (e.g., associated with reliability or SNR of the local measurement); the coefficients a_{nm} describe coupling strength between nodes n and m; τ_{nm} is propagation delay from node n to node m; $\eta_n(t)$ is the channel noise. Coupling coefficients a_{nm} may be associated with channel parameters, $a_{nm}^2 = p_m |h_{nm}|^2 / d_{nm}^2$, where p_m is the power transmitted by m-th node, h_{nm} is the fading coefficient, d_{nm} is the distance between nodes n and m. In general the coefficients a_{nm} may be asymmetric to take into account, e.g., different transmitting power at different nodes [10]. Due to radio-wave propagation loss in wireless communications the broadcasted signals decay with the distance, $a_{nm} = f(d_{nm})$, which results in local-only coupling among dynamical systems (nodes).

2.2 Fully Connected Network with Phase-Coupling

As an illustrative example first we consider a behavior of a globally connected network (1) with coupling function $h() = \sin()$ in absence of noise and delays in interactions. Dynamics of this networks is then described by well-known Kuramoto model for globally phase-coupled oscillators [4]

$$\dot{\theta}_n(t) = \omega_n + \frac{K}{N} \sum_{m=1}^{N} \sin(\theta_m(t) - \theta_n(t)), \quad n = 1...N \quad (2)$$

where $\omega_n = g_n(y_n)$ are local frequencies, $\theta_n(t) = x_n(t)$ are initial phases (cf.(1)). The equations (2) have a range of solutions from periodic to chaotic. For example, periodic solutions include: (i) phase sync (consensus) mode $\theta_n(t) = \theta^*(t)$ for all n; (ii) phase-lock (frequency sync) mode where all oscillators have waveforms of period T shifted by a phase $\theta_n(t) = \theta_0(t + nT/N)$; (iii) partial sync there both modes may co-exist.

The system (2) may be analytically tractable, and in the limit $N \to \infty$ there is a critical value of coupling strength K_c, such that for $K > K_c$ both frequency and phase sync appear in the system [4]. Following [4][5] we define a complex mean field for N oscillators with equal unit amplitude as

$$R(t) = \frac{1}{N} \sum_{n=1}^{N} e^{i\theta_n(t)} = re^{i\psi(t)} \tag{3}$$

Global coupling may be seen as the total mean field acting on a selected oscillator, then (3) may be rewritten as

$$\frac{d\theta_n}{dt} = \omega_n + Kr\sin(\psi - \theta_n), \quad n = 1...N \tag{4}$$

where r and ψ are mean-field amplitude and phase, respectively. If identical oscillators are all in phase-sync, then it results in just one oscillation with max mean field amplitude ($r = 1$), while random-phase oscillators show a chaotic behavior with minimum mean field amplitude ($r \to 0$). For this reason the mean-field amplitude r is also referred as the order factor.

3 Distributed Estimation

Given the system above the next step is to define a mapping of local decision/information onto the model (4). Local measurements may be mapped: (i) as a local initial phase; (ii) as a local initial frequency; (iii) as a data packet. The latter case assumes processing in base-band domain and considered elsewhere [16]. In Section 5 we show that cases (i)-(ii) may be implemented with low power consumption in analog domain without a need for high speed ADC.

As a practical example, we consider mapping of local measurements on radio frequencies (RF) of local nodes, local dynamical systems are presented by phase-lock loops. Local measurements (local frequencies) are modeled as random values taken from Gaussian distribution with variance σ_ω^2, a histogram of frequency distribution ($N = 50$) used in simulations is shown at Fig. 2. Provided that coupling strength K is large enough compared to frequency variations, the system evolves from quasi-chaotic (Fig. 3) to partial synchronization at $K = 0.5$ (Fig. 4), where nodes with close frequencies are frequency locked, resulting in growing mean-field which in turn attracts further staying apart (in frequency) nodes into the frequency lock at $K = 0.6$ (Fig. 5).

It may be shown that nodes with local frequencies $|\omega_n - \omega_m| > Kr$ can not be attracted to the frequency lock, it results in a partial frequency sync. But even in a case when all oscillators with different initial frequencies are synchronized, it results at best in phase-locking (constant phase difference) or frequency sync mode with frequency $\omega^* = \frac{1}{N} \sum_{n=1}^{N} \omega_n$, but not in phase sync mode (consensus), where phase differences are zero [5].

Taking into account propagation attenuation a_{nm} and reliability of local measurements $1/c_n$ in (1), it is easy to show (see Appendix) that the sync state is the weighted average, $\omega^* = \dfrac{\sum\limits_{n=1}^{N} c_n \omega_n}{\sum\limits_{n=1}^{N} c_n}$, which in channels with additive white

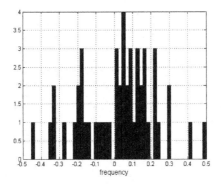

Fig. 2. Distribution of local measurements, $N = 50$

Gaussian noise corresponds to the maximum likelihood (ML) estimate obtained without dedicated fusion centers [6]-[10].

Note that if one of the local measurements (frequencies) has a high reliability (high SNR) and differs significantly from others nodes, this frequency will not be locked and the order factor r will have a lower value and significant variations. The change of the mean field r below a threshold may be used in CR system to indicate that one of nodes senses a strong signal in a certain frequency band which is not visible to the other nodes. In particular, this property may be used to resolve the hidden terminal problem.

3.1 Networks with Complex Topology

Fully-connected networks have the smallest average path length that facilitates a fast convergence. However, most of real large-scale networks (e.g., cellular wireless networks) are only sparse (or locally) connected. Convergence to a consensus on networks with difference sparse topologies has been studied in a number of papers [6]-[12]. In particular, recently it is shown that if the network is *connected*, i.e. there is a path between any pair of nodes, then local exchange of information among the nearby nodes is sufficient to reach a *global* consensus on the *average* of observable values without requiring any control node. A global consensus can be reached both with linear or non-linear coupling [6].

For a fully connected network (e.g., Fig. 6) with different local decisions $g_n(t)$ in (1), there is a transition to sync mode in the limit $N \to \infty$ when coupling strength $K > K_c^{(global)} = \dfrac{2}{\pi \hat{g} N}$, where $\hat{g}$ is the mode of the unimode distribution, $\hat{g} = \max_{N \to \infty} \{g_1, g_2, ..., g_N\}$ with normalization $\lim_{N \to \infty} \dfrac{1}{N} \sum_{n=1}^{N} g_n = 1$. For a connected network with local coupling the similar transition to sync takes place for coupling strength $K > K_c^{(local)} = \dfrac{2}{\pi \hat{g} \lambda_2}$, where λ_2 is the 2nd smallest eigenvalue of the Laplacian matrix derived from connectivity matrix $A = \{a_{mn}\}$.

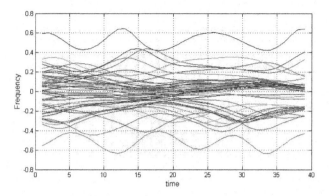

Fig. 3. Evolution of frequencies (local decisions) in time, K=0.3, quasi-chaotic mode

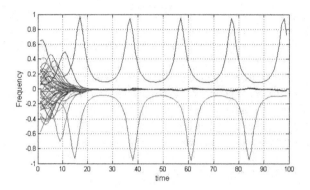

Fig. 4. Evolution of frequencies (local decisions) in time, K=0.5, partial sync

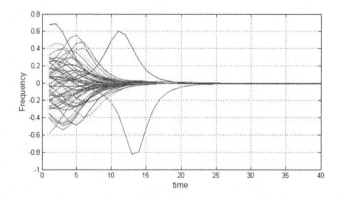

Fig. 5. Evolution of frequencies (local decisions) in time, K=0.6, sync mode

Dynamics of arbitrary connected networks may be analyzed using weighted Laplacian L_w of the graph: $L_w = BWB^T$ where B is the incidence matrix presenting interconnections of nodes with weighting matrix W [11][13]. For example,

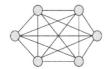   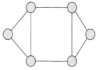

Fig. 6. Fully connected network   **Fig. 7.** Partially connected network

we may easily calculate that for uniformly weighted locally connected network at Fig. 7 the critical coupling strength is 6 times larger than for the globally connected network at Fig. 6. It means that results presented for fully connected networks may be applied for locally connected graphs (as long as they stay connected) with a properly selected coupling strength provided that the network parameters are known or upper bounded (e.g., by max Tx power).

4 Decentralized Synchronization

4.1 Packet-Based Transmission

The distributed estimation scheme outlined above recently attracts growing attention [9][10]. Unfortunately the scheme described in [9][10] can not used in practical wireless systems with packet transmission. In particular, the scheme [9][10] is based on continuous transmission which is impractical at least from the power consumption point of view and created interference to other systems. Besides, it also implies the full-duplex in Tx and Rx operations, which is problematic in practical systems due to a leakage from Tx into Rx circuitry at close frequencies (Tx power is much more than Rx power). A typical way to solve this problem is to use Tx/Rx switch and half-duplex regime to share the same antenna. However, the presence of half-duplex rises another problem of media access protocols which is not mentioned in [9][10]. Recall that due to interactions the system dynamics tunes phases and frequencies of oscillators, but does not control Tx/Rx switching intervals.

On the other hand, the correction term in (1) is actually the mean field formed by *simultaneously* transmitting nodes. It means that the whole system is first to be time-synchronized. Currently centralized synchronization is widely used in cellular wireless networks. However, centralized methods are sensitive to congestion problems and failures of control centers, these techniques are not suitable for network with changing topology and mobility. In the following we consider decentralized synchronization based on pulse-like coupling.

4.2 Self-synchronization with Pulse-Like Coupling

Collective behavior of pulse-coupled oscillators has been widely studied in physics and biology, e.g., [2][3]. Recently pulse-coupled methods are proposed for wireless sensor networks [15].

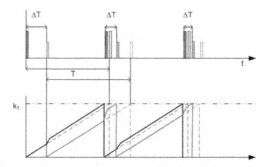

Fig. 8. Sync process for two nodes with pulse transmission

Let's consider the local dynamics of a node described by its state $x_n(t)$. We assume that $x_n(t)$ is a monotonically increasing function of time from some initial state to a threshold k_1, such that when the local state reaches the threshold, $x(t_k) > k_1$, the node transmits a pulse and returns to its initial state (forming a limit cycle trajectories similar to the phase-coupled oscillators case).

As an illustration, synchronization of two nodes with pulse transmission and linear behavior of local states is shown at Fig. 8. The upper and lower parts show pulse transmitting instants and dynamics of local states, respectively. In absence of interactions dynamics of states and transmitting moments are shown by dashed lines and not filled impulses, respectively. In presence of interactions the energy sensed during the silence period affects the states (solid lines) and drives firing moments to sync ($\Delta T \to 0$). If nodes are isolated (or the system in the sync mode), all nodes transmit pulses synchronically with the same period T.

One method to form local control signals to reach the global sync mode with impulse transmission is based on idealized assumption that that impulses are very short and not overlapped [15]. In this paper we extend this scheme for the practical packet-based transmission with possible overlapping. The purpose of (radio) packet based synchronization is two-fold: (i) to reduce/eliminate de-sync effects due to multipath propagation unavoidable in impulse transmission; (ii) to facilitate distributive estimation/control addressed in Section 3.

The main difference of packet-based transmission w.r.t. [14], [15] is that due to the noticeable duration of (radio) packet T_a the fired moments are delayed sequentially, which in turn results in delay accumulation and prevents the synchronization. To avoid delay accumulation one may use firing "in advance", which may be achieved by the proper lowing the firing threshold k_1 or adjusting the silent period duration $T_s = T - T_a$.

However, we found that these methods are suboptimal: our simulations of packet transmission show that in sync mode the packets are aligned with a jitter, which reduces sync accuracy. Fortunately, for distributed estimation/control described in the Section 3 we do not need precise synchronization: it is enough if at sync mode (radio) packets from different nodes are sufficiently long overlapped to create the effective mean field used further as a correction factor for the local estimation/decision.

At the same time in case of distributed estimation, each node must have a possibility to listen while other nodes are transmitting. A way to achieve it is to put (some of) simultaneously/synchronously transmitting nodes randomly, or according to some protocol, into the listening mode. Due to randomness, a number of active synchronously transmitting nodes is always less than in the case of continuous transmission; it increases the time to reach a global stable mode, but still preserves the convergence.

5   Proposed Scheme

A practical method to synchronize packet transmission in wireless networks without centralized control by using pulse-based coupling and tuning sensitivity and firing thresholds together with coupling strengths is described in [16].

We consider a wireless network with arbitrary topology and assume that all nodes have access to a common (radio) channel and exchange information with their neighbors by sending radio packets (waveforms) of duration T_a. A process of packet-based transmission is illustrated at Fig. 9. During silence period $t <$ T_s each node measures energy from other nodes at the common channel and compares it to a firing threshold k_1. If the threshold is reached, a node transmits a radio packet and resets its timer $t = 0$, otherwise the packet is transmitted at $t = T_s$. Note that in presence of other transmitting nodes the actual silence period for a given node is not constant.

When packets are in sync mode, they occupy the same time-slot and the measured value (hence, control signals) during silence period does not exceed the sensitivity threshold k_0.

A possible scheme with quasi-continuous time processing (Rx/Tx shaping filters are not shown) is depicted at Fig. 10. The suggested scheme consists of synchronization unit, estimator/detector unit and a control block. Methods of decentralized packet synchronization and distributed estimation are described above in Section 4 and Section 3, respectively. Similar to [15], we describe isolated local dynamics

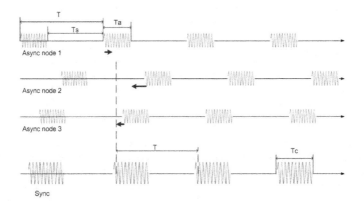

Fig. 9. Asynchronous and sync (lower part) packet-based transmission

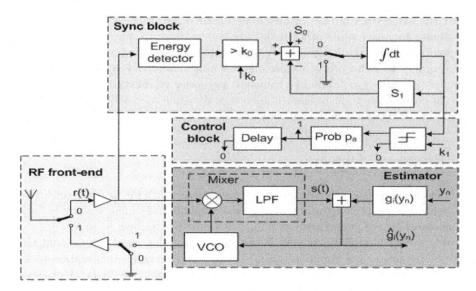

Fig. 10. Block scheme for distributed synchronization/estimation

by leaky integrate-fire model $\dot{x}_n(t) = S_0 - S_1 x_n(t)$, where S_0 is associated with accumulation speed and S_1 is a leakage factor. Possible implementation of "sync block" is sketched at Fig. 10, where a threshold k_0 controls noise sensitivity; S_0, S_1 and k_1 regulate coupling strength and the firing threshold. If the output signal from "sync block" is above the firing threshold, "control block" adjusts the instant of the next transmission; delay block inside of control unit sets duration of Tx mode (Tx mode: switches are at "1"; Rx mode: switches are at "0").

Functionality of all blocks are coordinated by the control block as follows. Each node is switching between *local* measurement mode (called as *silent* mode, max duration T_s) and *distributed* estimation mode (*active* mode, duration T_a). During *active* mode each node may be either in *transmitting* or in *listening* mode.

During silent mode each node (i) performs measurements on common channel to establish and maintain time synchronization and (ii) makes a decision (measurement) on parameters of interest (e.g., interference temperature) for distributed estimation/detection. In particular, during silent mode each node:

- Senses the common channel and makes updates for time synchronization;
- Makes local measurements and calculates (if needed) a local function $g_n(y_k)$.

During active mode a local state of n-th node, x_n, is updated with information from other modes and mapped onto a physical carrier (e.g., as oscillator frequency/phase or into a data packet). The control block at each node randomly, or according to a certain protocol, sets either active-transmitting mode or active-listening mode and keeps this setting during a given active mode.

- In case of active-transmitting mode the local information from n-th node is transmitted as a radio packet during $[\tilde{t}_k, \tilde{t}_k + T_a]$.

- In case of active-listening mode the state of n-th node is updated according to (1) and preserved as voltage controlled oscillator (VCO) parameter until the next active/transmitting mode.

For cognitive radio systems local measurements $y_n(t_k)$ $(n = 1, ...N)$ may present interference temperature at a given frequency band at different locations. In case of distributed control $y_n(t_k)$ may present a local decision, a planned action or/and a behavior strategy from the game theory perspective.

6   Conclusions

In this paper we outline the practical method to make distributed estimation in wireless networks without centralized control centers. We consider network nodes as local dynamical systems with impulse-like coupling to establish time synchronization among the transmitted packets combined with phase-coupling during packet durations to achieve distributed estimation/control.

The suggested method may be used for distributed spectral sensing and to resolve the hidden terminal problem.

References

1. Haykin, S.: Cognitive Radio: Brain-Empowered Wireless Communications. IEEE J. SAC 23(2), 201–220 (2005)
2. Hoppensteadt, F.C., Izhikevich, E.M.: Weakly Connected Neural Networks. Springer, NY (1997)
3. Strogatz, S.: Sync: The Emerging Science of Spontaneous Order. Hyperion NY (2003)
4. Kuramoto, Y.: Lec. Notes in Physics, vol. 30. Springer, NY (1975)
5. Acebron, J., Bonilla, L., Vicente, C., Ritort, F., Spigler, R.: The Kuramoto model: A simple paradigm for synchronization phenomena. Reviews of Modern Physics 73(1), 137–185 (2005)
6. Scherber, D., Popadopolus, H.: Distributed computation of averages over ad hoc networks. IEEE J. SAC 23(4), 776–787 (2005)
7. Olfati-Saber, R., Fax, A., Murray, R.: Consensus problems in networks of agents with switching topology and time-delays. IEEE Trans. Autom. Control 49(9), 1520–1533 (2004)
8. Olfati-Saber, R., Fax, A., Murray, R.: Consensus and Cooperation in Networked Multi-Agent Systems. IEEE Proc. 95(1), 215–233 (2007)
9. Barbarossa, S., Scutari, G.: Decentralized Maximum-Likelihood Estimation for Sensor Networks Composed of Nonlineary Coupled Dynamical Systems. IEEE Trans. on Signal Processing 55(7), 3456–3470 (2007)
10. Barbarossa, S., Scutari, G.: Bio-Inspired Sensor Network Design. IEEE Signal Processing Magazine 5, 26–35 (2007)
11. Dorogovtsev S.N., Goltsev A.V., Mendes J.: Critical phenomena in complex networks (2007) ArXiv:0705.0010v2

12. Papachristodoulou, A., Jadbabaie, A.: Synchronization in oscillator networks: Switching topologies and non-homogenous delays. In: IEEE Proc. Conf. Decision and Control (2005)
13. Barahona, M., Pecora, L.: Synchronization in Small-Word Systems. Phys. Review Letters, 054101, 89(5), 29 (2003)
14. Mirollo, R.E., Strogatz, S.H.: Synchronization of pulse-coupled biological oscillators. SIAM J. Appl. Math 50(6), 1645–1662 (1990)
15. Hong, Y.-W., Scaglione, A.: A Scalable Synchronzation Protocol for Large Scale Sensor Networks and its Applications. IEEE J. SAC 23(5), 1085–1099 (2005)
16. Nefedov, N.: Decentralized Synchronization/Estimation/Control in Cognitive Radio Networks. Nokia report NC62407, US.854.0079.U1 (2008)

Appendix

Let's consider a connected network with local coupling

$$\dot{x}_n(t) = g_n(y_n(t)) + \frac{K}{c_n} \sum_{m=1}^{N} a_{nm} \sin(x_m(t) - x_n(t)) + \eta_n(t) \tag{5}$$

Following [7][6], let's multiply each equation by c_n and sum over n,

$$\sum_{n=1}^{N} c_n \dot{x}_n(t) = \sum_{n=1}^{N} c_n g_n(y_n(t)) + K \sum_{n=1}^{N} \sum_{m=1}^{N} a_{nm} \sin(x_m(t) - x_n(t)) + \sum_{n=1}^{N} c_n \eta_n(t)$$

Thanks to the symmetry of coefficients a_{nm} and anti-symmetry of $\sin(x) = -\sin(-x)$, if the system is in sync mode, then

$$\dot{x}_n(t)_{t \to \infty} \to \dot{x}^*(t) = \frac{1}{c^*} \sum_{n=1}^{N} c_n \dot{x}_n(t) + \frac{1}{c^*} \sum_{n=1}^{N} c_n \eta_n(t) = w^* + v(t) \tag{6}$$

where $c^* = \sum_{n=1}^{N} c_n$

In other words, all state derivatives converge to a globally asymptotically stable and unique (constant) value $\dot{x}_n(t) \to w^*$ irrespective of initial conditions.

SICTA Modifications with Single Memory Location and Resistant to Cancellation Errors

Sergey Andreev, Eugeny Pustovalov, and Andrey Turlikov

St. Petersburg State University of Aerospace Instrumentation,
Bolshaya Morskaya street, 67,
190000, St. Petersburg, Russia
{corion,eugeny,turlikov}@vu.spb.ru

Abstract. In this paper we consider a cross-layer MAC-PHY technique that combines the collision resolution tree algorithm (TA) with the possibility of successive interference cancellation (SIC). The overview of the previous work shows that no simple protocol has been proposed to use a single memory location for the captured signal. We, consequently, propose two such protocols that demonstrate the throughput - implementation complexity trade-off. Further, we address the system operation during which interference cancellation errors are possible. We propose the third protocol that is resistant to cancellation errors. A simple yet effective technique is applied to address the throughput performance of all three protocols to demonstrate their superiority over known conventional TA protocols.

1 Introduction and Background

1.1 Conventional Random MAC Protocols

Cross-layer techniques enable telecommunication systems to achieve higher data rates than it is possible with the conventional OSI model interaction. In particular, the mutual collaboration of *physical* (PHY) and *media access control* (MAC) layers has proved to be very promising as MAC layer is practically a bottleneck of the modern telecommunication systems. A multiplicity of MAC protocols is known and analyzed [1], [2], thus leaving developers with a wide choice. Random MAC protocols are frequently used to cope with bursty traffic and to provide reasonably low packet delay even when a user population is high.

Broadly speaking, each random MAC protocol defines a *channel access algorithm* (CAA) and a *collision resolution algorithm* (CRA). The former arbitrates access to the shared broadcast channel, whereas the latter resolves the packet collisions (i.e. simultaneous transmissions of two or more data packets), whenever they arise. In ALOHA and ALOHA-based protocols, such as *diversity slotted aloha* (DSA), *binary exponential backoff* (BEB), *carrier sense multiple access* (CSMA) and others no particular CRA is specified. These protocols are generally easy to implement and once a collision occurs their underlying idea is to defer the subsequent packet retransmission to some future time in a 'hope' that the communications channel becomes idle.

S. Balandin et al. (Eds.): NEW2AN 2008, LNCS 5174, pp. 13–24, 2008.

By contrast, *tree algorithm* (TA) proposed in [3] and [4] defines a CRA to specifically address the collision resolution process and, consequently, to achieve higher performance. The performance of the original *standard tree algorithm* (STA) was enhanced by the *modified tree algorithm* (MTA). We refer to the STA and the MTA as to the *conventional* TAs in what follows. Each TA may incorporate one of three alternative CAAs, which are *gated* access, *window* access or *free* access.

According to the gated CAA the new data packets that arrive during the so-called *collision resolution interval* (CRI) are deferred. Once current collision is resolved, all the deferred packets are transmitted simultaneously to create the following collision and to start a new CRI. Window CAA is a generalization of the gated access scheme for the case when a new CRI is formed not by all the packets that arrive during the previous CRI, but by those arriving within a specified time window. The proper adjustment of this window may result in the higher performance of the TA. The gated and the window CAAs are together referred to as *blocked access* schemes. Finally, free CAA assumes the packet is transmitted immediately following its arrival.

1.2   Cross-Layer MAC-PHY Approaches

All conventional random MAC protocols assume that once a collision occurs no meaningful data packet could be recovered from it. However, for the wireless communications channel a *successive interference cancellation* (SIC) technique is known, which is a nonlinear type of a multiuser detection scheme, where users signals are decoded successively. More specifically, SIC first tries to detect and demodulate the strongest user signal currently present in the composite captured signal. After it is done, this signal contribution to the original signal is recreated and subtracted from it. A new composite signal is thus produced, which could be again the input for the iterative SIC procedure.

In [5] it was shown that SIC is capable of approaching the theoretical limits for an AWGN channel and its *cancellation error* was introduced and estimated. A cancellation error for a user is defined as its residual signal in the remaining composite signal after the subtraction of the recreated signal. The main reasons of the cancellation error are imperfect (amplitude and phase) channel estimation and incorrect bit decisions.

In [6] a concept of *successive interference cancellation in a tree algorithm* (SICTA) was first proposed to combine the advantages of SIC and conventional TAs. A new protocol was described and analyzed that adopts SIC to reuse the collision signals that are stored in a (potentially) unbounded memory. The main performance metrics considered were packet delay and *maximum stable through-put* (MST), which is defined as the highest possible (Poisson) arrival rate that still yields a finite packet delay with probability one. Note, that the STA and the MTA have the MST of 0.346 and 0.375, respectively, for binary tree and fair splitting (left and right tree branches are selected with equal probability). The proposed SICTA protocol has the MST of 0.693, which is twice that of the STA.

All known SICTA modifications theoretically required an unbounded memory storage for the received collision signals. By contrast, in [7] the performance of a novel TA with SIC property was investigated for which a single memory location suffices. Additionally, this protocol considered free CAA and was analyzed for a variety of the Markovian arrival processes and the error-free channel. The MST of the proposed protocol was shown to be 0.5698 for Poisson arrivals by using the additional control field/bit with separate feedback, indicating whether the packet is transmitted for the first time.

Motivated by the above work, we notice that the MST performance of original SICTA protocol [6] is yet to be evaluated under the single memory location assumption. Filling this gap results in two modifications of the original SICTA protocol as in [8]. First modification cancels only successful transmissions and, therefore, relies on a simpler SIC PHY scheme. Second modification cancels both collision signals and successful transmissions and, therefore, demonstrates the higher MST for the cost of a more difficult SIC implementation. Finally, following [9] we show that the introduced protocols suffer from a 'perpetual splitting' phenomenon when a probability of a cancellation error [5] is nonzero. We finally extend our model to account for the possible cancellation errors and modify the protocol operation to acquire robustness to the imperfect interference cancellation without severe tree truncation as in [9] and [10].

The rest of the paper is structured as follows. In Section II we formulate the assumptions of the reference information theoretical model and speculate on its variations in the aforementioned papers. Here we also describe the underlying idea of known protocols as well as of the proposed modifications. Section III conducts the MST analysis of the introduced protocols. In Section IV we compare the theoretical results and summarize the paper.

2   System Model and Protocols

2.1   Reference Information Theoretical Model

Following the multitude of works, e.g. [1], [2] and [11] we formulate a set of assumptions about the communications channel and the way users access it:

Assumption 1. The system time is slotted into equal slots. The duration of each slot is a unit of the system time, which is exactly the transmission time of one data packet. Each slot is assigned an integer nonnegative number and number t slot corresponds to the time interval of $[t, t+1)$. Hereinafter we refer to number t slot simply as slot t for the sake of brevity. Slot borders are known to all the users and each user is restricted to start its packet transmission only in the beginning of a slot.

Assumption 2. In each slot any and only one of the following error-free events may occur (channel-PHY feedback):

- Only one user transmits (**S** event - success);
- None of the users transmit (**E** event - empty);
- Two or more users transmit (**C** event - collision).

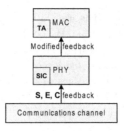

Fig. 1. Various feedback types

Assumption 3. When monitoring the channel activity the MAC layer of a user is notified of the channel event by the end of the current slot. The size of the PHY-MAC feedback (see Fig. 1) a user receives is subject to some variations below.

Assumption 4. There is an infinite user population, generating packets that are assumed to be unique. Each user is supplied with a buffer sufficient to store only one packet. The packet is stored from the instant of time it arrived into the system to the instant of time it is successfully transmitted. Packet inter-arrival times are assumed to be statistically independent random variables which are distributed exponentially with the mean value of $\frac{1}{\lambda}$. Thus, λ is the arrival rate of the new packets into the system. Notice, that infinite population provides a pessimistic estimation for a finite population system by considering each packet to be a virtual station.

2.2 Various PHY-MAC Feedback Types

The conventional STA with gated access requires only 'collision' - 'no collision' PHY-MAC feedback from the receiver (see **Assumption 3**) for its proper operation. New data packets that arrive during the previous CRI are transmitted in the first slot of the following CRI. If 'no collision' feedback is received for this initial slot, the CRI ends. Otherwise, each of the collided users flips a (biased) coin to choose the right subset with probability p and the left subset with probability $1 - p$ (see Fig. 2(a)). The procedure is repeated recursively until all the packets are received successfully. The example collision resolution tree in Fig. 2(a) comprises 7 slots, which is exactly the CRI length. For binary STA fair splitting with $p = 0.5$ is optimal and yields the MST of 0.346 [2], [12].

We notice that in the example collision resolution tree (see Fig. 2(a)) the collision in slot 5 is deterministic, since the collision in slot 3 is followed by the empty right slot 4, which means that all the collided users have chosen the left subset. In [3] and [4] a 'level skipping' was proposed to omit slot 5 and proceed directly to the next tree level. Therefore, the CRI length reduces to 6 slots (see Fig. 2(b)) instead of 7. The conventional binary MTA with gated access uses this idea and results in the MST of 0.375 for fair splitting. However, a biased splitting with $p = 0.582$ is optimal and yields the MST of 0.381 [12]. We finally notice that MTA requires the extended ternary 'success' - 'empty' - 'collision' PHY-MAC feedback to enable the discussed improvement.

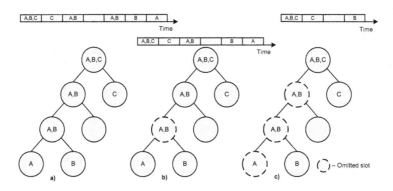

Fig. 2. STA (left), MTA (middle) and SICTA (right) examples

In the original SICTA protocol the receiver is supplied with the unbounded memory to store the collision signals. Consider the example in Fig. 2(c) where the CRI length is only 4 slots. The contents of the corresponding left slot is determined by canceling the interference from slot 1 after the successful reception of signal C in slot 2. Following the notation from [6], we denote the cancellation procedure as $\tilde{Y}_1 = Y_1 - X_C$. As slot 3 is empty the corresponding left slot is skipped following the rules of the MTA. Finally, the successful reception of signal B in slot 4 immediately yields the recovery of signal A by $\tilde{\tilde{Y}}_1 = \tilde{Y}_1 - X_B = Y_1 - X_C - X_B$.

As the entire left subtree of the STA is omitted by the original SICTA protocol, it results in the MST of 0.693, which is twice the MST of the STA. Additionally, SICTA requires extended k - 'empty' - 'collision' PHY-MAC feedback, where k is the number of successfully decoded packets plus the number of left slots identified as being empty after the SIC procedure.

Despite its high performance, SICTA is vulnerable to noise and imperfect interference cancellation in the error-prone communications channel. Indeed, suppose in Fig. 2(c) the last cancellation operation is degraded by the noise term N, that is $\tilde{\tilde{Y}}_1 = \tilde{Y}_1 - X_B + N$. If N is sufficiently large the signal A may not be restored and the collision resolution continues. Eventually, after A is transmitted successfully, the noise energy level may still be sufficiently high to lead receiver into believing there is another collision in the left slot. The protocol will require nonexistent users to further split until some external entity terminates it.

To overcome the above deadlock the SICTA/FS protocol was introduced to truncate the SICTA collision resolution tree after the first success [9]. For the operation of this protocol the ternary 'success' - 'empty' - 'collision' PHY-MAC feedback again suffices. However, during SICTA/FS analysis the **Assumption 4** has been replaced with the finite user population assumption, which considerably changes the system model and makes the derived MST value incomparable with those for SICTA and conventional TAs. (G)BEB-SICTA/FS protocols [13] also operate in the framework of this modified system model. Furthermore, SICTA/F1 protocol [10] for harsh wireless channel truncates the collision resolution tree after either first success or empty slot and, therefore, requires only 'collision' - 'no collision' PHY-MAC feedback.

2.3   Single Memory Location

Supplying the receiver with the unbounded memory storage for the collision signals is practically infeasible. Accounting for this fact, the question of the SICTA operation with the single memory location was first addressed in [7]. As we are interested in the similar investigation we extend the system model as follows.

Assumption 5. The receiver is able to store a single signal for which a single memory location is dedicated.

Importantly, in [7] each data packet was supplied with an extra field/bit with separate feedback, indicating whether the packet is transmitted for the first time. This again makes the derived MST incomparable with those for SICTA and conventional TAs as the system model has changed. Finally, in [8] two modifications of the FCFS protocol has been introduced, which conform to the derived system model and have the MST of 0.6048 and 0.6173, respectively. As mentioned above, the main drawback of the FCFS-based protocol is (potentially) infinite timer granularity [1], which may be difficult to achieve in the real systems.

Below we concentrate on the gated access SICTA with the single memory location. As in [8], we observe two implementation possibilities for the SIC operation. In the first straightforward scenario after the successful transmission has been detected and demodulated by SICTA, its contribution to the original signal is regenerated and subtracted from the overall received signal in memory (see Fig. 3(a)). Or, alternatively, a cancellation operation is only possible when a transmission is successful. We term the modification of SICTA that use this property SICTA with *success cancellation* (SICTA/SC). In some practical systems it may be also possible to cancel the collision signal from memory, even though the received signal is not detected and demodulated (see Fig. 3(b)). Clearly, this second implementation option impacts the performance of the protocol, which we term SICTA with *success and collision cancellation* (SICTA/SCC).

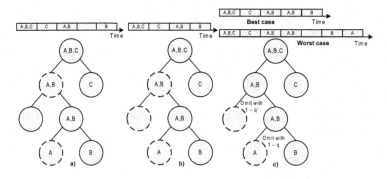

Fig. 3. SICTA/SC (left), SICTA/SCC (middle) and R-SICTA/SCC (right) examples

2.4   Presence of Cancellation Errors

We remind that in practical SIC schemes a cancellation error is possible [5], which is the residual signal in the remaining signal after the SIC procedure. For instance, after the cancellation of the signal A from the composite signal $X_A + X_B$ the resulting signal contains $\tilde{Y} = X_B + N_A$, where N_A is the residual signal A. After the subsequent cancellation of signal B we similarly obtain $\tilde{Y} = N_A + N_B$. If the $N_A + N_B$ energy level is sufficiently high the receiver incorrectly decides that the slot is not empty, but rather there is a collision between the nonexistent users. Below we assume that due to cancellation errors the interference is unsuccessful with some constant probability. That is, with this probability receiver obtains a meaningless signal after the next interference cancellation. In practice this probability could be derived as the worst-case estimate of the SIC operation.

Assumption 6. The interference cancellation is imperfect in a sense that after the successful signal is canceled from the composite signal the resulting signal contains a meaningless signal with probability q. Similarly, after the collision signal is canceled from the composite signal (following the second SIC implementation possibility) the resulting signal contains a meaningless signal with probability q'. We expect that in practice $q' \geq q$, that is, it is more difficult to cancel the collision signal than the successful signal.

The third protocol we propose is resistant to the imperfect cancellation of both types for the cost of its MST performance (see Fig. 3(c)). Basically, it refrains from skipping some collision slots like slot 3, whereas SICTA/SC and SICTA/SCC do. This allows avoiding 'perpetual splitting' phenomenon. We term this protocol *robust* SICTA/SCC (R-SICTA/SCC). Notice, that the rules for R-SICTA/SC that cancels only successfully received signals could be derived similarly, but this is left out of scope of this paper. In Fig. 3(c) the best case timeline corresponds to the case when both interference cancellation operations were successful, whereas the worst case timeline demonstrates both unsuccessful cancellations.

3   Performance Analysis

3.1   Example MTA Analysis

In order to demonstrate the approach that we use below to derive the MST of the proposed protocols, we firstly focus on the analysis of the MTA. Denote by τ the random CRI length. Formally, the conditional expectation $T_k = E[\tau | a$ collision of multiplicity k is resolved] gives the average CRI length for a collision of multiplicity k. In [3] it was shown that using the ratio $\frac{k}{T_k}$ the following bounds on the STA MST (R_{STA}) could be established:

$$\liminf_{k \to \infty} \frac{k}{T_k} < R_{STA} < \limsup_{k \to \infty} \frac{k}{T_k}. \tag{1}$$

The estimates for R_{STA} were summarized by [14] and the following refined figures were given:

$$0.34657320 < R_{STA} < 0.34657397. \tag{2}$$

As $\limsup\limits_{k\to\infty}\frac{k}{T_k} - \liminf\limits_{k\to\infty}\frac{k}{T_k} = 0.00000077$ we follow [14] to notice:

$$R_{STA} \cong \liminf_{k\to\infty}\frac{k}{T_k} \cong \limsup_{k\to\infty}\frac{k}{T_k} \cong \frac{ln2}{2}. \tag{3}$$

We denote the number of the collision resolution tree nodes in a STA by n. The number of successful, collision and empty slots during a CRI is denoted by n_s, n_c and n_e, respectively. As $n_s + n_c + n_e = n$, we use [11] to obtain:

$$n_s = k,$$
$$n_c = \frac{n-1}{2},$$
$$n_e = \frac{n+1}{2} - k. \tag{4}$$

Now we calculate the expected number of nodes (expected CRI length) in the MTA tree $E[m]$ by subtracting the omitted nodes (Fig. 4(a)):

$$E[m] = E[n] - \frac{1}{2}E[n_e] = \frac{3}{4}E[n] + \frac{k}{2} - \frac{1}{4}. \tag{5}$$

That is, $T_k^{MTA} = \frac{3}{4}T_k^{STA} + \frac{k}{2} - \frac{1}{4}$. Taking the $\liminf\limits_{k\to\infty}$ and $\limsup\limits_{k\to\infty}$ of (5) and accounting for (3) we establish:

$$R_{MTA} \cong \frac{1}{\frac{3}{4R_{STA}} + \frac{1}{2}} \approx 0.375. \tag{6}$$

3.2   SICTA/SC and SICTA/SCC

Here we consider perfect interference cancellation (**Assumptions 1 - 5** of the system model) and firstly derive the MST of the SICTA/SC protocol that cancels only successfully received signals. Analogously to the MTA analysis, we calculate the expected number of nodes (expected CRI length) in the SICTA/SC tree $E[m]$ by subtracting the omitted nodes (Fig. 4(a,b)) and accounting for (4):

$$E[m] = E[n] - \frac{1}{2}E[n_e] - \frac{1}{2}E[n_s] = \frac{3}{4}E[n] - \frac{1}{4}. \tag{7}$$

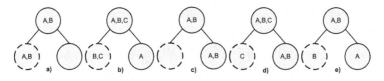

Fig. 4. Collision resolution tree examples for MST derivation

Using (3) the final expression for the MST of the SICTA/SC ($R_{SICTA/SC}$) is obtained as follows:

$$R_{SICTA/SC} \cong \frac{4}{3} R_{STA} \approx 0.462. \qquad (8)$$

Below we derive the MST of the SICTA/SCC protocol that cancels both successful and collision signals. We again calculate the expected number of nodes (expected CRI length) in the SICTA/SCC tree $E[m]$ by subtracting the omitted nodes (Fig. 4(a,b,c,d)) and accounting for (4). Notice that on average we have to add $\frac{1}{2}E[n_2]$ term, where n_2 is the number of collisions of size two during the CRI (Fig. 4(e)), as it was subtracted twice (Fig. 4(b,d)):

$$E[m] = E[n - n_e - n_s] + \frac{1}{2}E[n_2] = \frac{1}{2}E[n] - \frac{1}{2} + \frac{1}{2}E[n_2]. \qquad (9)$$

Following [14] we establish the bounds on the $\frac{E[n_2]}{k}$:

$$\limsup_{k \to \infty} \frac{E[n_2]}{k} < 0.721355,$$

$$\liminf_{k \to \infty} \frac{E[n_2]}{k} > 0.721340. \qquad (10)$$

Note that $\limsup\limits_{k \to \infty} \frac{E[n_2]}{k} = \liminf\limits_{k \to \infty} \frac{E[n_2]}{k} = \gamma$ with the accuracy of three decimal digits and $\gamma = 0.721$. Using (3) and (10) the final expression for the MST of the SICTA/SCC ($R_{SICTA/SCC}$) is obtained as follows:

$$R_{SICTA/SCC} \cong \frac{2}{\frac{1}{R_{STA}} + \gamma} \approx 0.5545. \qquad (11)$$

3.3 R-SICTA/SCC

We now consider imperfect interference cancellation (**Assumptions 1 - 6** of the system model) and derive the MST of the R-SICTA/SCC protocol that cancels both successful and collision signals in the presence of cancellation errors.

Remember (**Assumption 6**) that after the successful captured signal is canceled from the stored signal the resulting signal contains a meaningless signal with probability q and after the collision captured signal is canceled from the stored signal the resulting signal contains a meaningless signal with probability q' ($q' \geq q$). Below we calculate the expected number of nodes (expected CRI length) in the R-SICTA/SCC tree $E[m]$ by subtracting the omitted nodes (Fig. 4(a) with probability 1, Fig. 4(c,d) with probability $1 - q'$ and Fig. 4(e) with probability $1 - q$) and accounting for (4). We also account for n_2, that is, the number of collisions of size two during the CRI, to obtain the following:

$$E[m] = (\frac{1}{2} + \frac{1}{4}q')E[n] - \frac{1}{2} + \frac{1}{4}q' + \frac{k}{2} - \frac{1}{2}(q' - q)E[n_2]. \qquad (12)$$

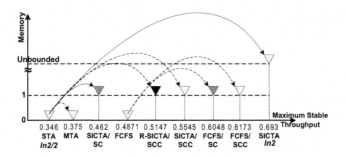

Fig. 5. MST comparison and relations of blocked access protocols

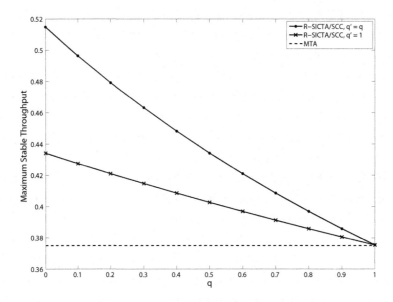

Fig. 6. R-SICTA/SCC MST for imperfect SIC

Using (3) and (10) the final expression for the MST of the R-SICTA/SCC ($R_{R-SICTA/SCC}$) is obtained as follows:

$$R_{R-SICTA/SCC} \cong \frac{4}{\frac{2+q'}{R_{STA}} + 2 - 2(q'-q)\gamma}. \tag{13}$$

Clearly, when $q' = q = 0$, $R_{R-SICTA/SC} \approx 0.5147$. We further emphasize two important special cases of the q' behavior (see Fig. 6). First is when $q' = q$ and $R_{R-SICTA/SCC} \cong \frac{4}{\frac{2+q}{R_{STA}}+2}$. Second is when $q' = 1$ and R-SICTA/SC protocol operates, for which $R_{R-SICTA/SC} \cong \frac{4}{\frac{3}{R_{STA}}+2-2(1-q)\gamma}$.

4   Comparison and Conclusions

The cross-layer combination of the PHY SIC technique and the MAC TA is prominent as it enables higher system throughput for the cost of moderate implementation complexity. A family of protocols that follow this approach is known, of which only the original SICTA has the provable MST of 0.693 in the infinite population model (see Fig. 5). Unfortunately, SICTA requires unbounded memory to store the captured collision signals that complicates its practical implementation. By contrast, some works are known to address similar protocols with a single memory location. However, neither addresses the behavior of the original SICTA protocol with a memory restriction.

We begin with the investigation of the single memory location SICTA that results in two distinct implementation possibilities. The former enables SIC only after the successful signal has been captured. The latter cancels both successful and collision signals to, consequently, achieve higher MST for the cost of a more complex PHY. We use a simple technique to derive the MST of the proposed protocols, which gives 0.462 and 0.5545 for the first (SICTA/SC) and the second (SICTA/ SCC) modification, respectively. In addition to their desirable complexity properties the novel protocols outperform all known conventional TAs (STA and MTA) and the SICTA/SCC has higher MST than that of the notorious FCFS (0.4871).

However, the same SIC technique may be used to modify the FCFS protocol itself. It is known that the similar modifications of the FCFS protocols with single memory location have the MST of 0.6048 and 0.6173, respectively. Despite the fact that the proposed protocols have lower MST, we notice that the practical implementation of the FCFS-based approaches is complicated due to the theoretically infinite precision of the system timer. Therefore, our proposals demonstrate a good balance between implementation complexity and guaranteed performance.

The practical performance of all the protocols that the SIC property is degraded by the imperfect interference cancellation. We extend our system model to account for the cancellation errors and introduce two probabilities. A meaningless signal is obtained after the cancellation of the successful signal with probability q and after the cancellation of the collision signal with probability q'. We notice that the discussed SICTA- and FCFS-based protocols that have the nonzero MST in the infinite population model suffer from a deadlock in the presence of cancellation errors. To improve the robustness of the proposed solutions, we modify SICTA/SCC to make it resistant to cancellation errors for the cost of its MST. The resulting modification (R-SICTA/SCC) has the MST of 0.5147 when $q' = q = 0$, which is still higher than that of the FCFS protocol. In addition, this protocol demonstrates satisfactory MST performance even for the high values of q and q' (see Fig. 6) and even in its worst-case behavior has the guaranteed MTA performance (0.375).

References

1. Bertsekas, D., Gallager, R.: Data Networks. Prentice-Hall, Englewood Cliffs (1992)
2. Rom, R., Sidi, M.: Multiple Access Protocols: Performance and Analysis. Springer, Heidelberg (1990)

3. Tsybakov, B.S., Mikhailov, V.A.: Free synchronous packet access in a broadcast channel with feedback. Problems of Information Transmission 14(4), 32–59 (1978)

4. Capetanakis, J.I.: Tree algorithms for packet broadcast channels. IEEE Transactions on Information Theory 25(4), 505–515 (1979)

5. Andrews, J., Hasan, A.: Analysis of cancellation error for successive interference cancellation with imperfect channel estimation. Technical report, EE-381K: Multiuser Wireless Communications (2002)

6. Yu, Y., Giannakis, G.B.: Giannakis. Sicta: A 0.693 contention tree algorithm using successive interference cancellation. Proceedings IEEE INFOCOM 3, 1908–1916 (2005)

7. Peeters, G.T., Van Houdt, B., Blondia, C.: A multiaccess tree algorithm with free access, interference cancellation and single signal memory requirements. Performance Evaluation 64(9-12), 1041–1052 (2007)

8. Van Houdt, B., Peeters, G.T.: Fcfs tree algorithms with interference cancellation and single signal memory requirements. In: International Workshop on Multiple Access Communications, vol. 1 (to be published, 2008)

9. Wang, X., Yu, Y., Giannakis, G.B.: A robust high-throughput tree algorithm using successive interference cancellation. In: IEEE Global Telecommunications Conference, vol. 6(28), pp. 5–10 (2005)

10. Wang, X., Yu, Y., Giannakis, G.B.: A deadlock-free high-throughput tree algorithm for random access over fading channels. In: 40th Annual Conference on Information Sciences and Systems, vol. 22, pp. 420–425 (2006)

11. Evseev, G.S., Turlikov, A.M.: A connection between characteristics of blocked stack algorithms for random multiple access system. Problems of Information Transmission 43(4), 279–345 (2007)

12. Massey, J.L.: Collision resolution algorithm and random access communications. Multiuser Communication Systems, G. Longo, CISM Course and Lecture Notes 265, 73–131 (1981)

13. Wang, X., Yu, Y., Giannakis, G.B.: Combining random backoff with a cross-layer tree algorithm for random access in IEEE 802.16. IEEE Wireless Communications and Networking Conference 2, 972–977 (2006)

14. Gyorfi, L., Gyori, S., Massey, J.L.: Principles of stability analysis for random accessing with feedback. NATO Security through Science Series: Information and Communication Security 10, 214–250 (2007)

An Improved OCDMA/OCDMA Overloading Scheme for Cellular DS-CDMA

Preetam Kumar and Saswat Chakrabarti

G S Sanyal School of Telecommunications
IIT Kharagpur, India
preetam@gssst.iitkgp.ernet.in, saswat@ece.iitkgp.ernet.in

Abstract. Overloading is a scheme to accommodate more users than the spreading factor N. This is a bandwidth efficient scheme to increase the number of users in a fixed bandwidth. One of the efficient schemes to overload a CDMA system is to use two sets of orthogonal signal waveforms (O/O). The first set is assigned to the N users and the second set is assigned to the additional users. An iterative multistage detection (IMSD) technique is used to cancel interference between the two sets of users. The interference cancellation receiver uses hard decisions (HDIC) or soft decisions (SDIC) to estimate the interference. In this paper, the BER performance of a new overloading scheme using scrambled orthogonal Gold code (OG/OG) sets is evaluated with SDIC receiver. When complex scrambling is not used, it is shown that OG/OG scheme provides 25% (16 extra users) channel overloading for synchronous DS-CDMA system in an AWGN channel, with an SNR degradation of about 0.35 dB as compared to single user bound at a BER of 10^{-5}. We have evaluated the overloading performance, when the two set are scrambled with set specific deterministic or random complex scrambling sequence. It is shown that the amount of overloading increases significantly from 25% to 63% (40 extra users) by using random complex scrambling for N=64. For deterministic (periodic) scrambling, the overloading percentage increases considerably to 78%. On a Rayleigh fading channel, an overloading of 100% is obtained at a BER of 5.10^{-4} with near single user user performance.

1 Introduction

The number of users supported in a DS-CDMA cellular system is typically less than spreading factor (N), and the system is said to be underloaded. Overloading is a technique to accommodate more number of users than the spreading factor N. This is an efficient scheme to increase users in a fixed bandwidth, which is of practical interest to mobile system operators. In fact this type of channel overloading is provisioned in the **3G** standard [1].

Among the approaches described in the literature, the most efficient ones use multiple sets of orthogonal codes [2]. The concept of overloading in a DS-CDMA system using two sets of orthogonal codes is explained with the help of Fig.1. For the first N users, the system allocates orthogonal codes drawn from the first set of N

S. Balandin et al. (Eds.): NEW2AN 2008, LNCS 5174, pp. 25–36, 2008.

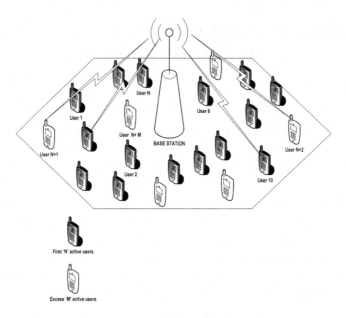

Fig. 1. Overloading concept in a single cell CDMA system

codes. When the number of intending users exceeds 'N', the excess users are accommodated in the system by providing suitable codes drawn from a second set of M codes. In this way, we are able to accommodate more number of users than the spreading length N (K>N), and the cell becomes overloaded.

The number of active users (K) in a conventional synchronous orthogonal CDMA environment is limited by the spreading factor N, which is WT where W is the transmission bandwidth and T is the duration of a symbol. When K exceeds N, the system becomes overloaded and the signatures are no longer orthogonal. This leads to multiple access interference (MAI). In an overloaded system, a conventional matched filter receiver is not optimal, due to the high level of MAI. Multiuser detection (MUD) is required in order to obtain a satisfactory performance of the users. Linear MUD's, such as the decorrelator, the minimum mean squared error detector or linear decision directed interference cancellation are devised to detect users in an underloaded system. The Maximum Likelihood (ML) detection is not an option because of its complexity that is exponential in the number of users. The nonlinear MUD's such as multistage parallel interference cancellation (PIC) and successive interference cancellation (SIC) [3], have good complexity- performance trade-off as compared to other MUD's. Hence these MUDs are suitable for overloaded systems.

Hence, orthogonal signatures (Walsh functions) are used in the downlink of IS-95 and UMTS mobile radio standards. Even in the uplink of UMTS, usage of orthogonal signatures has been advocated to realize multi-code channelization to increase the overall data rate. Also in the uplink, some systems that combine multicarrier modulation with CDMA can maintain orthogonality by inclusion of an appropriate cyclic prefix, and single-tap equalization [2].

It is interesting to note that several studies have been made in the recent past to understand, analyze and counter the detrimental effects of overloading. Almost all studies consider the uplink or reverse link and several studies suggest usage of appropriate multiuser detection (MUD) schemes at the base station   receiver. For example, a method of accommodating $K = N + M$ users in an N-dimensional signal space that does not compromise the minimum Euclidean distance of the orthogonal signaling has been presented in [4] for AWGN channel. A tree-like correlation coefficient structure of user signatures suitable for optimal multiuser detection has been proposed in [5]. In another approach, two sets of orthogonal codes which are orthogonal within the sets is introduced in [6]. In [6], the orthogonal sets are generated using Walsh-Hadamard (WH) codes, where the same WH code set is scrambled with set specific scrambling sequence (s-O/O). An iterative multistage detection technique has been proposed to cancel the interference between the two sets of user. In [7], it is shown that for uncoded BPSK modulated CDMA signal is shown that for uncoded BPSK modulated CDMA signal with N=64, an overloading of 11% can be achieved in an AWGN channel for s-O/O scheme. Another kind of receiver simplification is presented in [8], where   signals are divided into groups that are orthogonal to each other. A new overloading scheme using hybrid techniques has been proposed in [9], where the spreading codes and transmission modes are different for the two sets to increase the overloading performance. The attractive property of overloading scheme was the incentive to integrate a particular type of O/O, called quasi-synchronous sequences (QOS) [10], into cdma2000 standard [11].

To the best of our knowledge, the usage of orthogonal Gold codes has not been considered in any of the overloading schemes. In [12], a new method for generating different orthogonal sets of same length has been proposed. The new algorithm generates $(N-1)$ distinct, orthogonal sets of N sequences of length N. It has been shown that the peak value of crosscorrelation between different sets of same length is less than half the sequence length for $N \geq 32$. Such sequence sets would offer low intracell interference, when used in overloaded environment. Recently, the present authors have proposed a new overloading scheme using a set of Gold codes [13], which provides better performance than s-O/O scheme [7]. In this paper, we have evaluated the BER performance using orthogonal Gold code (OG/OG) sets with IMSD schemes. An efficient iterative multistage detection with soft decision interference cancellation is used to increase the amount of overloading.

This paper is organized as follows. In the next section, we describe the system model for the O/O overloading scheme. In section-3 we explain the IMSD operation and describe the process of iterative interference cancellation. Simulation results are presented and discussed in Section-4. Finally, we present the conclusion of this paper.

2 System Model

In the sequel we will consider the DS-CDMA system with processing gain N and the number of users K ($=M+N$). We assume that the channel is a nondispersive additive white Gaussian noise (AWGN) channel and that the different user signals are in perfect time synchronism. The signal $s_{u,k_u}(t)$ is the signature waveform of the k-th user in set-u, where $u \in \{1,2\}$, $k_1 \in \{1,2,3......N\}$ for set-1 and $k_2 \in \{1,2,3......M\}$ for set-2 users ($M \leq N$). Here N is number of users in set-1 and M number of users in set-2. The signature waveform may be expressed as:

$$s_{u,k_u}(t) = \sum_{j=1}^{N} s^{j}_{u,k_u} p_c(t - jT_c)$$ (2.1)

Here $s^{j}_{u,k_u} \in \{-1,1\}$, T_c is the chip duration and $p_c(t)$ is the real valued unit-energy rectangular chip pulse. All users signatures are normalized such that $\left\| s_{u,k_u}(t) \right\|^2 = 1$. We assume that all set-1 users are operational and hence N=Maximum number of users in set-1. Let us denote $\mathbf{S}_1$ and $\mathbf{S}_2$ as the signature matrices of the set-1 and set-2 users respectively. In this paper, we have considered two different orthogonal Gold code sets [12] for set-1 and set-2 users.

Let us denote $\mathbf{b}_1$ and $\mathbf{b}_2$ as the data matrices of the set-1 and set-2 users respectively. The data signal $b_{u,k_u}(t)$ of the k-th users in set-u, may be expressed as

$$b_{u,k_u}(t) = \sum_{l=-\infty}^{\infty} b^{l}_{u,k_u} p_{T_b}(t - lT_b)$$ (2.2)

where, data sequences $b^{l}_{u,k_u} \in \{-1,1\}$ are independent and identically distributed (i.i.d.) random variables taking values of +1 and -1 with equal probability. In (2.2), T_b is bit duration, N is the spreading factor and $p_{T_b}(t)$ is the rectangular pulse of the information data bits. Matrices $\mathbf{A}_1$ and $\mathbf{A}_2$ are diagonal matrices of received signal amplitudes for two sets of users.

The discrete-time matrix model of the received BPSK modulated CDMA signal after demodulating and chip-matched filtering is given as:

$$\mathbf{r} = \mathbf{r}_1 + \mathbf{r}_2$$
$$= \mathbf{b}_1\mathbf{A}_1\mathbf{S}_1 + \mathbf{b}_2\mathbf{A}_2\mathbf{S}_2 + \mathbf{n}$$ (2.3)

where

$$\mathbf{r}_1 = \mathbf{b}_1\mathbf{A}_1\mathbf{S}_1$$ (2.3)

$$\mathbf{r}_2 = \mathbf{b}_2\mathbf{A}_2\mathbf{S}_2.$$ (2.4)

The vector $\mathbf{n}$ is AWGN noise with zero mean, and variance equal to σ^2 .

When scrambling is used, the orthogonal Gold codes of both the sets are overlaid by a set-specific pseudo-noise (PN) sequence which is the same for all users within the set. In other words, we have $\mathbf{S}_1 = \dfrac{1}{\sqrt{N}}[\alpha_1\alpha_2........\alpha_N]$ and $\mathbf{S}_2 = \dfrac{1}{\sqrt{N}}[\beta_1\beta_2........\beta_M]$.

Let $\mathbf{P}_1 = (p_{11}, p_{22},.....p_{N1})^T$ and $\mathbf{P}_2 = (p_{12}, p_{22},.....p_{N2})^T$ designate the PN sequences overlaying the orthogonal Gold sequences in the two sets of users. In order to split the interference power evenly over the in-phase and quadrature components of the useful signal (irrespective of the carrier phase), we consider complex valued PN sequences: the chips p_{nu} takes their values from the set $C = \{\exp(j\pi/4), \exp(j3\pi/4), \exp(j5\pi/4), \exp(j7\pi/4)\}$. The scrambling sequence can be deterministic (periodic) or random. In periodic scrambling, the scrambling sequence randomly takes values form the set C and it is kept constant for all symbols. On the other hand, in random complex scrambling, it takes random complex values from the set C for each transmitted symbol.

In the next section, we explain iterative multistage interference cancellation receiver, which reduces the high level of interference due to overloading.

3 Iterative Multistage Detection

The received demodulated and chip sampled signal (2.3) is despreaded and we obtain soft outputs of the transmitted bits corrupted by multiple access interference (MAI) from other users and AWGN noise. In conventional matched filter detection, these outputs are fed to the decision device to make the hard decision of the transmitted information bits. In this work, iterative multistage detection (IMSD) technique is used to remove the MAI between two sets users. The basic principle of this receiver is to iteratively remove the estimated interference from each set due to the users of other set in multiple stages such that near single user performance is achieved. The interference power from set2-user (assuming that the useful signal power is normalized) is 1/N, and therefore the total interference power that affects set1-users is M/N. As long as M remains small compared to N, preliminary decisions can be made on the symbols transmitted by set1-users with some good reliability. But each of the set2-users gets an interference power of N (1/N) =1 from set1-users. Clearly the bit error (BER) performance will be poor for this set of users if detection is made prior to interference cancellation. As set1-users are detected with some good reliability, we can estimate the interference created from this set on set2-users. This estimated interference is removed from set2-uers before making the decision. Now in second iteration, interference from set2-users on set-1 are estimated form the first iteration outputs of set-1 and a more reliable set1 bits are obtained. This process continues till we get a near single user performance.

To explain the operation the following notations are used: $\hat{\mathbf{b}}_1^i$ and $\hat{\mathbf{b}}_2^i$ are decisions about set-1 & set-2 user data bits at i^{th} iteration respectively, $\mathbf{y}_1^i$ and $\mathbf{y}_2^i$ are set-1 and set-2 matched filter outputs at i^{th} iteration. At each iterative stage of the IMSD detector, the decision on the information bits are made according to the following expressions,

$$\hat{\mathbf{b}}_1^i = \phi\left(\mathbf{S}_1^T\left(\mathbf{r} - \mathbf{I}_2^{(i-1)}\right)\right) \tag{3.1}$$

$$\hat{\mathbf{b}}_2^i = \phi\left(\mathbf{S}_2^T\left(\mathbf{r} - \mathbf{I}_1^i\right)\right) \tag{3.2}$$

where,

$$\mathbf{I}_1^i = \hat{\mathbf{b}}_1^i \mathbf{A}_1 \mathbf{S}_1 \tag{3.3}$$

$$\mathbf{I}_2^i = \hat{\mathbf{b}}_2^i \mathbf{A}_2 \mathbf{S}_2 \tag{3.4}$$

are estimated Multiple Access Interference (MAI) on set-2 and set-1 users respectively.

In equations (3.1) and (3.2), $\phi(x)$ is the nonlinear decision function. For SDIC except for the last iteration, where we take hard decision, in other iterations several nonlinear decision functions can be used. We have used piecewise linear approximation of hyperbolic tangent and is defined as:

$$\phi(x) = \begin{cases} x/\theta & |x| < \theta \\ \operatorname{sgn}(x) & |x| \geq \theta \end{cases} \tag{3.5}$$

Here θ is selected to minimize the average BER.

4 Simulation Results

This section presents the Monte-Carlo simulation results of the proposed scheme with two IMSD techniques. The simulation has been carried out in MATLAB to evaluate the BER performance of the proposed scheme in an AWGN channel. Relevant simulation parameters are shown in Table 1. The value of the parameter θ is 0.5 for SDIC and it is fixed for all iterations. For all simulations, the system performance is evaluated by means of critical overload. We define the critical overload as the maximum achievable channel overload $\beta_{max} = (K_{max}-N)/N$ with interference cancellation, so that the SNR degradation for an average BER of 10^{-5} is less than 0.35 dB as compared to single user performance. It is a measure for the maximum acceptable channel overload, so that the system performance is degraded slightly as compared to the single user performance.

To increase the amount of overloading an efficient soft decision interference cancellation receiver is used as described in section 3. In Fig. 2, BER performance of this receiver at different overloading has been shown for N=64 at 28%, 25% and 22% overloadings. It is observed that 28% overloading cannot be achieved, with less than 1.0 dB SNR degradation at an average BER of 10^{-5}. If we reduce the overloading to 25%, the SNR degradation is about 0.35 dB as shown in Fig. 2 and we can ensure a BER of 10^{-5} for all users. So, we can obtain a *critical overload* of 25%, when the

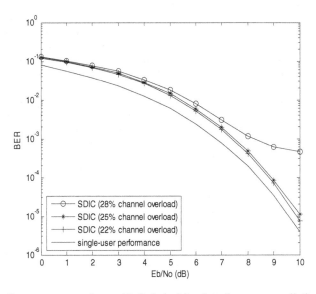

Fig. 2. BER performance comparison with Soft decision Interference cancellation (SDIC) with N = 64 at different values of overloading

Table 1. Description of some parameters relevant in Simulation

Parameters	*Specifications*
Transmission mode	Synchronous
Modulation/ Spreading	BPSK/BPSK
Spreading factor, N	64
Spreading codes	Orthogonal Gold codes
Power and phase of users	Equal
Type of Receiver	SDIC
Assumptions	Perfect chip, symbol and carrier synchronization

spreading factor N is 64. We have observed that the amount of critical load is only 19%, when the spreading length is reduced to 32.

The critical channel overload for s-O/O is 3% and 11% for N=32 and 64 respectively [7] for the same set of parameters. So there is a significant improvement in critical channel overload in OG/OG scheme as compared to s-O/O scheme.

In Fig. 3, the BER performance of OG/OG scheme with random complex scrambling is shown for N=32. Here, both the sets are scrambled by a set specific complex random scrambling sequence. We observe from the figure that, with complex scrambling the amount of overloading is 31%, with about 0.35 dB SNR degradation as compared to single user performance. In Fig. 4, overloading performance with periodic scrambling

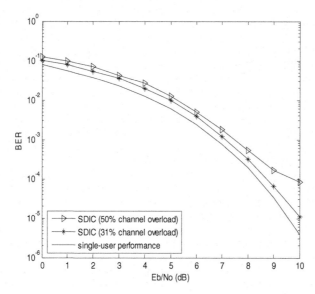

Fig. 3. BER performance comparison of OG/OG scheme with complex scrambling with SDIC recevier for N = 32

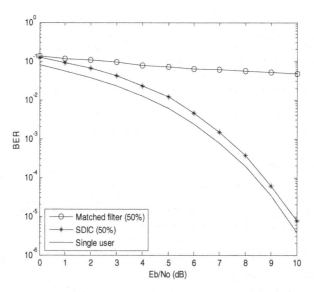

Fig. 4. BER performance comparison of OG/OG scheme with complex periodic scrambling with SDIC receiver for N = 32

is shown. It is interesting to observe that the overloading performance increases to 50% with less than 0.35 dB SNR degradation.

In Fig. 5, the BER performance with complex random scrambling is shown for N=64. Here we observe that we can support 40 extra users (63% channel overloading),

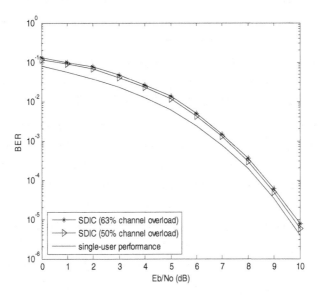

Fig. 5. BER performance comparison of OG/OG scheme with complex scrambling with Soft decision Interference cancellation (SDIC) for N = 64

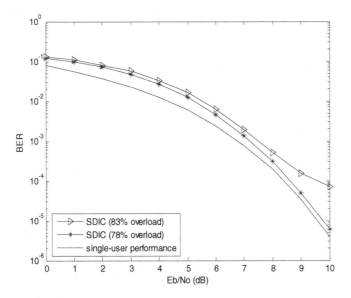

Fig. 6. BER performance comparison of OG/OG scheme with complex scrambling with Soft decision Interference cancellation (SDIC) for N = 64

with less than 0.35 dB SNR degradation as compared to single user bound. In Fig. 6, BER performance with periodic complex scrambling is shown. It is shown that with periodic scrambling, critical load increases to 78%. This is a significant amount of channel overloading, which can be obtained with complex scrambling. Hence, complex scrambling increases the amount of overloading significantly in an overloaded DS-CDMA system as compared to unscrambled OG/OG scheme [13].

5   Overloading Performance on a Rayleigh Fading Channel

We notice that the case of an AWGN channel is obtained by taking the received signal amplitude matrix, $A = I_k$. The Rayleigh fading channel model can be described by fading amplitudes generated according to $a_k = a_k^{(I)} + ja_k^{(Q)}$, where $a_k^{(I)}$ and $a_k^{(Q)}$ are independent zero-mean real Gaussian distributed random variables with variance $\sigma_{a_k^{(I)}}^2 = \sigma_{a_k^{(Q)}}^2 = 1/2$.

In order to compare the performance of these schemes, we define the critical overload as the maximum achievable channel overload $\beta_{max} = M_{max}/N = (K_{max}-N)/N$ with interference cancellation receiver, so that the SNR degradation as compared to a single user system at an average BER of 5.10^{-4} is less than 1 dB in an AWGN channel. It has to be emphasized that the receiver does not require any kind of user sorting to yield the desired overloading performance. As a consequence, this measure guarantees that the mean BER performance remains close to that of the ideal BER curve provided that $M < M_{max}$. It is worth noting that the BER performance in case of perfect interference cancellation is identical to the performance of a non-overloaded system where the users

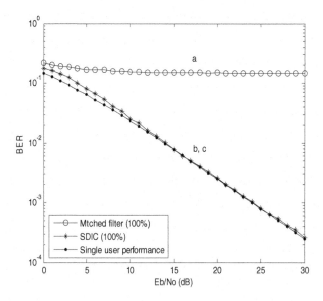

Fig. 7. BER performance of s-OG/OG scheme with complex scrambling over a Rayleigh fading channel for $N = 64$ for the following: a) Matched filter wiyh 100% overloading; b) SDIC with 100% overloading and c)Single user performance without overloading

are orthogonal, and also to the performance of a single-user system. The BER achieved by a single-user transmitting over a Rayleigh fading channel is given by

$$P_{e_b} = \frac{1}{2}\left(1 - \frac{1}{\sqrt{1 + N_0 / E_b}}\right) \tag{5.1}$$

Figure 7 shows the BER performance of scrambled OG/OG scheme, where a single set of orthogonal Gold code is used. The amount of overloading is fixed at 100% for N = 64. It is interesting to observe that the overloading increases considerably from 40% to 100% (64 extra users). The overloading for s-O/O [6] scheme is only 75% with complex scrambling with same set of simulation parameters. When we choose two different sets of orthogonal codes and complex scrambling, the achievable overloading is again 100%. This is a significant amount of overloading on a flat Rayleigh fading channel.

6 Conclusions

Efficient use of the available radio spectrum is an important requirement for future wireless communication. Overloading is an efficient scheme to increase the capacity of a DS-CDMA system. A new overloading scheme for DS-CDMA cellular system has been proposed in this work. This scheme is based on orthogonal Gold code sets. The BER performance of soft decision interference cancellation receiver has been evaluated through simulation. It is shown that this scheme with soft decision interference cancellation (SDIC) can overload the DS-CDMA systems by 25% at BER of 10^{-5} for N=64, with an SNR degradation of about 0.35 dB as compared to single user bound. The amount of overloading increases significantly from 25% to 78%, with periodic complex scrambling for set 1 and set 2 users for N = 64 with SDIC receiver. On a Rayleigh flat fading channel, we can obtain an overloading of 100% at a BER of 5.10^{-4} with complex scrambling and SDIC receiver.

References

1. Sari, H., Vanhaverbeke, F., Moeneclaey, M.: Multiple access using two sets of orthogonal signal waveforms. IEEE Commun. Lett 4, 4–6 (2000)
2. Kumar, P., Ramesh, M., Chakrabarti, S.: Overloading Cellular DS-CDMA: A Bandwidth Efficient Scheme for Capacity Enhancement. In: Rao, S., Chatterjee, M., Jayanti, P., Murthy, C.S.R., Saha, S.K. (eds.) ICDCN 2008. LNCS, vol. 4904, pp. 515–527. Springer, Heidelberg (2008)
3. Verdu, S.: Multi – user Detection. Cambridge University Press, Cambridge (1998)
4. Ross, J.A.F., Taylor, D.P.: Vector assignment scheme for M+N users in N-dimensional global additive channel. Electronics. Lett. 28 (1992)
5. Learned, R.E., Willisky, A.S., Boroson, D.M.: Low complexity joint detection for oversaturated multiple access communications. IEEE Trans. Signal Processing 45, 113–122 (1997)

6. Vanhaverbeke, F., Moeneclaey, M., Sari, H.: DS/CDMA with Two Sets of Orthogonal Sequences and Iterative – Detection. IEEE Commun. Lett. 4, 289–291 (2000)
7. Vanhaverbeke, F., Moenclaey, M.: Critical Load of Oversaturated Systems with Multistage Successive Interference Cancellation. IEEE VTC, 2663–2666 (2003)
8. Djonin, D., Bhargava, V.K.: New results on low complexity detectors for oversaturated CDMA systems. In: Proceedings of Globecom 2001, pp. 846–850 (2001)
9. Kumar, P., Chakrabarti, S.: A New Overloading Scheme for DS- CDMA System. In: National Conference on Communication, pp. 285–288. IIT Kanpur (2007)
10. Yang, K., Kim, Y.K., Kumar, P.V.: Quasi-orthogonal Sequences for Code-division Multiple-Access Systems. IEEE Trans. Inform. Theory 46, 982–993 (2000)
11. Physical Layer Standard for cdma2000 Spread Spectrum Systems, Realse B, TIA/EIA 3GPP2 C.S0002-B, January 16 (2001)
12. Donelen, H., O'Farrell, T.: Methods for generating sets of orthogonal sequences. Electronics Letters 35, 1537–1538 (1999)
13. Kumar, P., Chakrabarti, S.: A New Overloading Scheme for Cellular DS-CDMA using Orthogonal Gold Codes. In: Proceedings of IEEE Vehicular Technology Conference (VTC-Spring), Singapore, pp. 1042–1046 (2008)

Placement Algorithms for WiMAX Mesh Network

Salim Nahle and Naceur Malouch

Université Pierre et Marie Curie - Paris 6 - Laboratoire LIP6/CNRS,
104 avenue du Président Kennedy,
75016 Paris, France
{name.surname}@lip6.fr

Abstract. Recently standardized, WiMAX promises high data rates over long ranges. It defines two modes of operation Point-to-Multi-Point PMP and MESH. In the PMP mode, subscriber stations (SSs) connect to the base station (BS) in single-hop transmissions. The mesh mode on the other hand, allows direct communications between SSs. WiMAX mesh networks constitute a real solution for extending the coverage of the BS. They can be used for providing access into under-covered zones like rural areas and hard-to-wire areas. Several aspects affect the performance of the mesh, such as routing, scheduling and SS locations which when optimized, result in improved performance in term of capacity. In order to cover a specific area, we propose a placement algorithm that adequately places the SSs using routing and scheduling algorithms that we have previously proposed for the purpose of maximizing the capacity. Knowing that placement problems are NP-Hard, we design a heuristic with two variants and we show by simulations, that our algorithms, compared to the intuitive bottom-up approach, always find the smallest number of SSs, but also guarantee a required data rate.

Keywords: WiMAX, Mesh networks, Placement algorithms.

1 Introduction

IEEE 802.16 defines two modes of operation Point-to-Multi-Point PMP and MESH. In the PMP mode, subscriber stations (SSs) connect to the base station (BS) in single-hop transmissions. The mesh mode on the other hand, allows direct communications between SSs. They can transfer data between them and can be used to forward other SSs' traffic towards the BS. WiMAX mesh networks constitute a real solution for extending the coverage of the BS to ensure seamless and ubiquitous connectivity. They can be used for providing access into under-covered zones like rural areas and hard-to-wire areas. IEEE 802.16d [1] provides signalling control messages for both scheduling and mesh construction. Nevertheless, scheduling and routing algorithms have not been detailed and left open to the vendors. Two scheduling schemes were suggested: centralized and distributed. The former is used to control the access to network resources in a mesh tree that has been built using the *Network Entry* NENT and MSH-NCFG

S. Balandin et al. (Eds.): NEW2AN 2008, LNCS 5174, pp. 37–48, 2008.
© Springer-Verlag Berlin Heidelberg 2008

contol messages. In this case, each node gathers its children's requests and reports them along with its own in a MSH-CSCH:Request to its parent SS (called sponsoring node (SN) in WiMAX terminology). The BS which is the root of the mesh tree, gathers the requests, allocates resources locally and broadcasts the grants. Distributed scheduling however is performed in a distributed manner. Requests and grants are exchanged in the two-hop neighborhood in a collision-free manner using the control minislots of the MESH frame that are dedicated for distributed scheduling. In this work, we only consider the centralized scheduling.

We use WiMAX Mesh in the context of an on-board video surveillance system, for ensuring seamless and ubiquitous connectivity to trains, where it is assumed to have a BS connected to the backhaul wired network, and several SSs are used to extend its coverage all over the train trajectory (further details can be found on the BOSS project webpage [2]). Fig. 1 shows the scenario studied where a BS is deployed to provide connectivity but is unable to ensure coverage all along the train trajectory, thus we deploy for accomplishing full connectivity several SSs that form a mesh tree rooted at the BS. Centralized scheduling is adequate for this scenario, since the BS is given, and all the traffic passes by it. It is also better for ensuring QoS.

We identify 3 key issues that affect the performance of the mesh tree: SS locations, routing and scheduling. In a previous work, we have proposed an algorithm for the construction of WiMAX mesh trees which takes into account the burst profiles, and chooses high data rate paths [3]. A burst profile corresponds to coding and modulation techinques used on a link, and is carried in the mesh configuration messages. Later on, we proposed an improved version of this routing along with a scheduling algorithm that maximizes network utilization during resource allocation [4].

In this paper, we propose a placement algorithm that intelligently locates the SSs, and uses the proposed routing and scheduling algorithms. Knowing that the placement problem is NP-Hard, we propose a heuristic with two variants that reduce the number of SSs by maximizing the coverage of SSs being used at an iteration. The first one maximizes the area covered by the SSs chosen at each iteration. The second maximizes the distance on the train trajectory that is covered by the SS which is the closest to it. Futhermore, our algorithm ensures a specific data rate (expressed in Mbps) imposed by the video surveillance applications,

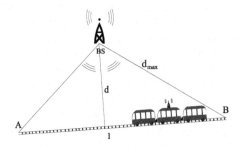

Fig. 1. Example scenario: Train trajectory and a located BS

which constitutes an additional input-constraint for the placement algorithm. It takes into account the different burst profiles defined in the standard. To the best of our knowledge, this is the first work that tackles SS placement in WiMAX mesh network, and takes burst profile into account for ensuring data rate. We show by simulations that our algorithm, compared to more intuitive placement algorithms, always finds the smallest number of SSs for covering the trajectory of a train, but also gaurantees the required data rate. Moreover, the algorithm is general enough so that it can be used to provide access in more complex areas. Here, the borders of the target area to cover will be considered as the target trajectory in our scenario.

The rest of the paper is organized as follows. In sections 2 and 3 we describe the used scheduling and routing algorithms respectively. We present our placement algorithm in section 4, and discuss its performance in section 5. Related work is discussed in section 6. Section 7 summarizes up the paper and highlights future work.

2   Fair Scheduling Algorithm

The scheduling algorithm must satisfy two objectives. First, it must gurantee the required data rate (r) by ensuring that the end-to-end data rate at each point on the trajectory is greater than r, $r_A \geq r$.

Second, the scheduling algorithm optimizes the utilization of the network resources by ensuring r with the minimum number of time slots. Assume α_{ij} is the proportion of the available minislot space dedicated for the wireless link between SS_i and SS_j. The values of α must be properly chosen in order to maximize the network utilization. Suppose as an example, that the train is at point A(x, y), on its railway which is 3 hops away from the BS, where each link supports a data rate r_i and $r_1 > r_2 > r_3$. It may be sufficient to allocate the same α for all the links to satisfy the first objective. Nevertheless this will result in resources wastage since the overall rate is bounded by the lowest rate r_3, and is $r/3$. Our scheduling ensures r by choosing the values of α for each route inversely proportional to the link rates. Hence, $\alpha_1 r_1 = \alpha_2 r_2 = \ldots = \alpha_n r_n = r$.

Note that $\alpha_1 + \alpha_2 + \ldots + \alpha_n$ must be less than *one* for every route, otherwise, there exists no feasible schedule that ensures r [4].

3   Routing

In IEEE 802.16 Mesh mode, Mesh Network Configuration (MSH-NCFG) and Mesh Network Entry (MSH-NENT) messages are used for advertisement of the mesh network and for helping new nodes to synchronize and join the mesh network. Further details on the parent choice in the mesh tree are not stated in the standard.

The routing algorithm used in this work is dedicated for the placement algorithm. We search to maintain r at each point of the trajectory by deploying a minimum number of SSs. Hence we do not seek the best-throughput route.

Routes are constructed in function of their distances(in kilometers). In a previous work, we showed that it is possible to increase network throughput by multiple hopping [5]. Several burst profiles were defined for the IEEE 802.16 physical layer. Each uses a modulation that offers a data rate for a certain range [6]. Hence, the inputs of the routing algorithm are first (r_i, R_i) of the used schemes where r_i is the data rate supported by a modulation scheme i and R_i is the maximum distance of a link so that i can be still used, and second the required rate r. For each SS, the routing algorithm searches among all the possible combinations, the route with the minimal number of hops that ensures r. Note that, the combinations are not that large even in worst case, since we only use six modulation schemes, hence six ranges, and the maximum number of hops allowed by the standard is seven, which is imposed by a 3-bits field in MSH-NCFG. The routing algorithm uses the scheduling defined in sec. 2 for determining the best scheduling obtained on a route. Among all the possible routes that ensures the required rate r, it selects the one with the minimum number of SSs and that with longest path. In other words, if 2 possible routes with n SSs exist, the first is $R_1, ..., R_n$, and the second is $R'_1, ..., R'_n$, it chooses the route with the longest aggregate distance $\sum_i^n R_i$, and hence we maximize the coverage of the selected SSs, since this routing algorithm will be used by the placement algorithm.

4   Placement Algorithm

As stated previously, finding the optimal locations of stations in a wireless network is NP-Hard. In this section, we propose an algorithm with two variants (hence 2 heuristic-algorithms) for minimizing the number of SSs needed to cover a certain trajectory while ensuring a data rate r all along this trajectory. The skeleton of the 2 algorithms is almost the same: We reduce the total number of SSs, by minimizing the number of SSs in each iteration. First, we consider the extremity points, each one alone, and we find the minimal number of SSs needed to ensure the required data rate (r). Second, we try to maximize the coverage of these SSs which is realized in two ways, resulting in 2 heuristics.

Next, we present two definitions for the purpose of algorithm description:

Definition 1. Root, of each iteration in the algorithm is one of the already placed SSs in the previous iterations or the BS. It is the closest SS (or the BS) to the farthest uncovered point.

Definition 2. Active zone, is the triangle formed upon joining the root with the two extremities of the remaining uncovered part of the railway. Initially, this zone is formed by connecting the BS to the two extremities of the path.

Our interest is in maximizing the coverage inside the active zone, since it contains all the uncovered path, and is rooted at the closest station to the farthest point. Hereafter, our algorithms will be called *Distant Point First*, or simply *DPF*, since at each iteration, we search the farthest uncovered point, and we choose the SS/BS which is closest to it as the root of the active zone, and we start by covering the farther extremity (Details of this procedure will follow next). The first algorithm referred to as *DPF1* maximizes the area of the polygon, formed by the extremity, the root and the SSs needed to cover this extremity. The second algorithm, referred to as *DPF2*, maximizes the coverage of the closest SS to the train trajectory.

4.1 Assumptions

The trajectory of the train as well as the location of the base station are supposed to be known which is a realistic assumption. We only consider the x and y coordinates, and have no constraints on SS possible locations. Nevertheless, the algorithm is intrinsicly extendible to take these constraints as inputs. We distinguish between different burst profiles by their data rates expressed in function of distance [6]. The BS and SSs operate on a single network channel.

4.2 Description of Distant Point First Algorithm

The main iteration of the algorithm is shown in fig. 3. We start each iteration by findig the root and the active zone. The root corresponds to the maximum uncovered distance, expressed as the following:

$$d_{max} = \max_{\forall A_i \in \rho_u} \left(\min_{\forall SS_j \in S} (d_{ij}, D_i) \right) \tag{1}$$

where A_i is a point on the trajectory of the train ρ, ρ_u is the uncovered part of this path. ρ_c is the part of the trajectory that has already been covered. Hence $\rho = \rho_u \cup \rho_c$. d_{ij} is the distance between A_i and a placed subscriber station SS_j. D_i is the distance between the base station and A_i. S is the set of already deployed SSs in previous iterations. The active zone is formed by connecting the root to the two extremities (Ext 1 and Ext 2 in the flow chart) of the uncovered part. Fig. 2 shows the deployment process for covering a train trajectory after the first main iteration. Different parameters of the algorithm appear in the figure, for instance D_3 is the distance between A_3 which is a point on the trajectory and the BS. ρ_c is the covered part after the first main iteration, and comprises two parts A_1A_2 and B_1B_2. ρ_u is the uncovered part after the first iteration A_2B_2.

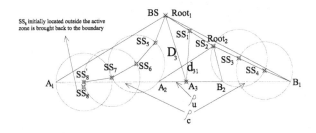

Fig. 2. Deployment example showing different parameters of the algorithm

The root of the first main iteration is the BS, and of the second one is SS_2, which is the closest SS to the farthest uncovered point A_3.

We proceed by selecting the extremity which is the farther from the root which is B_1 in the first main iteration and A_2 in the second. We apply the previously presented routing and scheduling to find the route and schedule that ensure the required rate which provide us the number of SSs needed and the distances between each two SSs. If the resulting rate is less than r, then we choose from S, another SS with lower number of hops and try to connect to it. In fact, in each iteration, due to multi-hopping, we try all the SSs in S that have fewer number of hops with the BS (this set of SSs is called S_{Root} in the chart), and if the number of SSs needed is inferior to that obtained from the root, we consider that SS for connecting the extremity instead of the root (This SS will be the temporary root for that extremity). S_{Root} in the second main iteration of fig. 2 contains, in addition to the root which is SS_2, SS_1 and SS_5. Note that the routing algorithm chooses, among the set of possible solutions (R_i, r_i) that satisfies r, the one with least number of SSs, and among the set of solutions

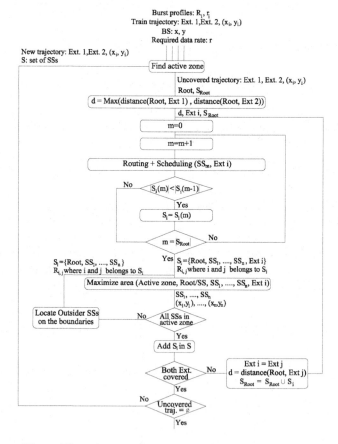

Fig. 3. Flow chart: Different steps of the algorithm

of the same number of SSs, we choose the one with longest aggregate distance $\sum_i^n R_i$ for maximizing coverage, this set is referred to as S_i.

Now, we use *DPF1* or *DPF2* to locate the SSs. The first maximizes the surface of the convex polygon formed by the root, S_i and the current extremity SS_i and if the number of SSs resulting from connecting to the root is more than the number of SSs resulting from connecting to one of the SSs or BS with smaller hop number, we add this temporary root to the polygon. Maximizing the area is not straightforward, hence we have developped an algorithm based on dynamic programming. We skip the details for space constraints.

The second maximizes the distance covered by the SS which is the closest to the railway by alligning the SSs with the root. This step replaces the *Maximize area* step in the chart. More precisely, the location of this SS is the intersection between the circle centered at the current extremity with the longest radius R_j, $j \in 1 \ldots n$, and the circle centered at the root with radius $(\sum_{i \neq j}^n R_i)$.

In both heuristic-algorithms, if one or more SSs were placed outside the active zone, we bring it back into the border to get full advantage of its coverage for maximizing the effective area, and we redeploy the other SSs $((n_i - 1)$ SSs). Then we consider the next extremity, and we add to the root and SSs with smaller number of hops, the recently deployed set of SSs (S_i), since an SS among them may be closer to the second extremity(expressed in the chart as $S_{Root} = S_{Root} \cup S_i$). This iteration is repeated until all the trajectory is covered.

4.3   Discussion and Other Approaches

Our algorithm starts by placing SSs from the extremity sides then moves down towards the trajectory (*side-down* approach). This allows to take into account the rate constraints since from the beginning of the placement algorithm, *end-to-end* rates can be evaluated. To illustrate this keypoint, we introduce intuitive approaches for placing SSs that start covering the railway from the bottom, and thus we call them *Bottom-Up* or simply *BU*. The first one, called *BU1*, computes the minimum number of SSs to cover the whole path, and then the SSs are moved up such as they are placed as far as possible from the trajectory towards the BS while keeping the trajectory covered. Next, SSs are placed on the intersections of the circles representing the transmission ranges until we hit the BS. Fig. 4(a) displays a deployment example using *BU1*, the range is fixed to $5km$. The second algorithm is a trivial one, that one may think of. It places enough number of SSs on the trajectory and connects the closest one to the BS through the minimum number of SSs as shown in fig. 4(b). It is worthy to note that these approaches can not intrinsically account for bandwidth constraints.

5   Evaluation

Hereafter, we study the performance of the placement algorithms. We start by presenting the feasible end-to-end data rate regions, then we compare the performance of DPF to BU Algorithms.

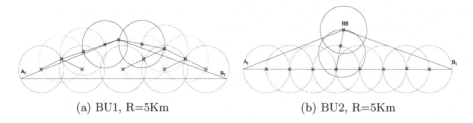

(a) BU1, R=5Km   (b) BU2, R=5Km

Fig. 4. Deployment example using BU algorithms

5.1 Study Region and Acheivable Rates

WiMAX mesh networks allow multi-hop communications, however, the number of hops being supported is limited to 7 (imposed by the 3-bits field in MSH-NCFG). Consequently the final range of the mesh network is limited to $7 \times R_{max}$ where R_{max} is the maximal range of a single-hop transmission. Figure 5, displays the feasible data rate region. It shows the possible normalized data rate in function of the BS distance from the trajectory d. The normalized data rate represents the supported end-to-end data rate for each distance, divided by the maximal possible datarate supported by the most robust modulation. For instance, according to [6], it corresponds to $64 - QAM$ 3/4 with 11 Mbps. Fig. 5(a) and fig. 5(b) correspond to 2 path lengths: $20km$ and $30km$ respectively. For each path, we change the location of the BS, by changing d and hence d_{max} (see fig. 1). The guaranteed end-to-end data rate is bounded by the rate of the farthest point from the BS (with d_{max}). The envelop corresponds to maximum normalized data rate of the farthest point (d_{max}) which is the bottleneck, calculated thanks to our routing and scheduling algorithms presented in sections 2 and 3. For instance if $d = 10$ km and the path length $l = 20$ km, then only 11% of the maximum WiMAX data rate can be achieved and hence can be requested. Beyond this value any placement algorithm will fail, thus in our next simulations we always choose topology and rate parameters in the feasible region.

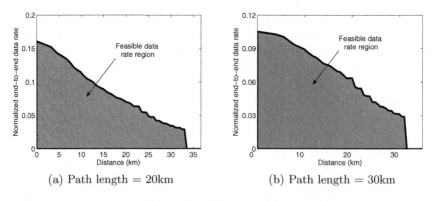

(a) Path length = 20km   (b) Path length = 30km

Fig. 5. Feasible data rate region

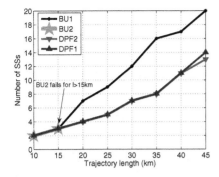 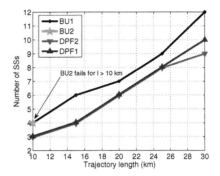

Fig. 6. Number of SSs needed (0.6 Mbps)   **Fig. 7.** Number of SSs needed (0.96 Mbps)

5.2   Performance of DPF Algorithms vs. BU Algorithms

Figures 6 and 7 displays the number of SSs needed to cover a trajectory in function of its length, for two values of r: $0.6Mbps$ and $0.96Mbps$ respectively. It is evident that this number increases as the length of the path increases. Nevertheless, it is clear that our algorithms $DPF1$ and $DPF2$ can find the smallest number of SSs, and have almost similar performance, except for the length $l = 45km$, where $DPF1$ results in fewer number of SSs than $DPF2$. $BU2$ fails to ensure r for $l > 15$ km and $l > 10$ km respectively.

In Figures 8 and 9, we compare the performance of $DPF1$ and $DPF2$ with $BU1$ and $BU2$ repectively. We find for each path length l the best ensured rate by $BU1$ and $BU2$ and the number of SSs to realize it, then we find for each length the number of SSs needed to guarantee the same rate using $DPF1$ and $DPF2$. This technique is used to force an inoffensive comparison against BU since otherwise BU fails to respect the constraint r in most cases. We run $BU1$ and $BU2$ for three fixed ranges R, $5km$ (high), $3.6km$ (medium) and $1.6km$ (small). In fact, the Bottom-Up approach can not account for link rates while placing nodes and the end-to-end rate is obtained only after the end of the placement algorithm. It is computed by dividing the rate that corresponds to the range R by the number of hops of the longest path. Thus we compare the performance of $DPF1$ and $DPF2$ to the best performance of $BU1$ and $BU2$. Note that the distance between the BS and the path d is fixed to $8.3km$ in these simulations.

Figures 8(a), 8(b), 8(c) show that $DPF1$ and $DPF2$ deploy in most of the cases lower number of SSs compared to $BU1$ despite the priority provided to the latter in this set of simulations. Moreover, $BU1$ can never ensure data rate for $l > 45km$ (fig. 8(b)) and $l > 10km$ (fig. 8(c)). For these longer paths, we plot the number of SSs obtained by $DPF1$ and $DPF2$ for having mimnimal coverage since no rate is imposed by $BU1$. Similarily, figures 9(a), 9(b), 9(c) show that $DPF1$ and $DPF2$ perform better than $BU2$ which can never ensure data rate for $l > 25km$ (fig. 9(b)) and $l > 10km$ (fig. 9(c)).

Fig. 10(a) and 10(b) show the guranteed end-to-end data rate in function of path length. In these simulations, we fix the distance between the BS and

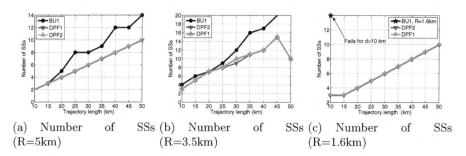

(a) Number of SSs (R=5km) (b) Number of SSs (R=3.5km) (c) Number of SSs (R=1.6km)

Fig. 8. BU1 vs. DPF1 and DPF2 in the case BU1 imposes its feasible rate

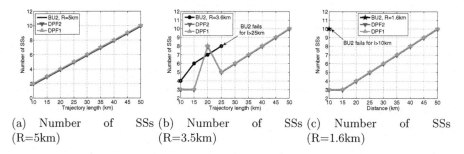

(a) Number of SSs (R=5km) (b) Number of SSs (R=3.5km) (c) Number of SSs (R=1.6km)

Fig. 9. BU2 vs. DPF1 and DPF2 in the case BU2 imposes its feasible rate

the trajectory d to $8.3km$ and we vary the length l. It is obvious that the rate achieved by our algorithms outperforms the rate obtained using *BU1* and *BU2*. *BU1*, with its best performance, can sometimes achieve comparable end-to-end data rate to our algorithms, but needs more SSs (especially for $R = 3.6km$).

As a conclusion, we have realized from figures 6 and 7 that *DPF* always find, compared to *BU*, the smallest number of SSs to cover a train trajectory while ensuring a data rate r. Moreover, *DPF* can always find smaller number of SSs even for ensuring data rates that are imposed by *BU1* and *BU2* as we have seen in figures 8 and 9. On the other hand, figure 10 illustrates that our algorithms can guarantee higher end-to-end data rates compared to BU1 and BU2, that may fail to gurantee full connectivity with rate r even for values of r and d chosen from the feasible data rate region.

6 Related Work

Placement algorithms have been given more attention in the context of cellular networks such as GSM or UMTS (for instance[7]). However, in these networks the base stations are supposed to be wired. In wireless mesh networks, the backbone is completely wireless with few wired Internet gatways. This renders the problem more challenging. On the other hand, several works treat the placement problem for wireless multihop networks [8], [9], [10], [11], [12] and [13].

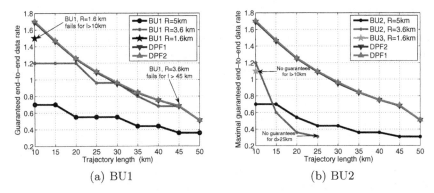

(a) BU1   (b) BU2

Fig. 10. DPF vs. BU: Guaranteed end-to-end data rate

Several algorithms were proposed in the context of wireless sensor networks [8], [9] and [10]. These algorithms are mainly based on virtual forces inspired from robotics. They require additional equipments to capture the additional sensing for virtual forces. In our work, non additional equipments are needed, nor software changes. Moreover, the short range and small capabilities (such as energy) of sensors significantly distinguish sensor networks from WiMAX mesh. Subscriber stations that are used do not pose these problems.

Srinivas *et al.* presented algorithms for constructing mobile network [13]. They rely on mobile backbone nodes to ensure the connectivity which is not the case in our study where SSs are supposed to be motionless. In addition, the same range is considered for all the backbone nodes, however, we use different ranges according to the burst profile on each link.

Robinson *et al.* studied depolyment factors for wireless mesh networks [12] and found that random networks are not suitable for large-scale mesh deployment. We carefully place the nodes so as to maximize their coverage. They also found that using multiple radios reduces network cost. In this work, we do not consider multiple radios, but we plan to explore them as a future work. Wang *et al.* proposed placement algorithms for mesh routers [11]. They assume fixed coverage and fixed capacity, however, the range and data rates change with distance and network state.

Nevertheless, to the best of our knowledge this is the first work that addresses SSs placement in WiMAX mesh networks. In our work, we rely on prior observations that multi-hopping may overperform single hops communications since more data rates are supported on shorter links ([3] and [5]). The deterministic nature of the channel access in WiMAX, helps reducing interference in the network, but also packet losses, which means that potential data rates can be used on different links. These rates are known from the burst profile information carried in the MSH-NCFG. Compared to Wi-Fi-based meshes, the data rates can not be reliable metric in those networks, since they strongly depend on the traffic and interference state of the network. For instance, in bursty conditions, and due to the random access to the channel, and the known hidden terminal problems, packet losses increase, which means that potential data rates can not be used.

7   Conclusion

We have proposed, throughout this paper a placement algorithm that intelligently finds SS locations, while using routing and scheduling algorithms that we have previously proposed for the purpose of maximizing the capacity. Knowing that the placement problem is NP-Hard, we have proposed two heuristics that maximizes the coverage area in each iteration. We have shown by simulations, that our algorithms always find the smallest number of SSs for covering certain zones when compared to intuitive bottom up approaches that naturally do not account for bandwidth constraint, but also guarantees a required data rate r which is chosen from the feasible rate regions that were presented in turn. Other algorithms often fail to ensure this r.

We are currently extending our algorithm, to account for frequency spatial reuse, multi-channels and multi-radios which can impove further the performance of WiMAX mesh networks. There, more realistic interference models shall be used, in addition to considering a set of candidate locations for the SSs that is provided by the operator.

References

1. IEEE 802 Standard Working Group: IEEE Standard for Local and Metropolitan Area Networks–Part 16: Air Interface for Fixed Broadband Wireless Access Systems. Standard 802.16d-2004, IEEE (2004)
2. BOSS project homepage (October 2006 - March 2009),
   http://www.celtic-boss.org/
3. Nahle, S., Iannone, L., Donnet, B., Malouch, N.: On the Construction of WiMAX Mesh Tree. IEEE Communication Letters 11(12) (2007)
4. Nahle, S., Malouch, N.: Joint Routing and Scheduling for Maximizing Fair Throughput in WiMAX Mesh Network. In: Proceedings of IEEE PIMRC (to appear, 2008)
5. Nahle, S., Iannone, L., Donnet, B., Friedman, T.: Investigating Depth-Fanout Trade-Off in WiMAX Mesh Networks. In: Proc. of 1st WEIRD Workshop (2007)
6. Betancur, L., Hincapié, R., Bustamante, R.: WiMAX channel: PHY model in network simulator 2. In: Proc. Workshop on ns-2: the IP network simulator (2006)
7. Amaldi, E., Capone, A., Malucelli, F.: Planning UMTS base station location: Optimization models with power control and algorithms. IEEE TRANSACTIONS ON WIRELESS COMMUNICATIONS 2(5) (2003)
8. Howard, A., Mataric, M.J., Sukhatme, G.S.: Mobile Sensor Network Deployment Using Potential Field: a distributed scalable solution to the area coverage problem. In: International Conf. on Distributed Autonomous Robotic Systems (2002)
9. Zou, Y., Chakrabarty, K.: Sensor deployment and target localization based on virtual forces. In: Proc. IEEE Infocom (2003)
10. Poduri, S., Sukhatme, G.S.: Constrained coverage for mobile sensor networks. In: Proc. IEEE ICRA (2004)
11. Wang, J., Xie, B., Cai, K., Agrawal, D.P.: Efficient Mesh Router Placement in Wireless Mesh Networks. In: Proc. IEEE MASS (2007)
12. Robinson, J., Knightly, E.: A Performance Study of Deployment Factors in Wireless Mesh Networks. In: Proc. IEEE Infocom (2007)
13. Srinivas, A., Zussman, G., Modiano, E.: Mobile Backbone Networks- Construction and Maintenance. In: Proc. ACM Mobihoc (2006)

On-Line Wireless Channel Modeling for Performance Control Purposes

Dmitri Moltchanov

Institute of Communication Engineering,
Tampere University of Technology,
P.O. Box 553, Tampere, Finland
moltchan@cs.tut.fi

Abstract. Performance control of applications running over wireless channels depends on ability to estimate statistical characteristics of wireless channels in real-time and represent them in terms of the model. Inherent time-varying dynamics of these characteristics pose a number of problems that have to be addressed before reliable estimates of the current channel state will be available. In this paper we consider the problem of on-line estimation of the state of the wireless channel based on bit error observations. The proposed approach is based on smoothing of the original error sequence which is shown to provide better description of the wireless channel for any given instant of time compared to conventional approaches. We also discuss applicability of the proposed approach in performance control of applications running over the air interface.

1 Introduction

Wireless channel characteristics are affected by many factors including atmospheric conditions, landscape, mobility of users, etc. In performance evaluation studies, wireless channels are conventionally modeled using covariance stationary processes. Nowadays, it is clear that stationary behavior rarely holds in practice. Since state of the wireless channel statistics may significantly change in time during an active session, different values of channel adaptation parameters are needed to provide best possible performance at any given instant of time. For wireless channels with dynamic adaptation of the protocol parameters to time-varying wireless channel conditions the performance control system has been proposed in [7]. Controllable parameters include the strength of FEC code, ARQ functionality, size of PDU at different layers, the rate with which traffic is generated. Authors demonstrated that for different wireless channel statistics different values of these channel adaptation mechanisms result in best possible performance. Note that in [7] authors assumed that the bit error process is locally stationary with abrupt changes.

While the approach originally proposed in [6,5] and adopted in [7] performs well when there are only abrupt changes in wireless channel statistics, performance of the system is expected to degrade when changes are gradual. This is mainly due to the biased estimation of statistical parameters of stationary

S. Balandin et al. (Eds.): NEW2AN 2008, LNCS 5174, pp. 49–60, 2008.

segments. Indeed, even local stationarity is restrictive assumption about wireless channel characteristics. The principal difference between modeling wireless channel characteristics using covariance stationary and local covariance stationary processes is in how many past observations are taken into account in estimating the current state of the wireless channel. The major shortcoming of both approaches is that authors assume that past behavior of wireless channel may sufficiently well describe its current behavior. When statistical characteristics vary in time, the current state may depend on few recent observations only. As a result, looking for accurate on-line estimator of the channel state we have to limit the effect of past behavior of the error process while still have enough observations to estimate statistics reliably.

In this paper assuming no a-priori information about stationarity of wireless channel statistics we propose a model to describe its behavior. Although the model is simple Markovian in nature, its parameters are allowed to change in time. To estimate the current state of the channel we propose to limit the number of past observations that are taken into account using either sliding window or exponential smoothing estimators. We compare performance of both approaches and demonstrate that any of these estimators perform better compared to conventional channel modeling using covariance stationary processes.

The rest of the paper is organized as follows. In Section 2 we briefly review related studies. Wireless channel statistics are considered in Section 3. The wireless channel model is also proposed there. Two approaches to estimate parameters of the model are proposed in Section 4. Numerical examples are shown in Section 5. Conclusions are drawn in Section 5.

2   Related Work

2.1   Propagation Models

To estimate performance of wireless channels, propagation models are often used. We distinguish between two types of propagation models. These are large-scale and small-scale propagation models. The former models focus on predicting the received local average signal strength (RLASS) when a mobile user moves away from the transmitter over large distances RLASS gradually decreases. Most large-scale propagation models assume that the RLASS decays as the power law function of distance between the transmitter and a receiver. The RLASS is usually computed by averaging the received signal strength over movements of 1 to 10 meters. Although it is known that during an active session a mobile user may move between areas with different RLASS neither outdoor nor indoor models take into account this behavior. They also do not take into account rapid fluctuations of the received signal strength.

Small-scale propagation models address one of abovementioned shortcomings of large-scale propagation models capturing fast fluctuations of the received signal strength over short time durations. Due to this, most performance evaluation studies of information transmission over wireless channels performed so far, were limited to small-scale propagation phenomenon. However, these models still fail

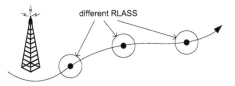

Fig. 1. A moving user experiencing different propagation characteristics

to predict the signal strength attenuation caused by movements over larger distances, e.g. between areas with different RLASS (see Fig. 1). Additionally, propagation characteristics experienced by a mobile node at a certain separation distance from the transmitter are always assumed to resemble characteristics of covariance stationary process. While this assumption is quite natural for stationary user, it does not hold when a user is on the move. As a result, stationary assumption about propagation characteristics does not hold in practice.

2.2 Bit Error Models

Conventionally, Markov models are used to capture characteristics of bit error process. In general, bit error models are divided into two categories. These are models, based on partitioning of the received signal strength and models, based on direct fitting of parameters. The former models are complex functions of small-scale propagation characteristics. The latter models implicitly takes into account stochastic properties of propagation environment at a certain separation distance from the transmitter. However, both kind of models assume that a user is stationary meaning that error rate remain constant in time. For comprehensive review of these models see [1].

Recently, it was demonstrated that covariance stationary property does not necessarily hold for wireless channel statistics. For example, authors in [4] found their GSM bit error traces to be non-stationary and proposed an algorithm to extract covariance stationary parts. They further used doubly-stochastic Markov process to model those parts separately. The modeling trace is finally obtained by concatenation. Among other conclusions, authors suggested that a given bit error trace can be divided into a number of concatenated covariance stationary traces. However, they did not provide any theoretical background for their segmentation procedure. Moreover, their approach is off-line in nature limiting application area of the model to off-line studies. In [6] authors observed local stationary properties for SNR statistics measured over IEEE 802.11b wireless channel under different mobility patterns. They applied exponentially-weighted moving average (EWMA) change-point statistical test to detect these segments and estimate their statistics in real-time. In [5] to isolate time instants at which mean of bit and frame error traces measured over IEEE 802.11b wireless channel change in time, similar approach has been adopted.

3   Wireless Channel Model

3.1   Wireless Channel Statistics

In this paper we use bit error observations of IEEE 802.11b wireless channel operating according to distributed coordination function (DCF) in DSSS mode at 11Mbps [2]. Setup of experiments and detailed statistical analysis can be found in [3]. Due to the limited space we consider only two bit error traces. We refer to these traces as Trace 1 and Trace 2. The whole number of bit error observations in both traces is $1E5$. According to setup of experiments this corresponds to approximately 25 transmitted frames where each frame was of 4096 bits.

To illustrate the time-varying nature of wireless channel characteristics we apply cumulative sum (CUSUM) statistics. Let $\{Y(t), n = 0, 1, \dots\}$ be the sequence of bit error observations. The following CUSUM statistics was firstly proposed by Page in [9]

$$C_Y(t) = \sum_{i=0}^{n} (Y(i) - \mu_Y), \tag{1}$$

where μ_Y is the mean of the process.

CUSUM statistics is plotted on chart. If during a period of time most of the values are greater that the mean of the whole trace the CUSUM statistics is increasing. Therefore, a segment of the CUSUM statistics with positive slope indicates a period where the values tend to be above the mean. Similarly, a segment with negative slope corresponds to a period of time where the values tend to be below the mean of the trace. A sudden change in direction of the CUSUM statistics indicates a sudden change in the mean. A period of time where CUSUM statistics remains relatively the same refers to a segment where the average did not change.

CUSUM statistics for considered traces are shown in Fig. 2. Note that the global bit error rate is 0.097 for trace 1 and 0.144 for trace 2. Firstly, we note that due to changing mean both traces are likely realizations of non-stationary processes. Indeed, bit error rate changes in time significantly for both traces. Moreover, note that although there are few abrupt changes in channel statistics, the majority of those are gradual.

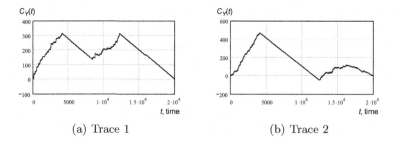

(a) Trace 1   (b) Trace 2

Fig. 2. Cumulative sum charts for bit error observations

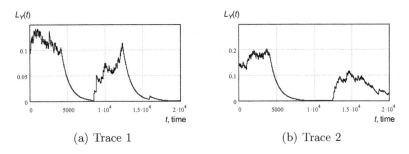

(a) Trace 1   (b) Trace 2

Fig. 3. EWMA charts for bit error observations

Absolute values of changing mean can be observed in Fig. 3, where exponentially smoothed statistics with weighting coefficient set to 0.001 is shown. Note that the bit error rate significantly varies. There are long duration of time during which errors do not occur. These time spans are identified by exponential decreases in Fig. 3. There are also long duration of time during which there are a plenty or bit errors observed at the air interface. As a result, performance experienced by applications significantly varies in response to time-varying characteristics of the bit error process. Performance control system aimed to provide best possible performance to applications at any given instant of time should appropriately react to both gradual and abrupt changes in bit error statistics. To do so, the current state of the wireless channel should be detected. It is important to note that stochastic behavior of other traces from [2] is similar.

3.2 Covariance Stationary Model

Assume that the error process is covariance stationary. In what follows we use the model based on direct fitting of parameters proposed in [8]. Let us denote the bit error process by $\{W(t), t = 0, 1, \dots\}$, $W(t) \in \{0, 1\}$. We model the bit error process using discrete-time Markov modulated process. When such process is allowed to have two states only, and at most single event is allowed in a slot, it reduces to switched Bernoulli process (SBP). To parameterize covariance stationary *binary* process, only mean and lag-1 autocorrelation coefficient have to be captured.[1] In [8] it was shown that there is SBP matching mean and lag-1 autocorrelation of covariance stationary bit error observations. It is given by

$$\begin{cases} \alpha = (1 - \gamma)\mu \\ \beta = (1 - \gamma)(1 - \mu) \end{cases} \quad \begin{cases} e_1 = 0 \\ e_2 = 1 \end{cases} , \quad (2)$$

where $f_{1,E}(1)$ and $f_{2,E}(1)$ are probabilities of error in states 1 and 2, respectively, α_E and β_E are transition probabilities from state 1 to state 2 and from state 2 to state 1, respectively, $K_E(1)$ is the lag-1 autocorrelation of bit error observations, $E[W_E]$ is the mean of bit error observations. The proposed model

---

[1] For covariance stationary binary process $\sigma^2 = f(\mu) = \mu(1 - \mu)$.

captures first- and second-order statistical characteristics in terms of error rate and autocorrelation function (ACF). Note that extension to the case of general finite-state Markov chain (FSMC, [10]) is straightforward. It is also important that the model in form of (2) was found to be appropriate when wireless channel characteristics are stationary in time.

4   Time Dependent Model

In what follows we assume that wireless channel statistics at any given instant of time is covariance stationary and can be well represented by the model given by (2). However, parameters of the model are allowed to change in time and depend on the current state of wireless channel statistics. In order to parameterize such model we have to estimate mean and and lag-1 autocorrelation coefficient such that only partial information about previous behavior of the error process is taken into account. The best possible way to represent the behavior of the error process is to somehow capture its current behavior. Not making any specific assumption about the stochastic process of interest it means that in estimating statistics of the process we have to take into account as few past observations as possible. However, in order to have reliable estimate of statistics we should have sufficient number of observations. As a result, there is a trade-off between the accuracy of the channel estimation and accuracy of the estimator itself. When the underlying process is non-stationary and experiences changes in its statistical parameters there is no guarantee that these estimators will be unbiased irrespective of the number of measurements taken into account. In order to estimate statistical characteristics we propose to use low pass filters based on sliding window and exponentially weighted estimators.

4.1   Sliding Window Estimator

The simple on-line estimator of the current values of channel statistics can be obtained using sliding window (SW) approach. Let $\{Y(n), n = 0, 1, \ldots\}$, $Y(n) \in \{0, 1\}$ be realization of the bit error process. The estimated value of the mean of the bit error process at time t, estimated with N observations back and denoted by $\hat{\mu}_{SW}(t, N)$, is given by

$$\hat{\mu}_{sw}(t, N) = \frac{1}{N} \sum_{j=t-N}^{t} Y(j), \qquad t = N, N+1, \ldots, \tag{3}$$

where parameter N is some constant determining the length of SW.

Observing (3) it is clear that $\hat{\mu}_{SW}(t, N)$, $t = N, N+1, \ldots$ are just sample mean with sliding window moving to the right direction each time a new observation becomes available. Therefore, for covariance stationary process

$$\hat{\mu}_{sw}(t, N) \stackrel{N \to \infty}{=} E[Y] = \frac{1}{N} \sum_{i=0}^{\infty} Y(i) \tag{4}$$

which is consistent, unbiased and effective point estimator of the mean.

The sliding window smoothing is usually applied to estimate average of time series as in (3). However, we apply this smoothing procedure to variance and lag-1 autocorrelation which are required for the model as follows

$$\hat{\sigma}_{sw}^2(t, N) = \frac{\sum_{j=t-N}^{t}(Y(j) - \hat{\mu}_{sw}(t, N))^2}{N - 1},$$

$$\hat{c}_{sw}(t, N) = \frac{\sum_{j=t-N-1}^{t-1}(Y(j) - \hat{\mu}_{sw}(t, N))(Y(j + 1) - \hat{\mu}_{sw}(t, N))}{(N - 1)\hat{\sigma}_{sw}^2(t, N)}, \qquad (5)$$

where to ensure that estimators are unbiased for small values of N we use $1/(N-1)$ multiplier instead of $1/N$. To get lag-1 autocorrelation autocorrelation coefficient $\hat{c}_{sw}(t, N)$ should be divided by $\hat{\sigma}_{sw}^2(t, N)$. Similarly to (3) estimators (5) converge to consistent estimators of $\{Y(n), n = 0, 1, \dots\}$ when $N \to \infty$.

Note that sliding window statistics applied to original observations allows to get a view of the trends existing in the data set. To parameterize sliding window estimators we have to determine the value of past observations that have to be taken into account. If the value of N is relatively small, e.g. less than 10, the sliding window statistics may significantly fluctuates. When the value of N is relatively big, e.g. greater than 1000 the history of the process may destructively affect its current state. Indeed, all the past observations including the most recent value in sliding window statistics are taken into account with the same weight reducing reactive properties of the model. Indeed, in those cases when state of the channel changes abruptly recent observations are much more important compared to previous behavior of the process. Since N last observations are taken into account in computation of the current value memory of sliding window statistics is limited at lag N.

4.2   Exponential Smoothing

The straightforward approach to assign more weight to recent observations is to decrease weights as the function of the distance between observation points. Exponentially or geometrically decreasing function is natural choice as it combines simplicity of analysis and the desired effect of weighting. For some stochastic observations of interest $\{Y(t), t = 0, 1, \dots\}$ exponentially smoothed (ES) statistics is given by

$$L_Y(t) = \gamma Y(t) + (1 - \gamma)L_Y(t - 1), \qquad (6)$$

where parameter $\gamma \in (0, 1)$ is some constant.

In (6) $L_Y(t)$ extends its memory not only to the previous value but weights values of previous observations according to constant coefficient γ. In (6) this previous information is completely included in $L_A(n - 1)$. To show it, let us rewrite $\{L_Y(t), t = 0, 1, \dots\}$ statistics recursively, starting from $L_Y(0) = Y(0)$

$$L_Y(0) = Y(0),$$
$$L_Y(1) = \gamma Y(1) + (1 - \gamma)Y(0),$$
$$L_Y(2) = \gamma Y(2) + \gamma(1 - \gamma)Y(1) + (1 - \gamma)^2 Y(0),$$
$$\dots \tag{7}$$

Since for any constant t the following holds

$$\gamma \sum_{i=0}^{t-1}(1 - \gamma)^i + (1 - \gamma)^t = 1, \tag{8}$$

it is easy to see that (7) converges to (6).

Let now $\{Y(t), t = 0, 1, \dots\}$, $Y(t) \in \{0, 1\}$ be realization of the bit error process. When exponential smoothing is applied to original observations the resulting statistics is called exponentially weighted moving average (EWMA). The value of EWMA statistic at the time t, is an estimator of the mean of $\{Y(t), t = 0, 1, \dots\}$, denoted by $L_\mu(t)$ and given by

$$\hat{\mu}_{es}(t, \gamma) = \gamma Y(t) + (1 - \gamma)L_Y(t - 1). \tag{9}$$

In fact, (9) defines a new stochastic process $\{\hat{\mu}_{es}(t, \gamma), t = 0, 1, \dots\}$, $\hat{\mu}_{es}(t, \gamma) \in [0, 1]$, as a function of initial observations and this process has different statistical characteristics compared to those of $\{Y(n), n = 0, 1, \}$. Given that $\hat{\mu}_{es}(0, \gamma) = E[Y]$ and $\{Y(t), t = 0, 1, \dots\}$ are observations of covariance stationary process, the mean of the process, $\{\hat{\mu}_{es}(t, \gamma), t = 0, 1, \dots\}$ is given by

$$E[\hat{\mu}_{es}(t, \gamma)] = E[Y](1 - (1 - \gamma)^t) + (1 - \gamma)^t E[Y], \tag{10}$$

that converges to $E[\hat{\mu}_{es}(t, \gamma)] = E[Y]$ as $n \to \infty$ meaning that (9) is unbiased estimator of the mean if the underlying process if covariance stationary.

The variance of $\{\hat{\mu}_{es}(t, \gamma), t = 0, 1, \dots\}$ is given by

$$\sigma^2[\hat{\mu}_{es}(t, \gamma)] = \sigma^2[Y] \left(\frac{\gamma}{2 - \gamma}\right)(1 - (1 - \gamma)^{2n}). \tag{11}$$

It is easy to see that when $n \to \infty$ (11) is approximated as

$$\lim_{n \to \infty} \left(\frac{\gamma}{2 - \gamma}\right)(1 - (1 - \gamma)^{2n} = \frac{\gamma}{2 - \gamma}. \tag{12}$$

Note that exponential smoothed statistics is not effective estimator of the mean since there is always stochastic factor biasing the value of $\sigma^2[\hat{\mu}_{es}(t, \gamma)]$ and we haver have $\sigma^2[\hat{\mu}_{es}(t, \gamma)] \overset{n \to \infty}{=} 0$. However, since observations may be realization of non-stationary process, we do not have any reason to require it.

To provide estimates of the lag-1 autocorrelation coefficient let us now define two stochastic process $\{\sigma^2(t), t = 0, 1, \dots\}$ and $\{c(t), t = 0, 1, \dots\}$ on top of initial observations as follows

$$\sigma^2(t) = (Y(t) - \hat{\mu}_{es}(t, \gamma))^2,$$
$$c(t) = \frac{\sum_{j=t-N-1}^{t-1}(Y(j) - \hat{\mu}_{es}(t, \gamma))(Y(j + 1) - \hat{\mu}_{es}(t, \gamma))}{(N - 1)\sigma^2(t)}. \tag{13}$$

Although exponential smoothing is often applied to initial observations leading to well-know EWMA statistics it can also be used to smooth other statistics. Using the following formulas we apply exponential smoothing to provide estimates of variance and lag-1 autocorrelation coefficient

$$\hat{\sigma}_{es}^2(t, \gamma) = \gamma \hat{\sigma}_{es}^2(t, \gamma) + (1 - \gamma)\hat{\sigma}_{es}^2(t - 1, \gamma),$$
$$\hat{c}_{es}(t, \gamma) = \gamma \hat{c}_{es}(t, \gamma) + (1 - \gamma)\hat{c}_{es}(t - 1, \gamma). \tag{14}$$

The reason to use exponentially smoothed statistics is as follows. Although, according to (9) and (14), the most recent value always receives more weight in computation of statistics, the choice of γ determines the effect of previous observations of the process on the current value. When $\gamma \to 1$ all weight is placed on the current observation, and exponentially smoothed statistics degenerate to initial observations. Contrarily, when $\gamma \to 0$ the current observation gets only a little weight. In this case most weight is assigned to previous observations. Note that large values of γ decrease reactive properties of the statistics. However, it also smoothes observations such that inherent stochastic variability does not affect the statistics significantly. The first value of exponential smoothed statistics is usually set to the expected value of observations or, if unknown, to its estimate. As a result, for on-line real-time implementation there should always be a certain warm-up period involving estimation of the expected value.

5 Numerical Examples

In this section we compare performance of the proposed channel estimation techniques with that of classical approach where the whole trace is considered to be covariance stationary and two-state Markov chain is used to model it. We firstly compare prediction properties of these approaches using the number of channel symbols that are incorrectly predicted and then turn our attention to prediction capabilities in terms of the frame delay and throughput.

Smoothed traces using MA and EWMA statistics for different N and γ are shown in Fig. 4. When N and γ are small both MA and EWMA statistics try to resemble original bit error observations. When N and γ increase reactive properties of both statistics get worse. As a result, we may expect that prediction performance worsens as N and γ increase. Indeed, we demonstrated that bit error statistics is characterized by time-dependent behavior. Thus, when N and γ increase more 'old' observations are taken into account in estimation of the current state of the wireless channel.

To provide comparison between considered models in terms of the number of correctly predicted symbols we used all three models to generate exactly the same number of observations as initial trace has. Then, we estimated deviation from the original trace using $\sum_{i=0}^{N-1}(Y(t) - X(t))^2$, where N is the number of observations, $\{Y(t), t = 0, 1, \ldots, N\}$ are original observations, $\{X(t), t = 0, 1, \ldots, N\}$ are observations generated by the model. Note that the proposed comparison gives the number of channel symbols that are incorrectly received. Results as

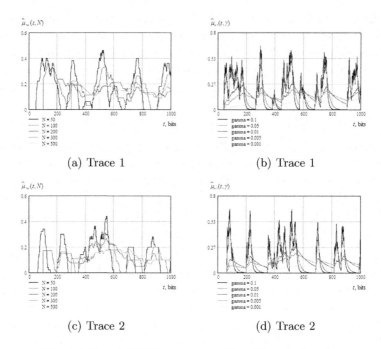

Fig. 4. MA and EWMA estimators for different N and γ

shown in Table 1. Both proposed models with any choice of N and γ outperform conventional approach. Also, as we expected, with increase of N and γ performance of both approaches degrades. Next, performance of EWMA statistics is better than that of MA statistics.

Consider how the proposed approaches can be used to optimize performance provided to applications running over wireless channels. Assume that we can choose different forward error correction codes (FEC) in response to changing channel conditions. We use $(255, 131, 18)$ and $(255, 87, 26)$ BCH codes. Performance parameters of interest are frame delay and throughput. Length of the frame is constant and equal to 255 bits. They arrive according to switched Poisson process $\{W_A(n), n = 0, 1, \dots\}$, $n = 255t$ with mean $E[W_A] = 0.5$, variance $\sigma^2[W_A] = 0.5$, and lag-1 autocorrelation $K_A(1) = 0.0$. There is also buffer that accommodates at most 40 frames. Automatic repeat request (ARQ) is enabled.

Table 1. Comparison between proposed approaches

Trace	MA			EWMA			Single model
Parameters	$N = 50$	$N = 200$	$N = 500$	$\gamma = 0.1$	$\gamma = 0.01$	$\gamma = 0.001$	–
Trace 1	1217	1223	1228	1016	1193	1255	1616
Trace 2	1755	1795	1814	1471	1709	1827	2447

We assume that a wireless channel is covariance stationary for 50% of time with bit error rate $E[W_E] = 0.08$ and lag-1 autocorrelation $K_E(1) = 0.0$. Then, parameters change to $E[W_E] = 0.02$, $K_E(1) = 0.0$. EWMA statistics with $\gamma = 0.1$ is used to estimate the current state of the wireless channel at the data-link layer operating with frame error statistics. Performance evaluation framework that we use here is described in detail in [7].

Results are shown in 2. To decrease delay experienced by frames and to increase throughput of the channel bit error rate $E[W_E] = 0.08$ requires $(255, 87, 26)$ FEC code while $E[W_E] = 0.02$ needs $(255, 131, 18)$ FEC code. The proposed EWMA statistics accurately estimated bit error statistics such that appropriate codes were chosen by the performance control system. Note that these results are even better than those obtained in [7] where the model specifically designed to detect abrupt changes in channel statistics was used. One can also see that fixed FEC codes lead to non-optimal performance of information transmission.

Table 2. Performance results: $E[W_A] = 0.5$, $\sigma^2[W_A] = 0.5$, $K_A(1) = 0.0$

Approach	$E[Q]$, slots	T, frames/s.
Optimal choice of the code	2.2172	61.7263
Fixed code (255,131,18)	13.6519	53.1967
Fixed code (255,87,26)	2.1281	43.4999

6 Conclusion

Estimating bit error statistics is important task in controlling performance of information transmission over wireless channels. However, inherent time-varying dynamics of wireless channel characteristics pose a number of problems that have to be addressed before reliable estimates of the current channel state will be available. In this paper we considered the problem of on-line estimation of the state of the wireless channel based on real-time observations of bit error pattern. Both proposed approach is based on smoothing of the original error sequence. Using numerical results we demonstrated that EWMA statistics outperforms MA statistics resulting in more accurate channel representation. Then, we adopted the proposed EWMA statistics as a channel estimator for performance control system proposed in [7]. Using frame delivery delay and throughput of the channel as performance metrics of interest we demonstrated that EWMA statistics provide reliable description of the wireless channel for any given instant of time.

References

1. Arauz, J., Krishnamurthy, P.: Discrete rayleigh fading channel modeling. Wireless Comm. and Mobile Comp. J., 413–425 (July 2004)
2. Khayam, S.: IEEE 802.11b error traces. Technical report, Michigan State University, (Accessed: 11.11 2004), http://www.egr.msu.edu/waves/people/Ali_files/

3. Khayam, S., Karande, S., Radha, H., Loguinov, D.: Performance analysis and modeling of errors and losses over 802.11b LANs for high-bitrate real-time multimedia. Signal Processing: Image Communication 18(7), 575–595 (2003)
4. Konrad, A., Zhao, B., Joseph, A., Ludwig, R.: Markov-based channel model algorithm for wireless networks. Wireless Networks 9(3), 189–199 (2003)
5. Moltchanov, D.: Monitoring the state of wireless channels in terms of the covariance stationary PDU error process. In: Proc. ICT2006, Funchal, Portugal (May 2006)
6. Moltchanov, D.: State description of wireless channels using change-point statistical tests. In: Braun, T., Carle, G., Fahmy, S., Koucheryavy, Y. (eds.) WWIC 2006. LNCS, vol. 3970. Springer, Heidelberg (2006)
7. Moltchanov, D.: Cross-layer performance control of wireless channels using active local profiles. Journal on Communications Software and Systems (JCOMMS) 3(3), 148–164 (2007)
8. Moltchanov, D., Koucheryavy, Y., Harju, J.: Simple, accurate and computationally efficient wireless channel modeling algorithm. In: Braun, T., Carle, G., Koucheryavy, Y., Tsaoussidis, V. (eds.) WWIC 2005. LNCS, vol. 3510. Springer, Heidelberg (2005)
9. Page, E.: A test for a change in a parameter occurring at an unknown point. Biometrika 42, 523–527 (1955)
10. Zhang, Q., Kassam, S.: Finite-state Markov model for Rayleigh fading channels. IEEE Trans. on Comm. 47(11), 1688–1692 (1999)

Multi-source Video Transmission with Optimum Perceptual Quality over Wireless Ad Hoc Networks

Pezhman Goudarzi and Mohammad Shahram Moin

Iran Telecom Research Center, Tehran, Iran
pgoudarzi@itrc.ac.ir

Abstract. Many optimization theoretic based rate allocation strategies have been developed for allocating some optimal rates to the competing users in wireless ad hoc networks. By considering different objective functions (such as congestion level, total packet loss and so on), the researchers propose some optimization framework by which the problem can be solved. Due to the rapid increase in the development of different video applications in such environments and the existence of difficulties in satisfying the pre-specified QoS limits, increasing the received video quality can be considered as an important and challenging issue. The quality of the received video stream is inversely proportional to the amount of distortion which is being imposed on the video stream by the network packet loss and the video encoder. The main objective of the current paper is to introduce an optimization framework in which by optimal rate allocation to some competing video sources, the aggregate distortion of the all of the sources be minimized. The simulation results verify the claims.

Keywords: QoS, PER, Distortion, Ad Hoc.

1   Introduction

optimization theory is an important tool for many rate allocation algorithms in wireline or wireless networks. Wireless ad hoc networks are computer networks in which the communication links are wireless. The network is ad hoc because each node is willing to forward data for other nodes, and so the determination of which nodes forward data is made dynamically based on the network connectivity. This is in contrast to wired network technologies in which some designated nodes, usually with custom hardware (variously known as routers, switches, hubs, and firewalls), perform the task of switching and forwarding the data. Ad hoc networks are also in contrast to managed wireless networks, in which a special node known as an access point manages communication among other nodes. Ad hoc networks can form a network without the aid of any pre-established infrastructure [1].

The requirements of a specific set of QoS parameters (delay, jitter, packet loss, etc) must be guaranteed for each real-time application. However, for most

S. Balandin et al. (Eds.): NEW2AN 2008, LNCS 5174, pp. 61–71, 2008.

real-time applications of wireless ad hoc networks, intrinsic time-varying topo-
logical changes provides challenging issues in guaranteeing these stringent QoS
requirements.

Due to dynamic nature of these networks, traditional routing protocols are
useless. So, special proactive/reactive multihop routing protocols such as DSDV
are developed. Some of these routing protocols introduce more than one feasible path for a source-destination pair. These category of routing algorithms are
called multipath routing algorithms [2]. Multipath routing scheme can reduce
interference, improve connectivity, and allow distant nodes to communicate efficiently [2]. In multipath routing, multiple multihop routes or paths are used to
send data to a given destination. This allows a higher spatial diversity gain and
throughput between source and destination nodes. On the other hand, it is obvious that inherent load balancing feature of the multipath routing algorithms has
the capability of reducing the congestion as well as increasing the throughput of
the user traffic in multi hop wireless ad hoc networks. Moreover, using multiple
paths between any source-destination pair can improve the important reliability
and availability features of the routing strategy.

Multipath routing can provide both diversity and multiplexing gain between
source and destination. However, multihop and multipath routing can also increase the total packet loss between the source and destination, especially if
there is congestion in the paths or if the bit error rate of the paths is high due to
the bad wireless link conditions (existence of high noise or interference levels).
Therefore, supporting multimedia data with stringent maximum loss requirement over multihop ad hoc networks with multipath routing can be considered
as an important and challenging research area.

Sending multimedia traffic over wireless ad hoc networks is a challenging issue
and many active research areas exist that all try to propose a solution to the
problem from different points of view.

In [3] a congestion-minimized stream routing approach is adopted. In [4] the
authors analyze the benefits of an optimal multipath routing strategy which seek
to minimize the congestion, on the video streaming, in a bandwidth limited ad
hoc wireless network. They also predict the performance in terms of rate and
distortion, using a model which captures the impact of quantization and packet
loss on the overall video quality.

Some researchers such as Agarwal [5], Adlakha [6] and Zhu [7] follow some
congestion-aware and delay-constrained rate allocation strategies. Agarwal et
al. in [5] introduce a mathematical constrained convex optimization framework
by which they can jointly perform both rate allocation and routing in a delay-
constrained wireless ad hoc environment. Adlakha et al. extend the conventional
layered resource allocation approaches by introducing a novel cross-layer optimization strategy in order to more efficiently perform the resource allocation
across the protocol stack and among multiple users. They showed that their
proposed method can support simultaneous multiple delay-critical application
sessions such as multiuser video streaming [6].

For multipath video streaming over ad hoc wireless networks, received video quality is influenced by both the encoder performance and the delayed packet arrivals due to limited bandwidth; Hence, Zhu et al. propose a rate allocation scheme to optimize the expected received video quality based on simple models of encoder rate-distortion performance and network rate-congestion trade-offs [7]. As the quality of wireless link varies, video transmission rate needs to be adapted accordingly.

In [4] setton et al. analyzed the benefits of optimal multipath routing on video streaming in a bandwidth limited ad hoc network. They show that in such environments the optimal routing solutions which seek to minimize the congestion, are attractive as they make use of the resources efficiently. For low latency video streaming, they propose to limit the number of routes to overcome the limitations of such solutions. To predict the performance in terms of rate and distortion, they develop a model which captures the impact of quantization and packet loss on the overall video quality.

In the current work, a similar approach such as [5] is being adopted by which a constrained optimization framework is introduced for optimal rate allocation to the real-time video applications. In [8], the authors do a similar optimization, but they take the average congestion of the overall network as the QoS criterion and minimize it to find the optimal solution for rate allocation on the available paths using simulations. In [9], the authors propose a distributed rate allocation algorithm which minimizes the total distortion of all video streams. Based on the subgradient method, their proposed scheme only requires link price updates at each relay node based on local observations and rate adaptations at each source node derived from rate-distortion (RD) models of the video. They show by simulation that their proposed scheme can achieve the same optimal rate allocation as that obtained from exhaustive search.

The presented work in this paper differs from that of [9] in that in our work, we have assumed that each video source may use multi-path routing for partitioning and transmission of the total video traffic. On the other hand, we have included the effect of the packet loss in the perceived video distortion. Our work differs from [5] and [8] in that we have used the total distortion as an objective Quality of Experience (QoE) measure in place of the QoS criterion used in [5] and [8]. In order to compute the total distortion, we have assumed that multiple video sources use the same wireless ad hoc medium for transmission and their associated distortions are additive [4]. On the other hand, the presented work differs from [5] in considering more than one (and possibly interfering) multipath-routed video sources which compete for the available bandwidth in a bandwidth limited wireless ad hoc network.

The paper's objective is to develop an optimal rate allocation framework bases on which, the overall distortion of all of the video sources is minimized. We also have used a penalty function approach for finding an iterative solution algorithm for the proposed constrained optimization problem such as those introduced in [10] and [11].

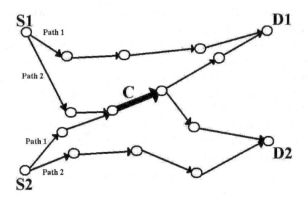

Fig. 1. Two competing multipath-routed video sources

The rest of the paper is organized as follows. In Section 2 the proposed optimization framework has been developed in detail. Section 3 is devoted to the numerical analysis and finally in Section 4 some concluding remarks are presented.

2   Proposed Optimization Framework

Consider the multihop wireless ad hoc network depicted in the Fig.1. Assume that there exist $\mathcal{N}$ video sources and the existing multipath routing protocol (e.g. DSDV), introduces n_k disjoint multihop paths between each source-destination pair (S_k, D_k) periodically $(1 \leq k \leq \mathcal{N})$. Each path is associated with a traffic flow and these multiplexed flows are aggregated in the destination node to produce the initial source-generated traffic stream. The number n_k is selected based on the assumption of availability of the current paths throughput information for the video source node S_k and the sufficiency of the aggregate estimated throughput for the traffic's minimum bandwidth requirements.

Each path j related to the source k contains $\mathcal{M}_{jk}$ wireless links from source to destination for $1 \leq k \leq \mathcal{N}$ and $1 \leq j \leq n_k$.

We assume a simple strong Line Of Sight (LOS) with BPSK signaling for node's wireless transmissions and also neglect the interfering effect of wireless transmissions between different paths [1]. We have used the Independent Basic Service Set (IBSS) setup (DCF mode) for implementing the MAC layer of the 802.11 WLAN standard which enforces the WLAN network in the ad hoc mode. It is assumed that BPSK DSSS is used in the physical layer. As the Bit Error Rate (BER) performance of the BPSK spread spectrum system in an AWGN environment is identical to that of conventional coherent BPSK system [12], it is sufficient to calculate the latter performance for evaluating the BER of the proposed system.

We also assume that the transmitted data is fragmented in equal length packets of length L bits enabled with FEC error correction capability up to M bits and this leads to the coding gain γ.

In the current paper, our objective is to minimize the total distortion associated with multiple video sources. Thus, a mathematical formulation must be presented that express the distortion of each video source in terms of its allocated rate. According to [13], this distortion is a function of the Packet Error Rate (PER) associated with each video source. In the following paragraphs, the packet error rate computation method is presented.

The Bit Error Rate (BER) of the link i in the j'th path of the k'th video source can be represented for a simple strong LOS propagation model with BPSK signaling as follows [1]:

$$b_{ijk} = Q\left(\frac{\eta_{ijk}}{\sqrt{r_{ijk}}}\right) \tag{1}$$

where:

$$\eta_{ijk} = h_{ijk} \cdot \sqrt{T_{ijk}} \qquad \forall k, j, i \in \mathcal{R}_{jk}$$

$$Q(y) \triangleq \frac{1}{\sqrt{2\pi}} \int_y^\infty e^{-x^2/2} dx$$

h_{ijk} is a physical constant, $\mathcal{R}_{jk}$ is the (nonempty) set of wireless links associated with the j'th flow of the k'th video source and T_{ijk} and r_{ijk} are the transmitted power and the total transmission rate associated with the i'th link in the j'th flow of the k'th source respectively. As it is said before, we assume that T_{ijk} is fixed during transmission and therefore do not depends on the transmission data rate r_{ijk}.

Assume that $\mathcal{R}_{jk}$ can be partitioned in two disjoint subsets. One subset is associated with those wireless links that are common between more than one video sources which we denote by $\mathcal{R}_{jk}^c$ (it is assumed that this subset is not empty) and the other set contains non-common wireless links which we denote by $\mathcal{R}_{jk}^{nc}$. So, we can write:

$$\mathcal{R}_{jk} = \mathcal{R}_{jk}^c \bigcup \mathcal{R}_{jk}^{nc} \qquad \forall j, k \tag{2}$$

We represent the set cardinality operator by $|.|$, so we have $|\mathcal{R}_{jk}| = \mathcal{M}_{jk}$. We also assume that $|\mathcal{R}_{jk}^{nc}| = \mathcal{O}_{jk}$ and thus we have $|\mathcal{R}_{jk}^c| = \mathcal{M}_{jk} - \mathcal{O}_{jk}$.

The r_{ijk} consists of two components: one is the traffic rate allocated to the j'th flow of the k'th source which is denoted by x_{jk} and another part is associated with the time-varying i'th link's cross (background) traffic a_{ijk}. Thus we have:

$$r_{ijk} = x_{jk} + a_{ijk} \qquad \forall k, j, i \in \mathcal{R}_{jk} \tag{3}$$

So, the available capacity (throughput) is denoted by e_{ijk} and is equal to $e_{ijk} = \mathcal{C}_{ijk} - a_{ijk}$. Where $\mathcal{C}_{ijk}$ is the capacity of the link i in the j'th path of the k'th video source.

In some cases (as is depicted in Fig.1), two or more multi-path video sources may compete for a common wireless link (in the Fig.1 this link is shown by bold line). Therefore, the available capacity of the common link must be shared between the competing flows in an optimal manner.

Assume that for each common link $i \in \mathcal{R}_{jk}^c$ there exists an associated set $\mathcal{S}_{jk}^i$ which represents the set of all ordered pairs $(path, source)$ that use the common link i in the path j of the source k (for example, in Fig.1, the path 1 of source 2 share the common link $\mathbf{C}$ with the path 2 of source 1). So the ingress and egress nodes associated with this common link are common between more than one flows.

For common links we assume that background traffic is composed only of those flows which are in $\mathcal{S}_{jk}^i$, i.e. we can write:

$$a_{ijk} = \sum_{(u,v) \in \mathcal{S}_{jk}^i} x_{uv} \qquad \forall k, j, i \in \mathcal{R}_{jk}^c \tag{4}$$

With the assumption of independent links' bit error rate, the total bit error rate along the j'th path of the k'th source can be calculated as follows:

$$\mathcal{B}_{jk} = 1 - \prod_{i=1}^{\mathcal{M}_{jk}} (1 - b_{ijk}) \qquad \forall j, k \tag{5}$$

The total PER of the j'th path of the k'th video source is composed of the congestion-related and Non congestion-related (wireless link) losses which we denote by $p_{j,k}^\mathcal{Q}$ and $p_{j,k}^\mathcal{R}$ respectively.

If the FEC induced error correction capability of a packet with length L bits is M bits ($M > 1$) and with the assumption of independent bit errors (lack of burst errors), the wireless link-related PER along the j'th path (flow) of the k'th source can be calculated as:

$$p_{jk}^\mathcal{R} = 1 - \sum_{m=0}^{M} \binom{L}{m} \mathcal{B}_{jk}^m (1 - \mathcal{B}_{jk})^{L-m} \qquad \forall j, k \tag{6}$$

Now we are in a position that must compute the congestion-related part of the PER.

First, assume that the end-to-end queueing delay of the j'th path of the k'th source can be represented with a random variable with the probability density function (pdf) $\beta_{jk}(t)$.

By adopting the same approach as in [5], it can be assume that congestion-related packet loss occurs when the end-to-end queuing delay of the j'th path of the k'th source exceeds a predetermined threshold Δ_{jk}. In mathematical terms the mentioned fact can be represented as follows:

$$p_{jk}^\mathcal{Q} = \int_{\Delta_{jk}}^{\infty} \beta_{jk}(t) dt \qquad \forall j, k \tag{7}$$

As in [5] simple M/M/1 queueing model and FIFO service discipline are adopted for the nodes. With the assumption of M/M/1 queueing model, the service time of each queue is an exponentially distributed random variable [14]. We also assume that these service times are independent. On the other hand, the end-to-end delay of each path j belonging to the video source k is equal to the sum of these independent random variables. Ignoring the source and destination nodes (hops), the total number of nodes in $\mathcal{R}_{jk}$, the number of non-common nodes in $\mathcal{R}_{jk}^{nc}$ and common nodes in $\mathcal{R}_{jk}^{c}$ would be $\mathcal{M}_{jk} - 1$, $\mathcal{O}_{jk} - 2$ and $\mathcal{M}_{jk} - \mathcal{O}_{jk} + 1$ respectively for each j, k.

Based on [15], for the nodes in $\mathcal{R}_{jk}$, the delay distribution (*pdf*) can be represented by *exponential* distribution as follows:

$$f_{ijk}(t) = \frac{e^{-t/\alpha_{ijk}}}{\alpha_{ijk}} \qquad \forall k, j, i \in \mathcal{R}_{jk} \tag{8}$$

where we can write as in [6]:

$$\alpha_{ijk} = \begin{cases} (e_{ijk} - x_{jk})^{-1} & \forall j, k, i \in \mathcal{R}_{jk}^{nc} \\ \left(\mathcal{C}_{ijk} - \sum_{(u,v) \in \mathcal{S}_{jk}^{i}} x_{uv} - x_{jk}\right)^{-1} & \forall k, j, i \in \mathcal{R}_{jk}^{c} \end{cases} \tag{9}$$

Thus the probabilistic distribution function of the end-to-end delay $(\beta_{jk}(t))$ is the convolution of all of these *pdf*'s [14].

The total PER related to the j'th flow of the k'th source can be simply shown that is equal to:

$$p_{jk} = 1 - (1 - p_{jk}^{\mathcal{R}})(1 - p_{jk}^{\mathcal{Q}}) \qquad \forall j, k \tag{10}$$

The total PER of the source-destination pair k with the assumption of independent path packet losses can be written as:

$$p_T^k = 1 - \prod_{j=1}^{n_k}(1 - p_{jk}) \qquad \forall k \tag{11}$$

According to [13], considering Mean Squared Error (MSE) criterion and assuming that the encoder and transmission distortions are uncorrelated, we can formulate the distortion of each video source k as follows:

$$\mathcal{D}_k = \mathcal{D}_{k0} + \frac{\theta_k}{\mathcal{R}_k + \mathcal{R}_{k0}} + \xi_k \cdot p_T^k \qquad \forall k \tag{12}$$

where:

$$\mathcal{R}_k = \sum_{j=1}^{n_k} x_{jk} \qquad \forall k \tag{13}$$

The empirical Rate-Distortion (R-D) model in [13] is used to represent the encoder distortion. $\mathcal{R}_k$ is the rate of encoded video and the parameters $\mathcal{D}_{k0}$, $\theta_k > 0$, and $\mathcal{R}_{k0}$ are calculated empirically from R-D curves. The scaling factor $\xi_k > 0$ depends on the encoding structure. In calculating (12), we have assumed a linear relationship between the total PER and the transmission distortion as mentioned in [13]. Based on the above facts, the formulation of the proposed total distortion minimization problem can be done as follows:

$$\min \mathcal{D}_T \triangleq \sum_{k=1}^{\mathcal{N}} \mathcal{D}_k \tag{14}$$

subject to:

$$\sum_{j=1}^{n_k} x_{jk} \geq x_{k,min} \qquad \forall k \tag{15}$$

$$0 \leq x_{jk} \leq \min_i (e_{ijk}) \qquad \forall j, k, i \in \mathcal{R}_{jk} \tag{16}$$

in which $x_{k,min}$ is the minimum required bandwidth for the k'th video source.

We must now remind our previous assumption that, the parameter n_k is assumed to be large enough such that the constraint (15) is met for all k.

We denote that the optimal solution vector of the system (14-16) as follows:

$$\chi^* \overset{\Delta}{=} \left(x_{11}^* x_{21}^* ... x_{n_1 1}^* ... x_{1\mathcal{N}}^* ... x_{n_\mathcal{N}\mathcal{N}}^* \right) \tag{17}$$

It is shown in [16] that under the following two assumption there exists a unique and optimal solution vector χ^* for the optimization problem (14-16).

$$0 \leq B_{jk} < \frac{1}{L} \qquad \forall j, k \tag{18}$$

$$x_{jk} < \min_i \left(e_{ijk} - \frac{2}{\Delta_{jk}} \right) \qquad \forall j, k, i \in \mathcal{R}_{jk} \tag{19}$$

Many iterative and optimal rate allocation algorithms have been proposed which lead to the optimal solution of constrained optimization problem (14-16) with the additional assumption in (19) [11]. From these algorithms we have selected the penalty function approach. A typical convex penalty function is convex with continuous positive second derivative.

It is shown in [11] that for solving the mentioned constrained optimization problem it is adequate to solve the following unconstrained one:

$$V(\chi) \overset{\Delta}{=} \mathcal{D}_T + \sum_k \int_0^{x_{kmin} - \sum_j x_{jk}} q(y) dy \tag{20}$$

The iterative gradient descent solution for solving the unconstrained problem (20) would be:

$$x_{jk}[n+1] = \left\{ x_{jk}[n] - \delta_{jk} \left. \frac{\partial V}{\partial x_{jk}} \right|_{x_{jk} = x_{jk}[n]} \right\}_{x_{jk}[n] \in \zeta} \qquad \forall j, k \tag{21}$$

where δ_{jk} is some positive and sufficiently small constant. Again, it must be stressed that the iterative algorithm in (21), allocates some rates to each path j of the video source k. By selecting the δ_{jk} parameters small enough, these allocated rates ultimately converge to the optimal solution vector χ^* [10].

3   Numerical Analysis

Consider a sample scenario which is depicted in the Fig.1. This scenario is consisted of two competing video sources $S1$ and $S2$ and each video source is routed through two disjoint path. Path 2 of the source 1 and path 1 of the source 2 are common in one wireless link. 16 nodes are randomly distributed in a $10^m \times 10^m$ area in this scenario. We have selected a simplified LOS propagation model for mobile nodes and the nodes mobility have been neglected by the assumption of a static network topology. The parameters η_{ij} and C_{ijk} are listed in the Tables 1 and 2 respectively (note that for example Path 21 in the Table 2 denotes path 2 of the source 1). $M = 2$, $N = 4$, $L = 1000$ and $x_{1,min} = x_{2,min}$ are selected to be 128^{Kbps}. Path 1 of the source 1 and path 2 of the source 2's cross traffics are selected to be CBR sources with rates 20^{Kbps} and 50^{Kbps} respectively and other links' cross traffics are being neglected. We assume that the path 1 and 2 of source 1 are consisted of 4 and 5 wireless links respectively. Also assume that the path 1 and 2 of source 2 are consisted of 5 and 4 wireless links respectively. Δ_{jk} parameter is assumed to be 5^{ms} for each j, k. δ_{jk} in iteration (21) is assumed to be 0.07.

In the Fig.2 the total distortion $(\mathcal{D}_T)$ of the proposed method is compared with an equal share scenario. In equal share scenario, equal rate allocation pattern exist for the two source i.e. path 1 of the source 1 and path 2 of the source 2 each achieve 103^{Kbps} and path 2 of source 1 and path 1 of source 2 each achieve the remaining $128^{Kbps} - 103^{Kbps} = 25^{Kbps}$. As the reader can verify, the total distortion of the proposed method is much less than that of the non-optimal equal share regime.

Table 1. Values of the parameters η_{ij}

η_{ij}	Link 1	Link 2	Link 3	Link 4	Link 5
Path 1	1	2	1.5	3	
Path 2	1	1	2	3	1.7
Path 3	4	2	1	1	2.1
Path 4	1.2	1.3	2	3.1	

Table 2. Values of the link capacities $(Kbps)$

C_{ijk}	Link 1	Link 2	Link 3	Link 4	Link 5
Path 11	120	110	120	120	
Path 21	60	60	50	60	60
Path 12	70	70	50	70	70
Path 22	110	110	100	110	

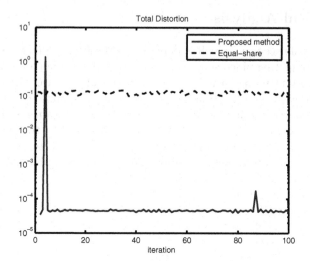

Fig. 2. Total Distortion comparison

4   Conclusions

In the current work, an optimization framework is introduced by which the rate allocation to each path of a multipath wireless ad hoc network can be performed in such a way that the total distortion of multiple video sources resulting from the network congestion and wireless environment can be minimized.

Main application of such algorithms is in rate allocation to those subsets of real-time traffics which require a minimum level of total distortion. As we have used a simple LOS propagation model for the mobile nodes and ignored the mobility, a more powerful algorithm which can support more general multipath fading propagation models and the mobility can be considered for future research.

References

1. Tonguz, O.K., Ferrari, G.: Ad Hoc Wireless Networks: A Communication-Theoretic Perspective. John Wiley, Chichester (2006)
2. Royer, E., Toh, C.-K.: A review of current routing protocols for ad hoc mobile wireless networks. IEEE Personal Communications 6, 46–55 (1999)
3. Setton, E., Girod, B.: Congestion-distortion optimized scheduling of video over a bottleneck link. In: IEEE 6th Workshop on Multimedia Signal Processing, 2004, pp. 179–182 (2004)
4. Setton, E., Zhu, X., Girod, B.: Congestion-optimized multi-path streaming of video over ad hoc wireless networks. In: IEEE International Conference on Multimedia and Expo, ICME 2004, vol. 3, pp. 1619–1622 (2004)
5. Agarwal, R., Goldsmith, A.: Joint rate allocation and routing for multi-hop wireless networks with delay-constrained data. Wireless Systems Lab, Stanford University, CA, USA, Tech. Rep (2004)

6. Adlakha, S., Zhu, X., Girod, B., Goldsmith, A.J.: Joint capacity, flow and rate allocation for multiuser video streaming over wireless ad-hoc networks. In: IEEE International Conference on Communications, ICC 2007, pp. 1747–1753 (2007)
7. Zhu, X., Han, S., Girod, B.: Congestion-aware rate allocation for multipath video streaming over ad hoc wireless networks. In: International Conference on Image Processing, ICIP 2004, vol. 4, pp. 2547–2550 (2004)
8. Setton, E., Zhu, X., Girod, B.: Congestion based multipath routing of multimedia data over ad hoc networks. In: IEEE International Conference on Multimedia and Expo, ICME 2004 (2004)
9. Zhu, X., Girod, B.: Distributed rate allocation for multi-stream video transmission over ad hoc networks. In: IEEE International Conference on Image Processing, ICIP 2005, vol. 2, pp. 157–160 (2005)
10. Kelly, F.P., Maulloo, A.K., Tan, D.K.H.: Rate control for communication networks: shadow prices, proportional fairness and stability. Journal of the Operational Research Society 49, 237–252 (1998)
11. Bertsekas, D.P., Tsitsiklis, J.N.: Parallel and Distributed Computation. Prentice Hall Inc., Old Tappan (1989)
12. Feher, K.: Wireless Digital Communication: Modulation and Spread spectrum Applications. Prentice Hall Inc., NJ (1995)
13. Stuhlmuller, K., Farber, N., Link, M., Girod, B.: Analysis of video transmission over lossy channels. IEEE Journal on Selected Areas in Communications 18, 1012–1032 (2000)
14. Papoulis, A., Pillai, S.U.: Probability, Random Variables and Stochastic Processes. McGraw Hill Inc., USA (2002)
15. Bertsekas, D., Gallager, R.: Data Networks. Prentice Hall Inc., USA (1987)
16. Goudarzi, P.: Minimum distortion video transmission over wireless ad hoc networks. In: 14th European Wireless Conference (2008)

Interference Aware Construction of Multi- and Convergecast Trees in Wireless Sensor Networks

Tomas Johansson, Evgeny Osipov, and Lenka Carr-Motyčkovà

Department of Computer Science and Electrical Engineering,
Luleå University of Technology
{Tomas.Johansson,Evgeny.Osipov,Lenka.Carr}@ltu.se

Abstract. In this paper we consider a problem of building a forwarding tree for multicast and convergecast traffic in short-range wireless sensor networks. Interference awareness and energy efficiency are the major design objectives for WSN protocols in order to maximize the network lifetime. The existing multicast algorithms aim at constructing low-energy cost trees. Adding interference-awareness, however, leads to increased throughput and further reduces the energy consumption by avoiding unnecessary retransmissions due to interference-induced packet losses. We propose a Localized Area-Spanning Tree (LAST) protocol for wireless short-range sensor networks. Unlike previous similar protocols, the LAST protocol reaches all the nodes in a given geographical area, rather than only specific individual nodes. When creating the tree, the protocol jointly optimizes the energy cost and the interference imposed by the structure.

Keywords: multicast, convergecast, wireless sensor networks, interference.

1   Introduction

Over the recent years wireless sensor networks (WSN) appeared as a unique networking environment with respect to routing amongst other aspects. In this paper we focus on constructing geographic based multicast and convergecast trees. In the context of wireless sensor networks, the converge cast traffic patterns are particularly important where data from a set of nodes at different locations should be transmitted to a single processing/control unit. The multicast communication is essential when the control unit sends a binary file with updated functionality to a group of nodes.

While multicast communications is a well researched area with multiple available solutions in the computer-oriented Internet, there are many challenging and yet unsolved problems in wireless sensor networks. Firstly, in comparison to address-centric computer networks it is difficult, and in many applications impossible, to assign hierarchically structured addresses to sensor nodes. Therefore, attribute-based and geographic based addressing is most common in WSNs. The second set of problems arises from specifics of wireless communications. The

S. Balandin et al. (Eds.): NEW2AN 2008, LNCS 5174, pp. 72–87, 2008.
© Springer-Verlag Berlin Heidelberg 2008

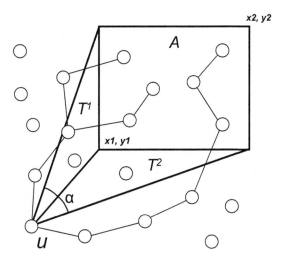

Fig. 1. The objective of LAST: Create an energy efficient multicast tree towards a geographic region. Energy cost is estimated as proportional to the area of the target area A in addition to the area of the two triangles T^1 and T^2.

major problem here is the broadcast radio transmission medium that introduces a phenomenon of cross-node interference when neighboring nodes transmit simultaneously. Most significantly, radio interference manifests itself in reduction of data flows throughput. In the context of energy constrained wireless sensor nodes, the interference dramatically reduces the network's lifetime[1] as well. The ultimate design objective for all WSN protocols, and multi- and convergecast in particular, is to minimize energy consumption in the network. This can be done either directly by aiming at keeping the communications related energy cost as low as possible, or indirectly by reducing the cross-link interference through topology control for minimizing the number of unnecessary retransmissions.

In this paper we introduce LAST, a Localized Area-Spanning Tree protocol. The main application of the algorithm is to request and collect data from all sensor nodes located in a specified area, as shown in Figure 1. When building the tree, the LAST protocol jointly optimizes the energy consumption and the induced interference of the tree. Overall, the protocol achieves better energy conservation in comparison to existing approaches. In short our protocol works as follows: each node is aware of its geographic position and the positions of its immediate neighbors. All nodes are also able to compute a distance based interference metric. When building the tree each node is given a task to cover a subregion of the target region by choosing one or several immediate neighbors that minimize: (a) a geographical distance towards the subregion; and (b) the energy consumption while going through a particular neighbor according to the interference metric.

---

[1] The lifetime of a network is defined as the time until the first node in the network runs out of energy.

Even though one of the objectives is to minimize the interference, creating a tree between the control unit and the target area rather than just a path would intuitively seem to increase the interference since the different branches would interfere with each other. However, the LAST algorithm takes both interference and energy cost into account when deciding whether or not to branch out at any given node. We define the angle α in Figure 1 to be u's *angle of visibility* of the target area A. If the angle of visibility is small, the algorithm is unlikely to introduce branches in the structure. On the other hand, if the angle is large, introducing branches would mean that the energy consumption of the tree would be reduced, while the branches would be largely separate from each other.

The remainder of the paper is structured as follows. Section 2 overviews the related work in the area of multicast topologies constrution algorithms. Section 3 present the network model and our assumptions, and in Section 4 we describe the LAST algorithm, including how interference and energy cost are measured and estimated. Section 5 presents a discussion as well as directions for our future work. We conclude the paper in Section 6.

2   Related Work

Multicast protocols can use both grid-based and tree-based structures to distribute the data from a sender to the receivers. In a tree-based structure, there is only one unique path between any sender-destination pair, while there may be several in a grid-based structure.

One example of an algorithm that creates a tree structure is GMP, [9] a distributed Geographic Multicast Protocol, where the protocol finds paths to destination nodes given their geographical coordinates. This is done by calculating virtual euclidian Steiner trees, which the actual paths are based on. The aim of the algorithm is to minimize the number of hops as well as the energy consumption.

GMR [7], the Geographic Multicast Routing protocol, is another example of tree-based multicast. The objective of this algorithm is to create a tree that reaches a set of nodes at certain given geographical coordinates. This is done by introducing the *cost over goodness* ratio, where the cost is the number of neighbors a node uses to forward the message, and goodness is the reduction of the total remaining distance to the destinations. The advantages of this algorithm is a low computation time as well as a lower number of transmissions than comparable algorithms.

In [5], the GMR protocol is extended into the multicast protocol HGMR (Hierarchical Geographic Multicast Routing). The network area is divided into cells, where one node in each cell is the designated access point for the cell. The hierarchical structure reduces the encoding overhead, while the protocol retains the forwarding efficiency of the GMR protocol.

MSTEAM, another multicast protocol with the goal to create a tree to a number of nodes with specified coordinates is presented in [2], where the first step is to create a minimum spanning tree over the target nodes. The next step is to create the actual multicast tree, and the decision when to branch is taken locally. Compared to similar protocols, the MSTEAM protocol performs well in simulations with regards to the energy efficiency.

Zeng et al. [10] introduce a grid multicast protocol, which unlike the previous algorithms is focused only on minimizing the energy cost, and not the hop count.

One common theme for the existing multicast protocols is that they focus on minimizing the energy consumption directly, but not indirectly through minimizing the interference of the resulting path structure.

In comparison to multicast, convergecast introduces the additional challenge of avoiding packet collisions when several sensors transmit data simultaneously. In [11], the authors introduce a block acknowledgment scheme in order to guarantee continuous packet forwarding.

3   Network Model and Assumptions

3.1   Model

A sensor network is modelled as an Euclidian graph $G = (V, E)$ with the vertices in V representing sensor nodes, and the edges in E representing communication links. The euclidian position (x, y) of the vertices in the graph corresponds to the physical position of the nodes in the euclidian two dimensional space, which means that the edge weight $w(u, v)$ represents the physical distance between nodes u and v. The energy cost to transmit a data unit from node u to its one-hop neighbor v is designated $c(u, v)$. Each node u has a maximum transmission range R_u.

In this paper, we discuss a nested structure of the target geographical region. The following notation is used: an area A can be divided into n subareas designated A_i^1, where $1 \leq i \leq n$. The subareas do not have to be disjoint, but the union of all subareas is identical to the original area A. The superscript indicates that we are at the first level below the original area A. Subarea A_i^1 can in its turn be divided further into m subsubareas A_{ij}^2, where $1 \leq j \leq m$. This process can be repeated any number of steps.

3.2   Assumptions

Throughout the paper, we make assumptions on the network as follows:

Bidirectional links. We assumed the sensor nodes to be equal. However, depending on the environment equal nodes in different locations will have different transmission ranges. This means that some of the links in the network will be unidirectional. In order to simplify the presentation of the algorithm, we will in this paper assume that the links in the network are bidirectional. However, the algorithm can easily be modified to handle unidirectional links as well.

Variable transmission power. We assume that all nodes can adjust their transmission power to any value from 0 to their maximum transmission power, depending on the desired transmission radius: when transmitting to node v, node u uses the lowest possible transmission power needed to reach v. A common path loss model says that the signal strength received by a node can be described as p/d^{α}, where p is the transmission power used by the sending node, d is the distance between two nodes, and α is an environment-specific path loss exponent.

Connectivity. We assume that the sensor network does not contain any nodes where packets can get stuck using multi-hop greedy geographic forwarding. In other words, for any node v in the network and any location l outside v's transmission range, v has at least one one-hop neighbor that is nearer to l than v is.

Topology knowledge. Each node knows only the positions of all its one-hop neighbors as well as the amount of signal loss when communicating with them. In reality, this information can be gathered by overhearing beacon transmissions between the neighboring nodes. We also assume that the nodes are global-topology agnostic. A detailed description of the information gathering is outside the scope of this paper.

Uniform distribution. In the beginning of the network lifetime, the nodes are assumed to be evenly distributed over the deployment area instead of clustered. This assumption holds until nodes start to run out of power.

4   Metrics and Algorithm

4.1   Metrics

Interference metrics. Unless the traffic patterns in a network is known in advance, the amount of interference can only be modelled on the properties of the network topology. Several different interference metrics have been proposed. In [1], the interference of a network is defined as the *maximum edge coverage* occuring in the network, where the *coverage ζ of an edge $e = (u,v)$* is defined as the number of nodes that are within distance $|u,v|$ to at least one of u and v, or more formally:

$$\zeta(e) = |\{w \in V, |v,w| \le |u,v| \cup |u,w| \le |u,v|\}| \tag{1}$$

and

$$MaxEdgeCoverage(G) = \max_{e \in E} \zeta(e)$$

Even though (1) implicitly assumes that the transmission area is circular for all nodes, this assumption is not necessary in order to compute the coverage metric. In the remainder of this paper, the coverage of a given link (u,v) is defined as the number of nodes that are affected by the transmission over that link, regardless of their distance to u and v.

In [8] the authors claim that defining interference as the maximum edge coverage in the network is problematic, since it is sender-centric, and since small changes in the network can drastically change the interference measure. In [3], we discuss another problem with the maximum edge coverage metric: it does not take the length of the paths in the graph into account. Removing edges with high interference, in order to reduce the interference of a graph, can cause the length of the paths between node pairs to grow indefinitely. Sending a message over a long path will then lead to a high total interference, even if each link in that path would result in low coverage. Based on these issues, we proposed to define the interference $TotI(P)$ of a path P to be the sum of the coverage of all edges in that graph.

$$TotI(P) = \sum_{e \in P} \zeta(e) \tag{2}$$

Unlike the previous metric, this metric reflects the total amount of interference when sending over a path. Removing edges with high coverage from the network is advantageous only if an alternative path with lower interference exists. The interference definition in this paper is similar to our previous definition: the total interference of a set of links is equal to the sum of the coverage of all links in the set.

Based on our assumptions on the topology knowledge, a node u can compute which neghbors will be affected when it communicates with a one-hop neighbor v. In other words, node u can compute the edge coverage of any link (u, v), and the interference $\zeta(u, v)$ can be computed locally.

Energy cost metrics. Since a node has no global topology knowledge, the energy cost cannot be computed directly. Based on our assumption on uniform distribution of nodes in the deployment field, we estimate the energy cost of a tree spanning a region as proportional to the area of the region, as illustrated in Figure 1.

Thus, the energy cost EC used in LAST to cover an area A from a given location is a function of the area of the A together with the area of the two triangles, as seen in the figure.

$$EC \propto A_{AREA} + T^1_{AREA} + T^2_{AREA}. \tag{3}$$

When comparing energy costs for covering different areas, the size of the areas can therefore be compared directly.

4.2   Localized Area-Spanning Tree

In this section, we present the operations of the LAST algorithm. The pseudocode in Listing 1 presents a high-level implementation of the algorithm. The algorithm is distributed; the tree is constructed starting from u, which selects one or several of its direct neighbors as child nodes in the tree. The child nodes in their turn choose their child nodes and so on.

Node u does not choose child nodes from all its one-hop neighbors, but only from those that are located closer to A than u itself. These neighbors form u's *candidate set* S_{CN}, as seen in Figure 2. The goal is to choose a subset of nodes from the candidate set to create a tree structure that jointly optimizes the interference and the energy cost. This is done by computing the interference- and energy metrics for each possible selection from the candidate set.

Listing 1. An overview of the LAST algorithm

TREECONSTRUCTION(A, u)
1 **if** this is the first received request
2   **then if** u is inside A
3   **then** BROADCASTREQUEST(A, u)
4   **else** $childNodes \leftarrow$ FINDCHILDREN(A, u)
5   **for each** $node$ **in** $childNodes$
6   **do** CONNECT$(u, node)$
7   TREECONSTRUCTION$(A_{sub}, node)$

FINDCHILDREN(A, u)
1 **for each** $S \in S_{CN}, |S| \leq 2$
2 **do for each** possible division of A into subareas
3   **do** calculate $EC(S)$
4   calculate $IC(S)$
5 **return** S with minimum $IC(S) + c * EC(S)$

RECEIVEREQUEST(A, u)
1 $v \leftarrow currentNode$
2 **if** v is inside A
3   **then if** this is the first received request
4   **then** CONNECT(u, v)
5   BROADCASTREQUEST(A, v)

Selecting child nodes. Assume that we, given the node u and the area A, have a subset $S \in S_{CN}$. Each node $v \in S$ is assigned to create a tree that covers the region A_v, which is a subarea of A. If the subset only contains one node, $A_v = A$. The interference cost of S is defined to be the total coverage of the edges from u to all nodes in S, as stated in (1):

$$IC = \sum_{v \in S} \zeta(u, v) \qquad (4)$$

The energy cost is the estimated total cost for each node v in S to cover its assigned subarea A_v, in addition to the energy cost to connect u with all nodes in S:

$$EC = \sum_{v \in S} c(u, v) + EC_v(A_v) \qquad (5)$$

The total cost of a given subset based on the amount of interference and the estimated energy cost is

$$IC + \beta * EC \qquad (6)$$

where the parameter β decides the relative importance of the interference and energy cost factors. The subset with the lowest cost is chosen by u, and the tree construction request is forwarded to those nodes.

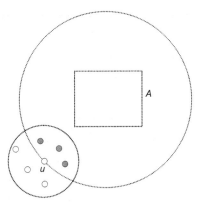

Fig. 2. The one-hop neighbors of u that are shaded are closer to the center point of A than u itself. They are therefore part of u's candidate set.

The chosen subset is the one that combines low interference with low energy cost. In some cases, these objectives may be contradictory: the best option to reduce the interference might be to have paths that take detours between the sender and the destination. However, this will lead to energy costs that are higher than if the energy-optimal path had been used. Depending on the scenario, the "value" of a decrease in energy cost in terms of interference will be different, and the factor c can be changed to reflect that.

Branching decision. If a subset with only one node is selected, that node is assigned to cover the entire area A. It is also possible for u to select several nodes simultaneously from S_{CN}, and assign each of them to cover a subsection of A, so that all the nodes together are responsible for covering the entire area A. As described in more detail in the next section, A can be divided either horizontally or vertically, and the subareas are always of equal size. This is illustrated in Figure 3. Each node in the subset becomes the starting point of a branch, and is assigned to cover one such subarea.

Eventually, the LAST process will reach node v that is inside the particular subarea that is to be covered. (An example of this can be seen in the figure, where v is inside A_1^1.) In that case, the algorithm will switch to the *broadcast mode*: the node v forms a connection to all its one-hop neighbors inside the area by broadcasting a request, which is forwarded by all the one-hop neighbors to their neighbors. If a node is reached by requests from several neighbors, it forms a connection to the node from which the first request is received. This is repeated until the tree covers the entire subarea.

Large candidate sets means that node u has to evaluate a large number of potential subsets. In order to reduce the computational load on individual nodes, u only considers subsets containing at most two nodes. In other words, the tree created by the LAST algorithm can branch into at most two branches at any given node. While this means that larger subsets will not be considered at all,

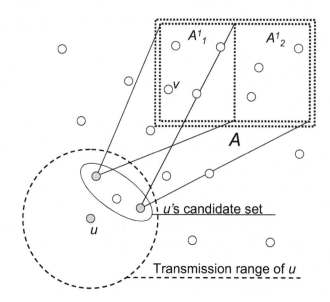

Fig. 3. Node u may use several nodes in its candidate set. Each of them will cover a subarea of region A.

the high number of branches would likely lead to a large amount of interference, which would make them unsuitable for the purpose of our algorithm.

For a candidate set S_{CN}, the number of unique non-empty subsets of a size between 1 and $|S_{CN}|$ is equal to $2^{|S_{CN}|} - 1$, which is no more than $2^{\Delta} - 1$, where Δ is the maximum degree of the network graph. However, the LAST algorithm is limited to no more than 2 branches at the same time, and correspondingly to subsets S such that $|S| \leq 2$. This means that the number of subsets considered by the algorithm is $O(\Delta^2)$.

Using brute-force calculation, we would have to compute the energy cost estimate and the edge coverage for all nodes in a particular subset in order to compute the energy cost and interference for that particular subset. Assuming a constant computation cost for calculating the estimated energy cost and the edge coverage for one node, the computation cost for one subset is constant. Thus, the total computation complexity for one node is $O(\Delta^2)$.

However, using this approach some of the calculations will be redundant. If node u is assigned to cover the region A, the two subareas that A will be divided into will be the same for any subset of S_{CN} of size 2. This means that for each node $v \in S_{CN}$, it is only necessary to estimate the energy cost to cover A and, at most, four different subareas of A (two for when A is divided horizontally and two more for when A is divided vertically). Since there will never be more than five different subareas to consider for each node, the total computation complexity for u to calculate the energy cost for all possible subsets will be $O(\Delta)$.

Division of target area. Depending on when and how the target area is divided, the topology and properties of the resulting tree will be completely different. This section discusses different aspects of choosing divisions that are more likely to result in trees with better performance.

The division of A can be done either horizontally or vertically, as is shown in Figure 4. Since different divisions will result in different trees, both the horizontal and the vertical divisions must be considered in order to evaluate all possible trees.

Since the areas and subareas will always be in form of rectangles, the communication overhead to describe their shape will be constant.

Packet losses during construction of the tree can lead to large parts of the tree never being constructed, which will result in only partial coverage of the target area. In order to reduce the effect of packet losses, the LAST algorithm creates subareas that overlap each other, as shown in Figure 5.

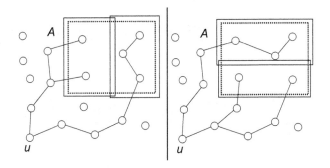

Fig. 4. In this case, node u has chosen a subset consisting of 2 candidate nodes, which means that A will be divided into two subareas. The resulting tree structures will look different, depending on whether A is divided horisontally or vertically.

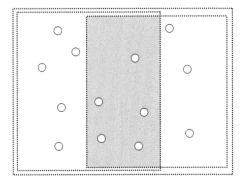

Fig. 5. The nodes in the gray area are covered by both the subareas. Thus, if the construction of a subtree covering one of the subareas fails, they will still be covered by the other subtree.

Minimum size of regions. As the tree constructed by the LAST algorithm branches, the regions to be covered by an individual branch become smaller. If the subareas are in the shape of thin rectangles, there is a high risk that the tree covering a subarea cannot connect all the nodes in the area without using nodes located outside of the area. This means that the energy cost of the tree will be higher, which is not desirable. A goal of the algorithm is therefore that neither the width nor the height of the subareas should be too small: an area A is not allowed to be divided so that the height or width of the resulting subarea falls below a given threshold.

No matter what the minimum size of a region is defined to be, it is not possible to guarantee that there exists a path between any two nodes in the region, if the path is required to be completely contained within the region. Figure 6 shows an example of this. Each of the nodes has a one-hop neighbor that is closer to the center node than they themselves are, so there are no stuck nodes. Still, the spiral topology can be of any size, which means that no region of any fixed size can be guaranteed to contain such a topology.

Target points. It is not certain that the best way, with respect to the energy cost, to construct a tree is to always aim for the center points of the target areas. Therefore, the LAST algorithm makes it possible to specify target points located anywhere in or on the edge of the areas.

When the LAST algorithm reaches a branching point, and the area to be covered is divided into two, the target points will be located on opposite corners along the split side that is facing the branching point. In the left part of Figure 4, the target points would be located at the two bottom left corners of the area A.

The method to estimate the energy cost described in 4.1 will generally result in trees that targets the center point c of A. The reason for this is that for a node

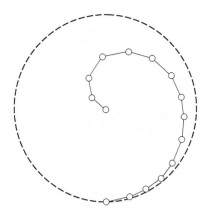

Fig. 6. Even though the network is assumed to have no stuck nodes, there is no limit on how much a path between two nodes might divert from the straight line between the nodes. This means that independent of the size and shape of a given area A, we cannot guarantee full connectivity within A using only the nodes inside A.

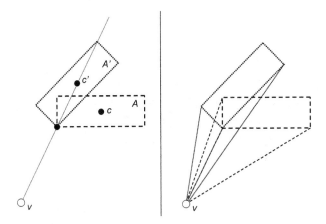

Fig. 7. When node v is assigned to enter the area A through point p_1, A is rotated and transformed into A'. (Note that the line through u and p_1 also goes through the transformed center point c'.) This results in a smaller triangle area and accordingly a lower estimation of the energy cost.

u, choosing the nodes that are closest to c as child nodes will typically result in a lower estimated energy cost compared to choosing other nodes. In order to model that the tree is supposed to enter A near to one of its corners, the estimation method has to be modified: The tree should target the specific corner instead of the center point of A, and subsequently cover A starting from that corner. The estimation method that accomplishes this works similarly to the previous described method, with the only difference that the area A is repositioned before the circle sector is constructed.

When entering an area A in a specific point p_1, the following computations are performed: the area A is rotated into A' so that a line that goes through v and p_1 also crosses the center point c' of A'. This can be seen at the left part of Figure 7. The actual location of A is indicated by a dashed rectangle, and the rotated position by a dotted rectangle. Note that the modified (dotted) circle sector covers a smaller area than the original (dashed) circle sector. This means that the estimation method indicates that a tree structure that enters A near p_1 will require less energy than a tree structure that would aim for A in general.

5   Discussion and Future Work

5.1   Degree of Overlap

In the case of packet losses, the construction of a branch might be interrupted, and a portion of the area A may not be covered. In order to remedy this, the subareas can overlap each other in order to provide some redundancy, as in Figure 5. This can lead to nodes being covered by several different branches, in which case a node only replies to the first branch it is contacted by.

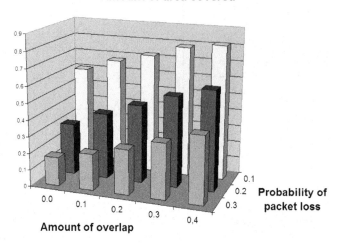

Fig. 8. Increasing the amount of overlap leads to a larger part of the area being covered

Determining the degree of overlap between different areas is a tradeoff between robustness and effectiveness. If an area is covered by several branches, the resulting tree structure will be ineffective with respect to both energy cost and interference. However, the impact of packet losses will be reduced.

In order to evaluate the results for different degrees of overlap, simulations were performed for different values of amount of overlap and probability of packet loss. For the purpose of the simulations, we assumed that the LAST algorithm would always create a tree that branches at four levels between the root node and the target area. This means that the tree has a total of $2^4 = 16$ leaves, where each leaf is responsible for covering a subarea of the target area.

Between each pair of branching points in the tree, there was a fixed probability of packet loss during the construction of the tree. When a packet is lost at some point, the construction of the entire subtree below that point is terminated.

In the simulations, the original area to be covered was always quadratical. At branching points, the area was always divided so that the resulting subareas would be as close to quadratical as possible. In other words, if the height was bigger than the width the area was divided horizontally, and vice versa.

The amount of overlap, as well as the probability of packet loss between two branching points, were varied in the simulations. For each possible combination, 500 simulation runs were performed. The results can be seen in Figure 8.

5.2   Implementation Issues

The case of non-uniform node distributions. Our estimation of the energy cost to cover an area from a specified node assumes that the sensor nodes are uniformly distributed over the area. If this is not the case, comparing different estimated costs may give incorrect results. However, to ensure a correct energy estimation it would be necessary to collect node density information over the entire network, which would be costly in terms of energy spent by the nodes.

Load balancing. In order to avoid overloading links and nodes near the root of the tree, it is not possible to use the links near the branches of the tree to their full capacity. As the root node starts the tree construction, it specifies the maximum possible bandwidth b. At the first branching point, each of the two branches can only transmit information at the maximum rate of $\frac{b}{2}$. At the next branching point, the maximum bandwidth is limited to $\frac{b}{4}$. This is repeated all the way to the leaves of the tree.

Branching penalties. If the sensor network is very dense, a higher number of branches may lead to increased interference that is not considered by our interference metric. In such a scenario, it might be desirable to avoid excessive branching in the tree structure even if our metric indicates that it is the best choice.

This can be controlled by adding an extra "penalty" value to (6) for all subsets with size 2, so that they have to perform significantly better than the subsets with only one member in order to be chosen.

5.3   Correctness of the Algorithm

The LAST algorithm is a recursive algorithm, with the base case that a branch has reached a node inside the subarea to be covered. If our assumptions of connectivity of the network holds, it can be shown that the following broadcast phase is guaranteed to eventually reach all the nodes in the subarea. For the algorithm to be proven correct, it has to be shown that it eventually terminates, and that the entire region A will be covered at that time. In this section, we present a simplified sketch of the proof.

We assume that, for any node u and any given point in the plane q outside that node's transmission range, that node has a neighbor that is closer to q than u itself. Thus, it can be shown that each node u in the network has a non-empty candidate set consisting of nodes that is closer to the region A than u itself. For every step in the tree construction algorithm, the node or nodes selected will be closer to the area to be covered than their parent node. Eventually, a node inside the area will be selected, and the base case has been reached.

If, at some point, two nodes in the candidate set are selected, and the tree splits into two branches, the task to cover the area A is divided over the branches so that the entire area will still be covered by all the branches taken together. The only way in which the algorithm will fail to create a tree that covers the entire specified region is if we take data losses into account. This has been discussed in more detail in Section 5.1.

5.4   Implementation

The LAST algorithm is currently being implemented in TinyOS operating system for sensor nodes. We extend the functionality of a combination of TinyLU-NAR [6] and GPSR [4] protocols. TinyLUNAR is the adopted to the specifics of sensor networks connection oriented routing scheme originally developed for mobile wireless ad hoc networks. The major property of LUNAR is simplicity of implementation in comparison to other protocols developed for MANETs. The most illustrative performance characteristic of TinyLUNAR is that it introduces only one byte overhead per data packet when forwarding on a multihop path.

In its turn GPSR [4], the Greedy Perimeter Stateless Routing is a geographic routing protocol proposed for wireless ad hoc and sensor networks that can be used to route data packets between any pair of nodes identified by their geographic coordinates. GPSR assumes that every sensor node is aware of its own location and the locations of its neighbors.

For the implementation of the LAST algorithm we combine the best features of the two protocols. Instead of broadcasting the route request messages in Tiny-LUNAR we carry them via GPSR to the target geographic region. Each node receiveing the RREQ message from GPSR establishes a backward label switching path. Upon the completion of the route establishment phase the messages are sent using the efficient forwarding mechanism of TinyLUNAR by this avoiding the overhead that otherwise would be introduced by packet forwarding directly over GPSR.

6   Conclusions

We presented a multicast tree formation algorithm for short-range wireless networks. Unlike previous algorithms that we are aware of, the LAST structure covers all the nodes in a geographic area rather than specific individual nodes. Previous algorithms have taken energy cost into account in order to prolong the lifetime of the network. In comparison to the existing approaches the LAST algorithm introduces interference as a metric when building multicast trees.

References

1. Burkhart, M., von Rickenbach, P., Wattenhofer, R., Zollinger, A.: Does topology control reduce interference? In: Proceedings of the 5th ACM Int. Symposium on Mobile Ad-hoc Networking and Computing (MobiHoc), pp. 9–19 (2004)
2. Frey, H., Ingelrest, F., Simplot-Ryl, D.: Localized minimum spanning tree based multicast routing with energy-efficient guaranteed delivery in ad hoc and sensor networks. Technical report, University of Southern Denmark, Ecole Polytechnique Federale de Lausanne, Universite des Sciences et Technologie de Lille (2007)
3. Johansson, T., Carr-Motyckova, L.: Reducing interference in ad hoc networks through topology control. In: Proceedings of The 2005 Joint Workshop on Foundations of Mobile Computing, pp. 17–23 (2005)
4. Karp, B., Kung, H.T.: Gpsr: greedy perimeter stateless routing for wireless networks. In: MobiCom 2000: Proceedings of the 6th annual international conference on Mobile computing and networking, pp. 243–254. ACM, New York (2000)
5. Koutsonikolas, D., Das, S., Hu, Y.C., Stojmenovic, I.: Hierarchical geographic multicast routing for wireless sensor networks. In: International Conference on Sensor Technologies and Applications, 2007. SensorComm 2007, October 14-20, pp. 347–354 (2007)

6. Osipov, E.: tinyLUNAR: One-byte multihop communications through hybrid routing in wireless sensor networks. In: Koucheryavy, Y., Harju, J., Sayenko, A. (eds.) NEW2AN 2007. LNCS, vol. 4712, pp. 379–392. Springer, Heidelberg (2007)
7. Sanchez, J., Ruiz, P., Stojmnenovic, I.: Gmr: Geographic multicast routing for wireless sensor networks. In: 3rd Annual IEEE Communications Society on Sensor and Ad Hoc Communications and Networks, 2006. SECON 2006 , vol. 1, pp. 20–29 (September 2006)
8. von Rickenbach, P., Schmid, S., Wattenhofer, R., Zollinger, A.: A robust interference model for wireless ad-hoc networks. In: Proceedings of the 19th IEEE International Parallel and Distributed Processing Symposium (IPDPS (April 2005)
9. Wu, S., Candan, K.S.: Gmp: Distributed geographic multicast routing in wireless sensor networks. In: ICDCS 2006: Proceedings of the 26th IEEE International Conference on Distributed Computing Systems, Washington, DC, USA, p. 49. IEEE Computer Society, Los Alamitos (2006)
10. Zeng, G., Wang, C., Xiao, L.: Grid multicast: an energy-efficient multicast algorithm for wireless sensor networks. In: Fourth International Conference on Networked Sensing Systems, 2007. INSS 2007, June 6-8, pp. 267–274 (2007)
11. Zhang, H., Arora, A., ri Choi, Y., Gouda, M.G.: Reliable bursty convergecast in wireless sensor networks. In: MobiHoc 2005: Proceedings of the 6th ACM international symposium on Mobile ad hoc networking and computing, pp. 266–276. ACM, New York (2005)

Optimum Resource Allocation for Amplify-and-Forward Cooperative Networks with Differential Modulation

Mohammadreza Rahmatpour and Vahid Tabataba Vakili

Dept. of Electrical Eng. Iran University of Science and Technology
mr.rahmatpour@ee.iust.ac.ir, vakily@iust.ac.ir

Abstract. The optimum resource allocation in relay networks is used to improve the error performance and increase the energy efficiency. In this paper, a two-dimensional resource allocation, i.e., the energy optimization and location optimization, is carried out based on the average symbol error rate (SER) for the system with and without a direct link. Differential modulation which bypasses the channel estimation at the transceiver is investigated using amplify-and-forward protocol for the system with multiple relays. We also show that the minimum error rate can be achieved via the joint energy-location optimization.

1 Introduction

Relay networks allow a source node to communicate with a destination node via a number of relay nodes in a wireless setup. By forming virtual antenna arrays in a cooperative manner, the spatial diversity gain can be achieved without imposing the antenna packing limitation [1], [2]. To reduce hardware complexity and communication overhead, the modulation schemes bypassing channel state information (CSI) have been adopted in relay systems. These systems are implemented by noncoherent modulations, differential modulations, and space-time coding (STC) techniques [3], [4], [11], [13].

Optimum resource allocation has recently emerged as an important research topic to improve the performance of relay networks (e.g. [5], [6], [8]). Though most of existing work on this topic focuses on the optimum energy allocation, relay location optimization is receiving increasing interests recently. For example, the optimum relay location is observed in [9],[12] with various protocols, and is considered jointly with the energy optimization in our recent work.

In this paper, we consider a relay network with multiple relays using differential modulation. The resource allocation of the relay system with a direct link (DL) and without a direct link (ND) is investigated using the amplify-and-forward (AF) protocol. The former evaluates a scenario where there is a direct link between the source node and the destination node, while the latter takes into consideration a situation where obstacles disable a direct transmission resulting in no direct link.

More importantly, the resource allocation problem is explored with the energy (power) optimization as well as the location optimization. Furthermore, the joint energy and location optimization is addressed. To enable the resource optimization, we first derive the average symbol error rate (SER) for reasonably high signal-to-noise ratio(SNR). The energy and location optimization is then carried out based on

S. Balandin et al. (Eds.): NEW2AN 2008, LNCS 5174, pp. 88–100, 2008.

this SER. We show that the optimum SER performance can be achieved by the joint energy and location optimization under the constraint of the total transmit energy and the fixed distance between the source and destination.

2 System Model

Consider a network setup with one source node (s), L relay nodes $\{ r_k \}_{k=1}^{L}$ and one destination node (d). The AF relaying protocol is considered, in which the relays amplify the signal received from the source and then forward it to the destination via rayleigh channels. We consider a time-division multiplexing (TDM), in which the transmission consists of two phases. In phase I, the source broadcasts the first symbol to all relays. In phase II, each relay amplify the signal and transmit it to the destination during distinct time slots.

A differential modulation is considered at both source and relay to bypass channel estimation. Specifically, with the nth phase-shift keying (PSK) symbol being as $s_n = e^{j2\pi c_n/M}$ $c_n \in \{0,1,...,M-1\}$, the corresponding transmitted signal from the source is $x_n^s = x_{n-1}^s s_n$ with the initial condition $x_0 = 1$.

In phase I, the encoded signal is broadcasted via a common rayleigh channel. The received signal at the kth relay and the destination are given by

$$y_n^{r_k,s} = \sqrt{E_s}\, h_n^{r_k,s} x_n^s + z_n^{r_k,s} \qquad , \qquad k = 1,2,......,L \qquad (1)$$
$$y_n^{d,s} = \sqrt{E_s}\, h_n^{d,s} x_n^s + z_n^{d,s}$$

where E_s is the energy per symbol at the source, and we denote the fading coefficients of channels between source and relays ($s-r_k$) and between source and destination ($s-d$) during the nth symbol as $h_n^{r_k,s}$ and $h_n^{d,s}$ and the corresponding noise component as $z_n^{r_k,s}$ and $z_n^{d,s}$, respectively.

Let $x_n^{r_k}$ denote the nth transmitted symbol from the kth relay, $k=1,2,...,L$. In phase II, the received signal at the destination corresponding to each relay is given by

$$y_n^{d,r_k} = \sqrt{E_{r_k}}\, h_n^{d,r_k} x_n^{r_k} + z_n^{d,r_k} \qquad , \qquad k = 1,2,......,L \qquad (2)$$

where E_{r_k} is the energy per symbol at the kth relay, h_n^{d,r_k} and z_n^{d,r_k} are the fading coefficients of channels between relays and destination (r_k-d) and the noise component at the destination, respectively.

In this paper, all fading coefficients are assumed to be independent and all noise components are independent and identically distributed (i.i.d) with $h_n^{i,j} \sim CN(0,\sigma_{i,j}^2)$ and $z_n^{i,j} \sim CN(0,N_0)$, $i,j \in \{s,r_k,d\}$. So, we can find the received instantaneous signal-to-noise ratio (SNR) between the transmitter j and the receiver i as $\gamma_{i,j} = |h_n^{i,j}|^2 E_j / N_0$ and the average SNR is $\overline{\gamma}_{i,j} = \sigma_{i,j}^2 E_j / N_0$.

At each of the relays, the received signal from the source is amplified

$$x_n^{r_k} = A_{r_k} y_n , \quad k=1, 2, ..., L \tag{3}$$

where A_{r_k} is the amplification factor. To maintain a constant average power at the relay output, the amplification factor is $A = (var\{ x_r(n)\})^{1/2} = (N_0 + \sigma_{s,r}^2)^{1/2}$ [13]. Then, we can represent the received signal at the destination corresponding to each relay node as

$$y_n^{d,r_k} = \tilde{h}_n^{d,r_k} x_n^s + \tilde{z}_n^{d,r_k} = y_{n-1}^{d,r_k} s_n + (\tilde{z}_n^{d,r_k})' \quad k = 1,2,....,L \tag{4}$$

$$\tilde{h}_n^{d,r_k} = \sqrt{E_s} \sqrt{E_{r_k}} A_{r_k} h_n^{r_k,s} h_n^{d,r_k} \tag{5}$$

$$\tilde{z}_n^{d,r_k} = \sqrt{E_{r_k}} A_{r_k} h_n^{d,r_k} z_n^{r_k,s} + z_n^{d,r_k}$$

and

$$(\tilde{z}_n^{d,r_k})' = \tilde{z}_n^{d,r_k} - \tilde{z}_{n-1}^{d,r_k} s_n \tag{6}$$

Therefore, according to the channels, the received signal is

$$y_n^{d,r_k} \sim CN(y_{n-1}^{d,r_k} s_n, \sigma_{h_k,eff}^2 \tag{7}$$

where the variance of the noise is given by

$$\sigma_{h_k,eff}^2 = 2N_0(E_{r_k} A_{r_k}^2 \sigma_{d,r_k}^2 + 1) \quad k = 1,2,..., \tag{8}$$

Similarly, the received signal at the destination corresponding to the source can be represented as

$$y_n^{d,s} = y_{n-1}^{d,s} s_n + (\tilde{z}_n^{d,s})' \tag{9}$$

with

$$y_n^{d,s} \sim CN(y_{n-1}^{d,s} s_n, 2N_0) \tag{10}$$

At the destination node, using the multi channel communication results in [10], the decision rule for $\tilde{s}_n = e^{j2\pi \hat{m}/M}$ can be obtained as (see[8],[13]):

$$\tilde{m} = arg_{m \in \{1,2,...,M-1\}} \max w_{d,s} l_m^{d,s}(y_n) + \sum w_{d,r_k} l_m^{d,r_k}(y_n) \quad (11)$$

where $w_{d,s}$ and w_{d,r_k} are combining weights given by $1/N_0$ and $2/\sigma_{h_k,eff}^2$, respectively, and $l_m^{d,i}(y_n) = \Re\{(y_n^{d,i})^* y_{n-1}^{d,i} s_n^m\}, i \in \{d,r_k\}$ with $s_n^m = e^{j2\pi m/M}$. It is assumed that the variances of the channel coefficients are available at the destination node. The decision rule for the system with no direct link can be obtained by only considering the summation term in Eq.(11).

3 Error Performance Analysis

In this section, we will derive the analytical expression of the error performance for the system described in the previous section. Under high SNR approximation, the symbol error rate (SER) for an L−relay system using differential binary phase shift keying (DBPSK) modulation is considered. Although the performance of a differential scheme has been analyzed in [8],[13], we will provide a simple and general expression for the average SER for the systems both with and with out a direct link.

At the destination, using Eq.(4), we evaluate the equivalent SNR from the source through the kth relay node as[13]

$$\gamma_{eq,r_k} = \frac{\gamma_{r_k,s} \gamma_{d,r_k}}{\bar{\gamma}_{r_k,s} + \gamma_{d,r_k} + 1} \quad (12)$$

Then, the relay system can be modeled as an equivalent multichannel system. In a L−relay system, the received SNR is

$$\gamma_{ND} = \sum_{k=1}^{L} \gamma_{eq,r_k} \quad (13)$$

and

$$\gamma_{DL} = \sum_{k=1}^{L} \gamma_{eq,r_k} + \gamma_{d,s} \quad (14)$$

For the system with no direct and with a direct link, respectively. Using Eqs. (13) and (14), we formulate the average SERs as follows:

The conditional error probability is determined as following: [10]

$$P_{e,ND}(\gamma_{ND}) = \frac{1}{2^{2L-1}} e^{-\lambda_b} \sum_{n=0}^{L-1} b_n \gamma_{ND}^k \quad (15)$$

where γ_b is given as

$$b_n = \frac{1}{n!} \sum_{k=0}^{L-1-n} \binom{2L-1}{n}$$

(16)

According to the average of $P_2(\gamma_{ND})$ over the fading channel statistics given by $p(\gamma_{ND})$ in [10] and assuming $\bar{\gamma}_{eq,rk} > 1$, $P_{e,ND}$ is easily shown to be:

$$P_{e,ND} = \frac{1}{2^{2L-1}} \sum_{n=0}^{L-1} \binom{n+L-1}{L-1} \sum_{k=0}^{L-1-n} \binom{2L-1}{k} \prod_{k=0}^{L} \left(\frac{1}{1+\bar{\gamma}_{eq,r_k}} \right)$$

(17)

where $1/(1+\bar{\gamma}_{eq,r_k})$ is SER for relay link.

According to [13] the PDF of γ_{eq,r_k} is determined as following:

$$p(\gamma_{eq,r_k}) = 2\frac{\bar{\gamma}_{s,r}+1}{\bar{\gamma}_{s,r}\bar{\gamma}_{r,d}} exp(-\frac{\bar{\gamma}_{eq}}{\bar{\gamma}_{s,r}})K_0(\beta\sqrt{\gamma_{eq}}) + \frac{2}{\bar{\gamma}_{s,r}\bar{\gamma}_{r,d}}\sqrt{\frac{\bar{\gamma}_{eq}(1+\bar{\gamma}_{s,r})\bar{\gamma}_{r,d}}{\bar{\gamma}_{s,r}}} exp(-\frac{\bar{\gamma}_{eq}}{\bar{\gamma}_{s,r}})K_1(\beta\sqrt{\gamma_{eq}})$$

(18)

where $\beta = 2\sqrt{\frac{1+\bar{\gamma}_{s,r}}{\bar{\gamma}_{s,r}\bar{\gamma}_{r,d}}}$, $K_0(.)$ and $K_1(.)$ denotes the zeroth-order and first order modified Bessel function of second kind. and SER for relay link is

$$P_r = \int_0^\infty e^{-\gamma_{eq}} p(\gamma_{eq}) d\gamma_{eq} = \bar{\gamma}_{r,d}^{-0.5} e^{0.5/\bar{\gamma}_{r,d}} W_{-0.5,0}(1/\bar{\gamma}_{r,d}) + (1+\bar{\gamma}_{s,r}) e^{0.5/\bar{\gamma}_{r,d}} W_{-0.5,0}(1/\bar{\gamma}_{r,d})$$

(19)

where $W_{\lambda,\mu}(.)$ denotes the Whittaker function and is equal to:

$$W_{\lambda,\mu}(z) = \frac{e^{-z/2} z^k}{\Gamma(1/2-\lambda+\mu)} \times \int_0^\infty t^{-\lambda-1/2+\mu}(1+\frac{t}{z})^{\lambda-1/2+\mu} e^{-t} dt$$

(20)

P_r can be written as:

$$P_r = \left[\frac{1}{\bar{\gamma}_{r_k,s}} + \frac{1}{\bar{\gamma}_{d,r_k}} ln(\bar{\gamma}_{d,r_k}) \right]$$

(21)

As a result at high SNR, the average SER of an L-relay AF system using DBPSK signaling with no direct link in which all L-relay Links are mutually independent can be approximated as:

$$\bar{P}_{e,ND} \approx C(L) \prod_{k=1}^{L} \left[\frac{1}{\bar{\gamma}_{r_k,s}} + \frac{1}{\bar{\gamma}_{d,r_k}} ln(\bar{\gamma}_{d,r_k}) \right]$$

(22)

And in the same way for a system with direct link:

$$\overline{P}_{e,DL} \approx C(L+1)\frac{1}{\overline{\gamma}_{d,s}} \prod_{k=1}^{L} \left[\frac{1}{\overline{\gamma}_{r_k,s}} + \frac{1}{\overline{\gamma}_{d,r_k}} ln(\overline{\gamma}_{d,r_k}) \right] \qquad (23)$$

Where $\overline{P}_{e,ND}$ and $\overline{P}_{e,DL}$ represent the average SER for the system with no direct link and with a direct link ,respectively, and

$$C(L) = \frac{1}{2^{2L-1}} \sum_{n=0}^{L-1} \binom{n+L-1}{L-1} \sum_{k=0}^{L-1-n} \binom{2L-1}{k} \qquad (24)$$

Notice that these expressions are similar to the average SER of the coherent system in [2].

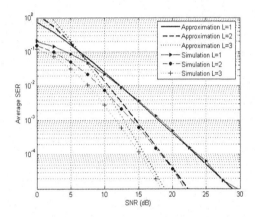

Fig. 1. SER comparison between approximation versus simulation (DL,SNR(dB), $\overline{\gamma}_{d,s} = \overline{\gamma}_{r_k,s} = \overline{\gamma}_{d,r_k}$)

Fig.1 shows the approximated and simulated SER when $L=1$, 2 and 3 for the system with a direct link. The figure confirms that the diversity is in direct proportion to the number of relays, and demonstrates that the approximations are pretty tight compared with the simulations, especially when SNR is high and L is small.

4   Optimum Resource Allocation

In this section, we will investigate the effects of resource allocation on the SER performance. To perform the optimization in this sections, we make use of the relationship between the average power of channel fading coefficient $\sigma_{i,j}^2$ and the inter-node distance $D_{j,i}$ as follows:[7]

$$\sigma_{i,j}^2 = C.D_{j,i}^{-v} \qquad i,j \in \{ s,r_k,d \}, \qquad (25)$$

Where v is the path loss exponent of the wireless channel and C is a constant which we set it to 1 without loss of generality. To express the energy constraint by SNR, let us define the total SNR $\rho = E/N_0$, the transmit SNR at the source node $\rho_s = E_s/N_0$, and the transmit SNR at the relay nodes $\rho_{r_k} = E_{r_k}/N_0$.

4.1 Energy Optimization

For each source, relay and destination node locations ($D_{s,r}$ and $D_{r,d}$), and the total SNR per symbol ρ, we will determine the optimum energy allocation ρ_s and ρ_{r_k} which minimize the average SER in Eqs .(22) or (23) by constraint:

$$\rho = \rho_s + \sum_{k=1}^{L} \rho_{r_k} \qquad (26)$$

By using Eq.(25), the average received SNR at the relay and destination can be expressed in terms of the transmit SNR as:

$$\bar{\gamma}_{r_k,s} = \rho_s \sigma_{r_k,s}^2 = \rho_s D_{s,r_k}^{-v} \qquad (27)$$

and

$$\bar{\gamma}_{d,r_k} = \rho_{r_k} \sigma_{d,r_k}^2 = \rho_{r_k} D_{r_k,d}^{-v} \qquad (28)$$

Let us first consider the system with no direct link. The optimum energy allocation which minimizes the average SER can be achieved by applying the first order conditions. By treating the approximated SER, $\bar{P}_{e,ND}$ as a function of ρ_s and ρ_{r_k}, we find the first order conditions for the optimum solution

$$\frac{\partial \bar{P}_{e,ND}}{\partial \rho_s} - \lambda = 0 \text{ and } \frac{\partial \bar{P}_{e,ND}}{\partial \rho_{r_k}} - \lambda = 0 \qquad (29)$$

$$\rho - (\rho_s + \sum_{k=1}^{L} \rho_{r_k}) = 0 \qquad (30)$$

Where λ is the Lagrange multiplier .Eqs.(29) give us

$$-\frac{\partial \bar{P}_{e,ND}}{\rho_s} \sum_{k=1}^{L} \frac{\rho_{r_k} \sigma_{d,r_k}^2}{\rho_{r_k} \sigma_{d,r_k}^2 + \rho_s \sigma_{r_k,s}^2 \ln(\rho_{r_k} \sigma_{d,r_k}^2)} - \lambda = 0 \qquad (31)$$

$$-\frac{\partial \overline{P}_{e,ND}}{\rho_{r_j}} \frac{\rho_s \sigma^2_{r_j,s}[\ln(\rho_{r_j}\sigma^2_{d,r_j})-1]}{\rho_{r_j}\sigma^2_{d,r_j}+\rho_s\sigma^2_{r_j,s}\ln(\rho_{r_j}\sigma^2_{d,r_j})}-\lambda=0 \tag{32}$$

Where $j =1, 2, ..., L$. However, the summation and logterms make it difficult to find the optimum solution. Therefore, we consider an idealized L-relay system with all relays located at the same distance from the source and the destination; that is, $D_{s,r_k} = D_{s,r}$ and $D_{r_k,d} = D_{r,d}$ $\forall k$. It is then reasonable to assign equal energies to all relays $E_{r_k} = E_r$ (equivalently $\rho_{r_k} = \rho_r$), $\forall k$. Then the first two equations of the first order conditions can be represented as:

$$-\frac{L\overline{P}_{e,ND}}{\rho_s} \frac{\rho_r\sigma^2_{d,r}}{\rho_r\sigma^2_{d,r}+\rho_s\sigma^2_{r,s}\ln(\rho_r\sigma^2_{d,r})}-\lambda=0 \tag{33}$$

$$-\frac{L\overline{P}_{e,ND}}{\rho_r} \frac{\rho_s\sigma^2_{r,s}[\ln(\rho_r\sigma^2_{d,r})-1]}{\rho_r\sigma^2_{d,r}+\rho_s\sigma^2_{r,s}\ln(\rho_r\sigma^2_{d,r})}-L\lambda=0 \tag{34}$$

Also, we have

$$\rho-(\rho_s+L\rho_r)=0 \tag{35}$$

With Eqs.(33) and (35), we arrive at

$$\sigma^2_{r,s}[\ln(\rho^o_r\sigma^2_{d,r})-1]\rho^{o2}_s + \rho^o_r\sigma^2_{d,r}\rho^o_s - \rho\rho^o_r\sigma^2_{d,r}=0 \tag{36}$$

From this equation, we can find the solution for ρ^o_s and, correspondingly, ρ^o_r. However, the logterm in Eq.(36) makes us to use a complicated closed-form solution analytically but, we could resort to a numerical search using Eq.(36).

Similarly, for the system with a direct link under the idealized L-relay assumption, we have

$$(L+1)\{\sigma^2_{r,s}[\ln(\rho^o_r\sigma^2_{d,r})-1]\rho^{o2}_s + [\rho^o_r\sigma^2_{d,r}-\frac{\rho\sigma^2_{r,s}}{L+1}\ln(\rho^o_r\sigma^2_{d,r})]\rho^o_s - \rho\rho^o_r\sigma^2_{d,r}=0 \tag{37}$$

As mentioned earlier, we could resort to a numerical search to find the optimum solution. In Eq.(37), the $\rho\sigma^2_{r,s}\ln(\rho^o_r\sigma^2_{d,r})/L+1$ term mainly affects the energy allocation compared with the system with no direct link. This effect is obvious especially when the relays are located close to the source.

The optimum energy allocation obtained from the numerical search for the systems both with and without a direct link is plotted in Fig.2. We consider the total SNR value of $\rho = 15db$ and a path loss exponent of $v = 4$ with various L values. A one-dimensional setup is considered; that is $D_{s,r} + D_{r,d} = D_{s,d}$ with $D_{s,d} = 1$. In the

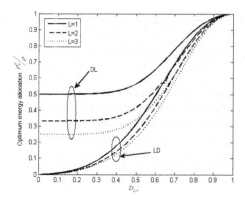

Fig. 2. Optimum energy allocation (DL and ND, $\rho = 15dB, v = 4$)

system with no direct link, for all L, the optimum energy allocation at the source increases as the relays move toward the destination. However, for the system with a direct link, a uniform energy allocation is optimum when the relays are located close to the source. this is because when a direct link exist ,and the relays are located close to the destination, much of energy is assigned at the source to assure that a transmitted signal can reach the relays; it is the same as in the system with no direct link.

4.2 Location Optimization

For any given transmit energies at the source and relay nodes (E_s and E_r , or equivalently ρ_s and ρ_r), we will determine the optimum location of the relays, $D_{s,r}$, which minimizes the average SER in Eqs .(22) or (23) while satisfying $0 < D_{s,r} < D_{s,d}$.

For the given energies at the source, the SER of a direct link is independent of the location, since $D_{s,d}$ is a fixed value with $1/\overline{\gamma}_{d,s} = D_{s,d}^v / \rho$. Hence, notice that the location optimization can be achieved without considering the direct link, i.e., the result of location optimization is the same for both the systems with and without a direct link.

The optimum location can be found by using the SER as a function of distance , $D_{s,r}$ and $D_{r,d}$, and solving the first order conditions:

$$L\overline{P}_{e,ND} \frac{vD_{s,r}^{v-1}\rho_r}{D_{s,r}^v\rho_r + D_{r,d}\rho_s \, ln(\, \rho_r D_{r,d}^{-v} \,)} - \lambda = 0 \qquad (38)$$

$$L\overline{P}_{e,DL} \frac{vD_{s,r}^{v-1}\rho_r [ln(\, \rho_r D_{r,d}^{-v} \,) - 1]}{D_{s,r}^v\rho_r + D_{r,d}\rho_s \, ln(\, \rho_r D_{r,d}^{-v} \,)} - \lambda = 0 \qquad (39)$$

And $D_{s,d} = (D_{s,r} + D_{r,d})$ Then, we have the following solution:

$$v(D^o_{s,r})^{v-1} \rho_r - v(1 - D^o_{s,r})^{v-1} \times \rho_s \{ ln[\rho_r (1 - D^o_{s,r})^{-v}] - 1 \} = 0 \quad (40)$$

We can find the optimum solution for $D^o_{s,r}$, and then $D^o_{r,d}$. Again, the logterm and path loss exponent make it difficult to find the closed-form solution. We could resort to a numerical search for the optimum solution.

4.3 Joint Energy and Location Optimization

So far, we have been focusing on the energy optimization and location optimization separately. Now let us consider the joint optimization which satisfies both the energy and location optimization. Mathematically, the analytical solution can be obtained by using Eqs. (22) and (23), for the systems with no direct link and with a direct link, respectively.

The joint optimization can be achieved by finding a common solution which minimizes the SER both the energy and location distribution. However, the analytical solution can not be easily obtained as we have seen even with an idealized case in the previous section. In general, the global optimization can be obtained. Specifically, for any given energy distribution, find the optimum location. Then, optimize the energy distribution based on the updated location. These steps proceed until convergence. The final solution provides the global optimum value which minimizes the SER.

5 Numerical Search and Discussions

Next, we will discuss the benefits of optimization including the joint energy-location optimization. Figs.3 and 4 verify the benefit of energy optimization and location optimization, respectively. We consider $\rho =15$dB and $L=1$, 2, and 3 for the system with no direct link.

In Fig.3, for the system without energy optimization, a uniform energy allocation is used; that is, $\rho_s = \rho_r = \rho /(L+1)$ at any $D_{s,r}$. From Fig.3, we observe that the energy-optimized system outperforms the unoptimized system as we expected. The figure shows that the unoptimized systems have the minimum SER almost at the midpoint, which agrees with the results in [2], [9]. However, the minimum points of the energy-optimized SER curves move toward the destination when $L> 1$.In Fig.4, for the system without location optimization, we place the relays at the midpoint of the source-destination link.

Similar to the energy optimization case, the location-optimized system outperforms the unoptimized system. The curves in Fig. 4 also exhibit more flatness compared with the ones in Fig. 3, and the optimum SER can be achieved with assigning more energy to the source. Similar trends are observed in the system with a direct link. Fig. 5 depicts the contour of SER for the relay system with no direct link when $L = 3$, $\rho = 15$dB. The vertical line and horizontal line correspond to the SER of the system with uniform energy allocation and mid-distance allocation, respectively, i.e., unoptimized systems.

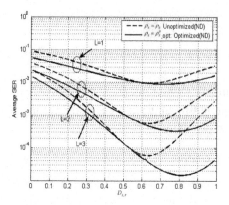

Fig. 3. SER comparison between relay systems with and without energy optimization (ND, $\rho = 15dB, v = 4, DBPSK$)

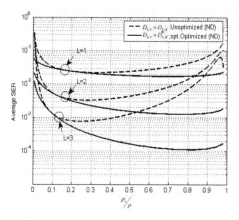

Fig. 4. SER comparison between relay systems with and without location optimization (ND, $\rho = 15dB, v = 4, DBPSK$)

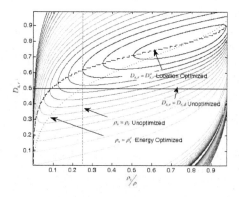

Fig. 5. The SER contour (ND, $\rho = 15dB, v = 4, DBPSK$)

We also plot the lines for the energy optimization and location optimization, which correspond to the minimum values along the $D_{s,r}$ axis and ρ_s / ρ, respectively. Fig. 5 shows that the optimum location allocation slowly changes compared with the optimum energy allocation.

Notice that the cross point of two optimizations is the global minimum SER of the system, accordingly this point corresponds to the joint energy and location optimization. It is interesting that we cannot achieve the minimum SER without any optimization, which is different from the case with the decode-and-forward (DF) protocol in [7], Fig. 5 shows that we can achieve the minimum SER by locating the relays close to the destination with more than half of the total transmit energy at the source. This is due to the fact that the SER decreases by reducing the effect of noise and variance on the amplification factor and allowing the source to transmit signal with high SNR.

6 Conclusions

In this paper, we investigated the optimum energy distribution and the optimum location of relays in a wireless system with multiple relays employing differential modulation. Two-dimensional optimization is studied on the basis of minimizing the average SER, which we derived for the amplify-and-forward relaying protocol with reasonably sufficiently high SNR. Our analysis shows that the optimum energy allocation depends on a direct link, but the optimum location allocation does not. Our simulations and numerical results show that both the energy and location optimizations provide remarkable SER advantages. We have also shown that the minimum SER can be achieved by the joint energy-location optimization, and that the minimum SER cannot be obtained without appropriate resource allocation.

In comparison with [7], in which optimum energy and location was investigated for arbitrary number of relays employing differential demodulation with decode-and-forward cooperative protocol, on base of an upper bound on the average SER, we could find that in both approaches, simulation and numerical results confirm that both the energy and location optimizations provide considerable SER advantages. In addition, they also confirm that the location optimization may be more critical than the energy optimization and the system SER is more sensitive to location distribution than to the energy optimization. In other words, the differential relay system with uniform energy distribution can achieve near optimum SER by appropriately choosing the relay location; while a system with relays located at the midpoint between the source and the destination cannot approach the optimum SER even with optimized energy distribution. According to [7], minimum SER with optimum energy allocation in DF is achieved when relays are near to source but in AF protocol, minimum SER with optimum energy allocation is achieved when relays are near to destination.

References

1. Laneman, J.N., Wornell, G.W.: Energy-efficient antenna sharing and relaying for wireless networks. In: Proc. of WCNC, vol. 1, pp. 7–12 (2000)
2. Ribeiro, A., Cai, X., Giannakis, G.B.: Symbol error probabilities for general cooperative links. IEEE Trans. on Wireless Communications 4(3), 1264–1273 (2005)

3. Chen, D., Laneman, J.N.: Modulation and demodulation for cooperative diversity in Wireless systems. IEEE Trans. on Wirelss Communications 5(7), 1785–1794 (2006)
4. Cho, W., Yang, L.: Distributed differential schemes for cooperative wireless networks. In: Proc. of ICASSP, vol. 4, pp. 61–64 (2006)
5. Annavajjala, R., Cosman, P.C., Milstein, L.B.: Statistical channel knowledge-based optimum power allocation for relayig protocols in the high SNR regime. Journal on Selected Areas on Communications 25(2), 292–305 (2007)
6. Boyer, J., Falconer, D.D., Yanikomeroglu, H.: Multihop diversity in wireless relaying channels. IEEE Trans. on Communications 52(10), 1820–1830 (2004)
7. Cho, W., Yang, L.: Energy and location optimization for relay networks with differential modulation. In: Proc. of ICASSP, vol. 3, pp. 153–156 (2007)
8. Himsoon, T., Su, W., Liu, K.J.R.: Differential modulation for multinode amplify-and-forward wireless relay networks. In: Proc. of WCNC, vol. 2, pp. 1195–2000 (2006)
9. Li, H., Zhao, Q.: Distributed modulation for cooperative wireless communications. IEEE Signal Processing Magazine 23(5), 30–36 (2006)
10. Proakis, J.: Digital Communications, 4th edn. McGraw-Hill, New York (2001)
11. Wang, T., Yao, Y., Giannakis, G.B.: Non-coherent distributed space-time processing for multiuser cooperative transmissions. IEEE Trans. on Wireless Communications 5(12), 3339–3343 (2006)
12. Yu, M., Li, J., Sadjadpour, H.: Amplify-forward and decode-forward: The impact of location and capacity contour. In: Proc. of MILCOM, vol. 3, pp. 1609–1615 (2005)
13. Zhao, Q., Li, H.: Performance of differential modulation with wireless relays in Rayleigh fading channels. IEEE Communications Letters 9(4), 343–345 (2005)

LBS Position Estimation by Adaptive Selection of Positioning Sensors Based on Requested QoS

Renato Filjar, Lidija Bušić, Saša Dešić, and Darko Huljenić

Ericsson Nikola Tesla d.d. Krapinska 45, 10 000 Zagreb, Croatia
{renato.filjar,lidija.busic,sasa.desic,
darko.huljenic}@ericsson.com

Abstract. With increasing attractiveness of location-based services (LBS), the need for consistent establishment and deployment of the LBS Quality of Service (QoS) hierarchy is strongly demanded. The position estimation is in the heart of every location-based service. Thus, LBS QoS is primarily concerned with position estimation performance, including position estimation errors and response time, achieved by either single position sensor, or a combination of several position estimation sensors and methods. Common LBS QoS establishment approach consists of either "as-is" (i. e. no-guarantee) or "best-effort" (again no-guarantee, but with some concern) approach. The proposed new solution starts with generic description of LBS QoS and methods for its deployment. As the result a method emerges that utilises position estimation by adaptive selection of positioning sensors based on requested QoS.

Keywords: Location Based Services (LBS) QoS Positioning Sensor Selection.

1 Introduction

Determination of the most appropriate Location-Based Services (LBS) Quality of Service (QoS) level is identified as one of the major challenges in the LBS development [1]. So far, various LBS QoS determination approaches have been utilised, that usually offers very limited or no guarantees to the end users.

Since the quality of position estimation sets of the foundation of the LBS QoS, it has been a matter of considerable research to rationalise the usage of position estimation and network resources in order to provide reasonable QoS to a wide number of users [2, 3].

Here we propose system architecture and an algorithm for the LBS position estimation by adaptive selection of positioning sensors based on requested QoS that enables provision of reasonable LBS QoS for particular service, and decreases the over-stretching of user and network resources.

2 Problem Description

Location-Based Services, as a group of telecommunication services, combines robust and accurate positioning with geospatial (location-related) content and telecommunication

S. Balandin et al. (Eds.): NEW2AN 2008, LNCS 5174, pp. 101–109, 2008.

networks in order to enable provision of location-related content and services to users [2, 3]. In an analogy to the other telecommunication services, the provision of LBS is driven by various levels of Quality of Service (QoS), i. e. different levels of position estimation accuracy and response time, according to requirements of particular service [1]. Accurate and robust estimation of the user position is the foundation concept of every LBS, regardless of its accuracy [2, 3]. Therefore, the parameters describing the LBS QoS are identified and established in relation to position estimation performance and its distribution throughout the elements of network architecture in support of the LBS provision [1]. Existing LBS-related industrial standards [4, 5] define the following key parameters of LBS QoS:

- Horizontal accuracy,
- Vertical accuracy,
- Response time (time between the request for position determination and the position estimate delivery).

Apparently, different position estimation methods will provide various levels of position estimation accuracy (Fig. 1). Satellite position estimation methods (based on GPS, Glonass, Galileo, GNSS and other satellite positioning systems) have already been identified as the most accurate and reliable, within their limits of operation, and thus assumed to be foundation position estimation methods for LBS development. Other methods (mostly mobile network-based, such as CellID, TOA or E-OTD) are assumed to be assistant methods, or methods of the second choice, due to their inferior QoS compared with satellite positioning methods.

Fig. 1. Different position estimation methods provide different position estimation accuracy (i. e. Quality of Service)

An adequate level of QoS is set for every LBS service. This means the highest level of position estimation performance is not compulsory for less demanding LBS, such as finding the nearest petrol station. On the other hand, the critical applications, such as notifications of car accidents, require either best-effort or guaranteed high-level QoS [4].

Position estimation process is based on readings (measurements) conducted on various signals propagating through space. A dedicated device, aimed to perform such measurements and eventually provide the results in a form of either raw signal measurements or as an initial position estimate, is called a position estimation (or positioning) sensor. Positioning sensor is considered an entity deployed within the system architecture for the purpose of either position estimation using particular positioning method, or provision of the additional (assistance and augmentation) data needed for particular positioning method. Various devices may be considered positioning sensors, including, but not limited to: GPS/Galileo/Glonass/GNSS sensor, sensor for WLAN-based positioning and network positioning sensors, (comprising various network-based positioning methods, such as Cell ID or E-OTD), augmentation and assistance service for positioning methods (such as A-GPS/GNSS and differential GPS/GNSS).

The position of a (mobile) user can be estimated either from single positioning sensor measurements, or from the combination of more positioning sensor estimates. Although the satellite positioning method (GPS, Glonass, Galileo, GNSS) is widely accepted as the most important and elementary LBS position estimation method [1, 5], it cannot be considered a single solution for all LBS positioning needs. However, applicability of satellite positioning systems in critical environments (in-doors, urban and mountainous areas) may be significantly reduced by either degraded position estimation performance (large positioning errors) or complete absence of positioning signal (coverage not available) [6, 7].

In order to tackle disturbing effects on satellite positioning systems, other positioning methods (especially those network-based) should be considered for utilisation in synergy with satellite positioning systems. Such a combination increases the general position estimation performance when both solutions (satellite and network positioning) are available, or provides continuation of position estimation service (although with reduced QoS, especially positioning accuracy) when satellite positioning becomes temporarily un-available. In general, the synergy between satellite positioning and other positioning methods provides more stable and consistent, and less vulnerable LBS QoS compared with the case of satellite positioning system being used alone.

Deployment of combination of several position estimation methods yields the position estimate and positioning error (QoS) estimate, which are obtained in a process called positioning sensor fusion. In positioning sensor fusion, satellite positioning, network-based positioning and other (optional) methods act as the single sources of position estimates (basic positioning sensors and methods). A particular integration method (Kalman filter, neural network, particle filter etc.) should be deployed in order to combine the outcomes of single sources of position estimates, according to the statistical nature of positioning sensor fusion process [6]. Integration methods provide best available position and QoS estimate, which are usually more accurate than those achieved by single position estimation method. In the present concept of positioning sensor fusion, position and QoS estimates are obtained using the come-upon state of methods and resources (position estimation on the as-is basis). Existing concept does not utilise the network resources efficiently, since much effort can be put into provision of the best possible QoS (provision of positioning assistance and augmentation, deployment of map-matching algorithm, utilisation of computation-demanding algorithm on network elements) for a service that does not require it. In addition, the end-users are usually being charged for this inefficiency.

3 Fundamental Concept

A method for position estimation achieved through adaptive selection of positioning sensors for Location-Based Services (LBS) based on requested QoS is addressed here, with the aim to provide reasonable, appropriate and satisfactorily level of the LBS QoS for particular service in question. The method is based on harmonisation of the particular user and service QoS preferences and capabilities, which provides the LBS QoS level determined by the position estimation performance of the selected positioning sensors. Position and LBS QoS estimates are considered the outcomes of the method for position estimation based on requested QoS. Position estimate is expressed by mandatory (latitude and longitude) and optional (height above the sea level, velocity and azimuth) elements. The LBS QoS estimate is assumed to be expressed by horizontal and vertical position estimation error, and response time. The position estimate is transferred to the LBS application, while the LBS QoS estimate is expected to be utilised internally within the proposed method.

The main concept within the proposed method calls for adaptive selection of positioning sensors, which will be used in position estimation in order to provide the level of LBS QoS both satisfactorily for the end user and not too demanding for the equipment and network resources. The adaptive selection is achieved through the ability of proposed method to suit different conditions of positioning environment (availability of particular positioning sensors and their performance, LBS QoS demands for particular service etc.).

The existence and availability of the following sub-sets of general profiles are necessary prerequisites for proposed method's implementation:

- LBS user profile,
- LBS service profile.

The LBS user profile describes user preferences and provides the list of supported positioning sensors for the particular LBS service invocation. User preferences relates to the choice of preferred positioning method and the willingness to pay for a dedicated positioning process.

LBS service profile determines the minimum requested LBS QoS for actual service in question. It defines the acceptable level of horizontal and vertical positioning accuracy, as well as the response time between the request for positioning and position estimate delivery.

The system architecture in support of the proposed method for position estimation achieved through adaptive selection of positioning sensors for LBS based on requested LBS QoS is presented on Fig 2. The system architecture consists of the client, the LBS application server, the positioning server, and the set of positioning sensors.

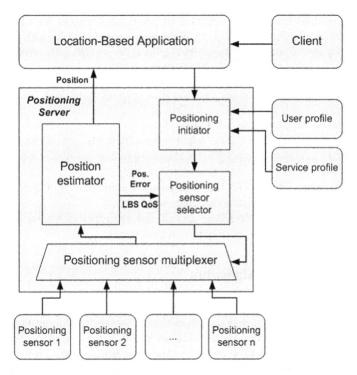

Fig. 2. System architecture in support of the method for adaptive selection of positioning sensors for LBS based on requested QoS

The client serves as the invocator of the LBS service. The Location-Based Application (LBA) handles the LBS invocation requests and position-related content and service provision to the end-users.

Positioning server is in the heart of the system, providing the following dedicated functionalities:

- Positioning sensor multiplexer, for reading the outputs of positioning sensors,
- Positioning initiator, for accepting LBS service initiation requests, and LBS QoS requirements harmonisation based on existing User and Service LBS profiles,
- Positioning sensor selector, for making decisions about activation of particular positioning sensors based on requested LBS QoS,
- Position estimator, aimed to perform positioning sensor fusion, and estimation of position and LBS QoS (position estimation error).

Position and LBS QoS estimation is obtained using different estimation methods that utilise one or the combination of the following:

- Choice of the basic positioning method,
- A method for positioning sensor,
- Choice of Positioning Assistance and Augmentation provision,
- Activation of the context matching algorithm.

Presented method for adaptive selection of positioning sensors for LBS based on requested QoS is an iterative process in which every iteration results with position and LBS QoS estimates. Control parameter for the iteration process is the congruence between initially requested and estimated LBS QoS.

Application of the method for adaptive selection of positioning sensor for LBS brings considerable benefits, with the list not limited to the following:

- Provision for guaranteed LBS QoS.
- Network resources are used much more efficiently, due to introduction of suitable and reasonable LBS QoS provision, instead of the best effort approach.
- End users are charged for the initiated service and delivered QoS more properly, since the service provides guaranteed QoS.
- A service will be delivered to an end user even though his/her initial set of active sensors does not satisfy QoS requirements.

4   Adaptive Selection Algorithm

This chapter presents the positioning sensor adaptive selection algorithm for proposed method. It is based on the current state-of-the-art in mobile communication technologies, especially in relation to up-to-date standardisation in the area of the 3G [5].

The algorithmic representation of the method for adaptive selection of positioning sensors for LBS based on the requested QoS is shown on Fig 3. The aim of the method is to provide position and positioning QoS estimates based on the delivery of suitable and reasonable QoS level for LBS in question, as described in the previous chapter. The algorithm is implemented within the elements of the positioning server.

The iterative process is embedded in positioning procedure of the LBS as follows. After the LBS client initiates particular LBS service, the first iteration of positioning sensor selection is conducted based on requirements for LBS QoS for the service in question, LBS user profile and the set of available positioning sensors. The result is the first set of position and LBS QoS estimates based on obtained list of available positioning sensors. If the obtained position and positioning error estimates failed to reach minimum requirements for service in question, or one or more positioning sensors fail to provide their readings, the iterative procedure follows in which a new set of positioning sensors is to be chosen, and the process repeats until at least the minimum requested LBS QoS is reached, or until all reasonable combinations of positioning sensors are examined without satisfying LBS QoS results. Finally, the whole system architecture is set to its default state, determined by User Profile parameter values, thus maintaining the initial user preferences (initial choice of activated positioning sensors, for instance).

The method starts with the QoS requirements harmonisation and available position estimates collection, both initiated immediately after invocation of the LBS service. The parameters of the LBS user profile and LBS service profile are acquired and combined in order to define the requested LBS QoS for particular service invocation.

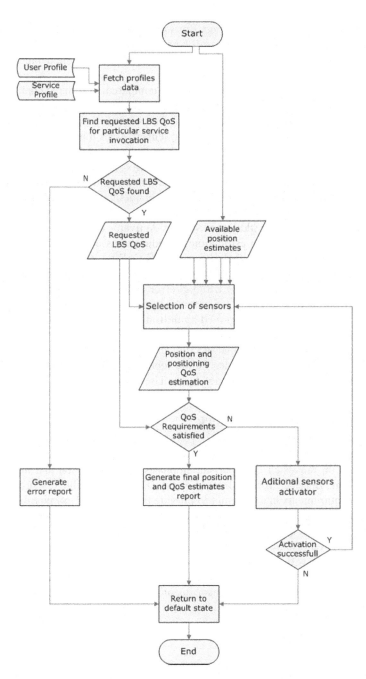

Fig. 3. Sensor selection algorithm supporting the method for adaptive selection of positioning sensors for LBS based on requested QoS

If the requested LBS QoS is determined for the invoked service, the resulting set of requested LBS QoS parameter values will be forwarded for further processing. If not, an error report will be delivered, which concludes with return to the default state. The default state is determined by user profile.

If the requested LBS QoS is established, it is compared with the available position estimates provided by already active positioning sensors in order to select the first-iteration line of active sensor(s), as seen from the point of view of the requested LBS QoS. Chosen first-iteration line of positioning sensors gives the first-iteration position and positioning error estimation. If those satisfy the requested LBS QoS set earlier, obtained first-iteration position estimate is to be delivered to the LBS application as the final result. Otherwise, the activation of additional positioning sensors (including assistance and augmentation to basic positioning methods, and possible fusion of various positioning sensors) is to be performed.

The main purpose of the Additional Sensors Activator element is to involve more positioning sensors in position estimation procedure, thus allowing for improvement of the actual LBS QoS towards the requested LBS QoS for particular service invocation. The choice of the additional sensors should be based on the list of supported positioning sensors and dedicated LBS QoS error analysis, which ultimately leads to iteration convergence. The LBS QoS error analysis and decision to activate certain positioning sensors should be performed by utilisation of the appropriate optimal control algorithm. Iterations continue until either the requested LBS QoS is reached and position and positioning QoS estimations delivered, or the inability to fulfil requested QoS generates the appropriate service failure report. Finally, the whole system architecture is set to default state, determined by the User Profile.

5   Conclusion and Future Work

Proposed method for the LBS position estimation by adaptive selection of positioning sensors based on requested QoS puts an emphasis on introduction and utilisation of the User and the Service Profiles. Instead of exposing the available resources to over-demanding requirements in an attempt to provide the best LBS QoS possible, the proposed method decreases the maximum LBS QoS available to reasonable level that satisfies the end-user, at the same time allowing for more reasonably efficient utilisation of LBS-related resources and ability to serve much more end-users.

Further research will focus on development of advanced positioning sensor selection algorithms based on advanced use of User and Service Profiles, and on achieving the satisfying levels of positioning sensor fusion performance.

Acknowledgements

The work described in this paper was conducted under the research project New architecture and protocols in converged telecommunication networks (No. 071-0362027-2329), approved by the Ministry of Science, Education and Sport, Republic of Croatia.

References

1. Filjar, R., Busic, L.: Enhanced LBS Reference Model. In: Proc. of NAV 2007 Conference, p. 8. Westminster, London (2007)
2. Kuepper, A.: Location-based Services: Fundamentals and Operation. John Wiley & Sons Ltd, Chichester (2005)
3. Steiniger, S., Neun, M., Edwardes, A.: Foundations of Location Based Services (lecture notes). Department of Geography, University of Zurich, Switzerland (accessed on 22 March 2006) (2008), http://www.geo.unizh.ch/publications/cartouche/lbs_lecturenotes_steinigeretal2006.pdf
4. 3GPP TS 22.071 V7.4.0. Location Services (LCS); Service Description; Stage 1 (2008).
5. 3GPP TS 23.271 V7.7.0. Functional stage 2 description of Location Services (LCS) (Rel 7) (2008)
6. Parkinson, B.W., Spilker Jr., J.J.: Global Positioning System: Theory and Applications, vol. I. AIAA, Washington (1996)
7. Filjar, R., Desic, S., Huljenic, D.: Satellite Positioning for LBS: A Zagreb Field Positioning Performance Study. Journal of Navigation 57, 441–447 (2004)
8. Doucet, A., de Freitas, N., Gordon, N.: Sequential Monte Carlo Methods in Practice. Springer+Business Media, Inc, New York (2001)
9. Schiller, J., Voisard, A.: Location-Based Services. Elsevier, Inc., San Francisco (2004)

Low Latency Cross Layer Handover Scheme in Proxy Mobile IPv6 Domain*

Geunhyung Kim

Department of Visual Information Engineering,
Dong-Eui University, Pusan, Korea
geunkim@deu.ac.kr

Abstract. In IP-based wireless networks, minimizing handover latency with few packet loss is one of the most important issues. To achieve this goal, host-based and network-based fast or localized mobility management solutions have been proposed. Proxy Mobile IPv6(PMIPv6) avoids tunneling overhead over the air and supports mobility for hosts without host involvement. However, the basic performance of PMIPv6 for handover latency and packet loss is not different from that of Mobile IPv6. In this paper, we propose an enhancement for PMIPv6 to reduce the *packet reception latency* and to minimize *packet loss* for both intra-local mobility anchor(LMA) and inter-LMA handover by pre-establishing bidirectional tunnel between MAGs within an administrative domain. As a result, we found that the proposed scheme, though it requires additional signaling messages to establish the bidirectional tunnels, guarantees lower packet reception latency and fewer packet loss than other recent approaches without erroneous movement prediction.

Keywords: Ubiquitous Mobile Network, Mobile IPv6, Proxy Mobile IPv6, Seamless Handover, Packet Reception Latency.

1 Introduction

Since wireless networks such as IMT-2000, wireless LAN, and Mobile WiMAX introduce real-time multimedia services such as voice over IP and interactive streaming, ubiquitous roaming support for real-time multimedia traffic in an access independent manner becomes increasingly important. Mobile IPv6 (MIPv6) [1] specifies a host-based global mobility management scheme for an IPv6 node to maintain network connectivity of a node as it roams around the Internet. MIPv6 performs three procedures sequentially to support handover: a movement detection, an address configuration, and a location update. These procedures have been shown to result in a long handover latency which strongly degrades the performance of packet transmission of mobile nodes(MNs). In order to reduce handover latency of MIPv6, IETF has standardized two schemes: hierarchical MIPv6 (HMIPv6)[2] and fast handover for Mobile IPv6 (FMIPv6)[3].

---

* This work was supported by Dong-eui University Grant.(2008AA204).

Despite of lots of efforts, MIPv6 has not widely deployed, due to its heavy specification. That is, MN is required to implement complex specification which is difficult to deploy and maintain on small portable terminals with battery. Because of the deployment problem of MIPv6, network service providers and vendors have provided their own network-based mobility mechanisms within the localized area. However, these proprietary solutions are not interoperable with each other. Therefore, IETF recognized the need for standardization of network-based mobility services and IETF NetLMM (Network-based Localized Mobility Management) Working Group has discussed the problem statement [4] and goal [5] of network-based localized mobility management and has recently adopted Proxy Mobile IPv6 (PMIPv6) [6] as a single promising proposal for localized mobility standard.

PMIPv6 aims at supporting IP mobility management for all hosts irrespective of the presence or absence of MIPv6 functionality. This protocol does not require the MN be involved in the layer 3 (L3) signaling required for mobility management. In a PMIPv6 domain, a mobile access gateway (MAG) acts as a proxy for the MN, that is, it tracks the MN's movement and sends mobility signals to the MN's local mobility anchor (LMA) on behalf of the MN. In a PMIPv6 domain, MN feels as if it is always at its home link despite the change of access router, since the MN's home network prefix (HNP) is hosted on every access link.

Like other mobility protocols, PMIPv6 suffers from packet loss or latency during handover to other access networks. To communicate efficiently for real-time applications on the move with PMIPv6, packet loss and handover latency should be minimized [7]. PMIPv6 mobility scope is classified into three cases: intra-link mobility, local mobility and global mobility, each referring to handover within one LMA, handover between adjacent MAGs within one LMA, and handover between adjacent LMAs respectively [4]. Among three mobility cases, this paper mainly focuses on the local mobility and global mobility within an administrative domain.

The L3 handover latency in PMIPv6 is caused by the network access authentication latency, location update latency, and address configuration latency [6]. These latencies are not appropriate for real-time applications and throughput sensitive applications [7]. Therefore, several handover optimization approaches [10] [11] have been introduced in IETF to reduce handover latency. These approaches borrow some ideas from FMIPv6 protocol to enhance PMIPv6 protocol by improving L3 handover performance and work in two operation mode: predictive and reactive mode. Both are based on the prediction made by link-layer information of future events. As a result, this prediction may sometimes be wrong. In addition, the allocated buffer space constitutes useless overhead if the prediction is wrong since packet forwarding for smooth handover performed in accordance with prediction.

In this paper, we propose an alternative low latency seamless handover scheme that can reduce the handover latency of MN and the packet loss by sending context information with MN's HNP from current MAG to adjacent MAGs, establishing bidirectional tunnel between current MAG to adjacent MAGs before the MN moves adjacent MAGs, and buffering packets on the current MAG until L3 handover finishes. In addition, the proposed mechanism reduces packet loss

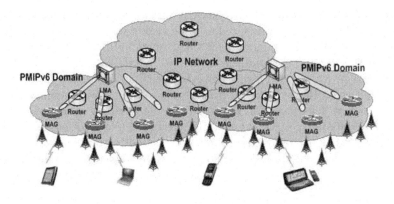

Fig. 1. A Functional Architecture of PMIPv6 Domain

caused by deferring transmission of deregistration message to LMA which results in the loss of packets that correspondent node (CN) transmitted.

The remainder of this paper is organized as follows: In the next section, we first give an overview of PMIPv6 in terms of benefit and problem, and discuss related works on fast handover for PMIPv6. In section 3, we present the proposed handover scheme to reduce the packet reception latency and packet loss. We discuss performance in section 4. Finally, we conclude the paper and give future work in section 5.

2   Related Work

2.1   PMIPv6 Overview

In this section, we investigate the PMIPv6, being standardized in IETF as the network-based mobility management, in terms of its operations. PMIPv6 aims at providing the localized mobility management solution without the MN's participation in any L3 mobility management related signaling procedures. To meet the goal, PMIPv6 introduces new entities such as the MAG and the LMA. The latter has the functional role of a home agent(an anchor point for the MN's HNP) for the MN in the PMIPv6 domain, such as assigning MN's HNP and managing the MN's reachability state. The former conceals the roaming information from the MN by emulating MN's home link properties and performs the mobility management related signaling on behalf of the MN. PMIPv6 gives L3 handover solution for wireless access networks like MIPv6, but it does not require any protocols on the MN for the L3 mobility management. Fig. 1 gives the functional architecture of PMIPv6 domain.

The PMIPv6 protocol operation consists of five phases from network attachment to data transmission as shown in Fig. 2. In the first phase, MN requests an access authentication to MAG through L2 access, when MN attaches an access network that belongs to PMIPv6 domain. Then, the MAG, on receiving the authentication request of the MN, requests authentication with regard to

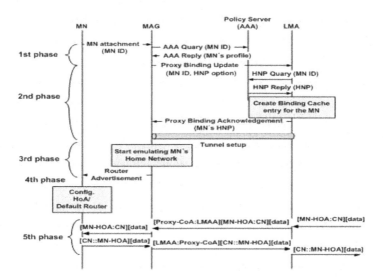

Fig. 2. Network Attachment Procedure in PMIPv6 Domain

the MN's ID to the policy server and retrieves the MN's profile (e.g., LMA address, and other address configuration parameters) if the MN is authorized for mobility service. The second phase is the location update, in which the MAG sends *Proxy Binding Update (PBU)* message to the LMA in order to register the current attachment point of the MN. The LMA registers the MAG address in regard to the MN and sends *Proxy Binding Acknowledgement (PBA)* message with the MN's HNP. It also creates the binding cache entry and sets up its endpoint of the bidirectional tunnel to the MAG. The MAG, on receiving the *PBA* message, sets up its endpoint of the bidirectional tunnel to the LMA and also sets up the data path for the MN's traffic. At this point the MAG has all the required information for emulating the MN's home link. In the third phase, the MAG emulates the MN's home interface on the access interface. Hence It sends *Route Advertisement(RA)* message to the MN on the access link advertising the MN's HNP as the hosted on-link-prefix. In the forth phase, the MN attempts to configure its interface either using stateful or stateless address configuration model after receiving the *RA* message on the access link.

Once the address configuration is completed, the MN has a valid address from its network prefix at the current attachment point. The serving MAG and LMA also have proper routing states for handling the traffic sent to or from the MN using an address from its HNP. Finally, the MN communicates with CNs in the fifth phase.

PMIPv6 does not use any mobility stack on the MN but rather uses the proxies on the access routers to help perform the mobility functions, such as the binding updates to the LMA. PMIPv6 focuses on extending MIPv6 to achieve mobility due to three main reasons. The first reason is that MIPv6 is a very mature mobility protocol for IPv6. Secondly, PMIPv6 allows re-using the MIPv6 home agent functionality to provide mobility to hosts without requiring any additional mobility management protocol. As the third reason, numerous MIPv6 enhancements can be re-used.

2.2 Benefit of PMIPv6

PMIPv6 is one of network-based localized mobility management solutions which can avoid both tunneling overhead over the air as well as hosts' involvement in mobility management. PMIPv6 has several benefits on the top of those of the previous works on the localized mobility management(LMM) as follows, since they are also the most important goals for IETF NetLMM Working Group[5].

– *Reduction in handover latency*: PMIPv6 can reduce the IP handover latency by limiting the mobility management(localized location registration) in the PMIPv6 domain.
– *Reduction in signaling overhead:* The mobility related signaling overhead over the air can be alleviated in the PMIPv6 since it can avoid tunneling overhead over the air and as well as the remote Binding Updates either to the Home Agent(HA) or to the Correspondent Node(CN) by adopting the MAG as a proxy mobility agent.
– *Provision for location privacy:* Keeping the MN's Home Address (MN-HoA) fixed over a PMIPv6 domain and enabling the MAGs to perform the proxy role of the MN minimize the chance that the attacker can infer exactly where the MN is located.

Beyond these benefits, supporting for heterogeneous wireless link technologies, supporting for unmodified MNs, and supporting for IPv4 and IPv6 have been identified as design goals for a possible IETF NeTLMM protocol [5]. The PMIPv6 has been developed based on the IETF NeTLMM's goal, it has potential to be deployed for real-time multimedia applications successfully in the near future. However, PMIPv6 should be complemented, since it has several problems to be ironed out, that we discuss in the next section.

2.3 Problem of PMIPv6

Even though PMIPv6 can reduce the handover latency, PMIPv6 is not enough to provide seamless mobility for real-time applications such as VoIP and mobile IPTV. The intra-LMA handover procedure of PMIPv6 consists of an L2 connection establishment and an L3 connection establishment that involves a network access authentication, a location registration with LMA, and an address configuration. We show the intra-LMA handover procedure and handover latency of PMIPv6 in Fig. 3.

As shown in Fig. 3, when MAG detects the MN's detachment, the MAG deregisters binding information on LMA. It incurs that LMA deletes binding and routing state for that MN and drops packets headed for that MN. Until new MAG re-registers the MN's attachment on the new link, all packets sent to

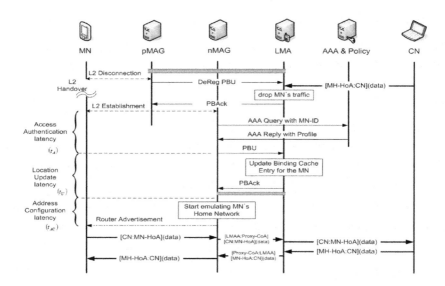

Fig. 3. Intra-LMA Handover Latency of PMIPv6

that MN are dropped. Therefore, the amendment of a point of time when MAG deregisters binding information on LMA is inevitable for seamless handover. In addition, MAG can send RA message to MN after receiving *PBA* message from the LMA. That is, MN confirms its home network prefix and knows new default routers after MAG receives MN's HNP from LMA. This procedure, address configuration, causes additional handover latency and packet loss.

2.4 Related Fast Handover Mechanisms

Since it is inevitably experience packet loss or delivery latency before it receives its HNP advertisement from the newly attached MAG when a MN moves to a new access network in PMIPv6 domain, several fast handover mechanisms have been introduced to reduce handover latency.

In [8], buffering packets at LMA during handover is proposed to avoid packet loss. It does not scale well because LMA must have buffers for all MNs currently associated with it. In [9], IAPP(Inter-AP Protocol) is adopted to reduce in authenticating and obtaining MN's profile. Still, it may experience on-the-fly packet loss between LMA and previous MAG. Moreover, since [8] and [9] buffer for incomming packet at the LMA, forwarding delay increases as the distance between LMA and MAG becomes longer. In [10] and [11], combining PMIPv6 with FMIPV6 is introduced. These schemes reduce packet loss and incoming packet delay significantly. However, it inherits potential risks of erroneous movement and out-of-order packets delivery problem from FMIPv6.

3 Proposed Scheme

In this paper, we consider a PMIPv6-based mobile system described in Fig. 4. In PMIPv6 specification, all packets originated from or sent to MNs within LMA are routed through LMA. When the number of MN in LMA increases, the traffic volume concentrating into the LMA increases dramatically. Therefore, we consider that a PMIPv6-based mobile system consists of several LMAs distributed geographically or topologically within an administrative domain, in order to resolve the scalability problem caused by traffic concentration. Since LMAs are in an administrative domain, we assume that LMAs share the profile information and can exchange messages with other LMAs.

When MN moves to new MAG in an LMA while preserving communication with a CN, L3 handover procedure starts only after L2 handover procedure comes to end. In the proposed mobile system, three handovers are defined: L2 handover, L3 handover within a LMA (intra-LMA handover) and L3 between LMAs (inter-LMA handover). We assume that each MAG has interface with connection to a distinct set of wireless access point. Since MAG is assigned IP address included in subnet prefix hosted at LMA and the P2P link between MN and MAG does not changes when MN moves within an MAG (L2 handover), L3 handover procedure is not required.

The distinct property of PMIPv6, that an MN's HNP is always hosted on the access link where it is anchored, is a foundation of our proposed mechanism. That is, if an MN enters in PMIPv6 domain at first and it is authenticated by the first attached MAG, then it can not receive other network prefix advertisements but its already assigned HNP from new MAGs, no matter how its prefix is obtained (e.g., through policy server or previous MAG). As discussed the problem of PMIPv6 before, if an MN does not receive its HNP as soon as its attachment point changes, packet loss would be happen. On the contrary, if new MAG knows some portion of the profile, especially HNP, of newly attached MNs in advance, new MAG can advertise MN's HNP as soon as MN attaches on its access network.

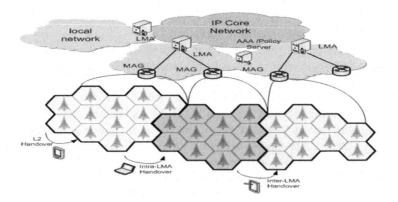

Fig. 4. Proxy Mobile IPv6-based Mobile System

In this paper, we define adjacent MAGs as the candidate MAGs to which the MN can handover and determine by L2 information. Adjacent MAGs of current MAG are geographically and administratively continuous with it. We assume that each MAG knows adjacent MAGs. When an MN moves and L2 event and information show the MN may move out the MAG, MAG sends MN's HNP to adjacent MAGs in advance in order to perform address configuration quickly after MN's handover. We can optimize PMIPv6 handover latency by sending MN's HNP to adjacent MAGs in advance. When an MN moves to a new access network, it can receive its HNP advertisement as soon as it attaches on the new access link. After this fast router advertisement, an MN is ready to receive packets from or send to LMA though new MAG. Even though fast router advertisement, the packet loss happens because an expeditious deregistration of binding on LMA.

To overcome this problem, we modify PMIPv6 signaling sequence in order that LMA keeps binding to previous MAG until new MAG receives PBA message from LMA. To prevent packet drop or buffering during this handover period, the tunnel between previous MAG to new MAG is used to receive packets from the previous MAG and to send outgoing packets in vice versa. Handover latency consists of an L2 handover, a network access authentication, a location update, and an address configuration. Among these procedures, we can skip network access authentication because adjacent MAGs have already obtained the mobile node's profile before handover or performed network access authentication prior to handover. Therefore, we can reduce packet reception latency and outgoing packet delivery latency since the MN receives RA message as soon as it attaches on new MAG.

For intra-LMA handover, 1) When current MAG detects MN's movement to another MAG by L2 *Link Going Down* event, current MAG establishes tunnels to adjacent MAGs. It also starts buffering packets of ongoing session for the MN and announces MN's HNP and ID. 2) After the new MAG detects MN's attachment, it sends FBU message to previous MAG to trigger transmission of buffed packets and sends PBU message to LMA. Previous MAG, on receiving FBU message, sends De-Reg PBU message to LMA concurrently. If adjacent MAGs does not receive L2 signal notifying MN's attachment, they remove pre-established tunnel and context information, such as MN's ID and HNP by timeout. 3) LMA, on receiving new PBU message for already bounded MN, sends PBA message to new MAG. LMA sends packets through bi-directional tunnel established between LMA and new MAG. 4) previous MAG, on receiving PBA message for De-Reg PBU message, transmits remaining packets quickly to new MAG and sends FBAck message indicating the buffer is empty and removes bidirectional tunnel. This signaling procedure also resolve out-of-order packets caused by different transmission paths from LMA to new MAG, since new MAG knows the end point of data transmission from previous MAG by receiving FBAck message which identifies the data packet on the buffer has forwarded to new MAG.

For inter-LMA handover, 1) when MN moves to new MAG within other LMA, previous MAG establishes bidirectional tunnel with new MAG in new LMA

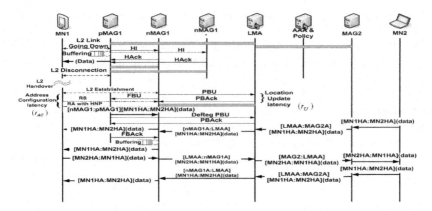

Fig. 5. Proposed signaling flow for Intra-LMA Handover

before L2 connection is disconnected. 2) When new MAG detects MN's attachment, it sends proxy BU to new LMA to establish bidirectional tunnel. And then, new LMA sends BU to previous LMA to establish bidirectional tunnel. 3) Previous MAG, on receiving FBU message, send De-Reg PBU message to previous LMA to remove the tunnel between previous MAG and previous LMA. By this procedure, proposed scheme has low latency handover latency performing the address configuration as soon as L2 handover and results in no packet loss.

4   Performance Analysis

In this section, we analyze numerically the proposed scheme, MIPv6, PMIPv6, and FPMIPv6 in terms of handover latency similar to that of [12] [13] [14].

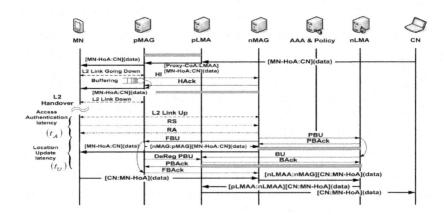

Fig. 6. Proposed signaling flow for Inter-LMA Handover

4.1   Handover Latency

With handover latency, we specify three latencies such as link switching latency[1], IP connectivity latency, and location update latency. The link switching latency (t_L) is due to L2 handover. The IP connectivity latency (t_I), which reflects how quickly an MN can send IP packets after L2 handover, is due to new IP address configuration and movement detection after L2 handover. An MN can send packets after IP connectivity latency affected address configuration latency. The location update latency (t_U) is the latency in forwarding IP packets to MN's new location.

In discussion on handover latency, we consider two scopes: when an MN sends packets and when the MN receives packets after L2 handover. The IP connectivity latency reflects how quickly an MN can send IP packets after L2 handover. To consider how quickly an MN can receive IP packets after L2 handover, we specify the *packet reception latency* (t_P), the period from the starting point of L2 handover to when an MN receives packets for the first time after L2 handover.

To analyze the handover latency, let us define the following parameters.

t_{MAG} : latency of an IP packet delivery between MAGs
t_{LMA} : latency of an IP packet delivery between LMAs
t_{ML} : latency of an IP packet delivery between MAG and LMA
t_{MN} : latency of an IP packet delivery between MAG and MN
t_{MH} : latency of an IP packet delivery between MAG and HA

In basic MIPv6, the *packet reception latency* t_P is $t_L + t_I + t_U$ and IP connectivity latency The address configuration time (t_{AC}) is the interval from the time an MN detects its movement to the time it assigns its interface with a new CoA (Care of Address) based on the prefix of the new MAG including AR(Access Router). In this step, an MN generates a new CoA and performs DAD procedure taking 1 sec to a minimum to determine if its new CoA does not conflict with others. It is obstacle to real-time service provisioning in the MIPv6 environment. The location update latency (t_U) is $2(t_{MN} + t_{ML})$ [12].

In Fig. 7, (a) represents the timing diagram of intra-LMA handover and (b) represents that of inter-LMA handover of PMIPv6. In intra-LMA handover, MN can send and receive packets after bidirectional tunnel , since the MAG emulates the MN's home interface on the access interface. The *packet reception latency* in intra-LMA handover is $t_L + t_A + t_U + t_{MN}$ and the packet reception in inter-LMA handover is $t_L + t_I + t_U + t_{MN}$ where the IP connectivity latency is $t_{MD} + t_{AC}$ and the location update latency is $t_{BU} + t_{nLM} + t_{New}$. t_{BU} is the round-trip delay between an MAG and an LMA and t_{nLM} is the round-trip delay between new LMA(nLMA) to previous LMA(pLMA).

Fig. 8 shows the timing diagram corresponding to the proposed scheme. The proposed scheme improvement *packet reception latency* preponderantly as described in the previous section. The *packet reception latency* is $t_L + 2t_{MAG} + t_{MN}$.

---

[1] Link switching latency is the same as L2 handover latency.

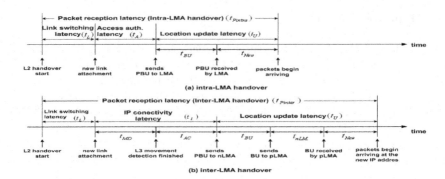

Fig. 7. Timing diagram in basic PMIPv6

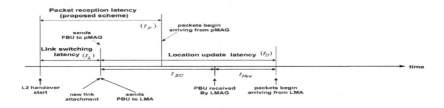

Fig. 8. Timing diagram in Proposed Scheme

Table 1. Handover Performance Comparison

Feature	PMIPv6	FPMIPv6 [11]	Proposed Scheme
Packet loss	$(t_L + t_I + t_U)$R	depends on prediction	No
Reception latency	$t_L + t_I + t_U + t_{MN}$	$t_L + 2t_{MAG} + t_{MN}$	$t_L + 2t_{MAG} + t_{MN}$
Outgoing latency	$t_L + t_I + t_U$	$t_L + t_{MN}$	$t_L + t_{MN}$
note	R : transmission rate	prediction error exits	robust for any mobiity

The *packet reception latency*, *packet loss*, and *outgoing delivery latency* of basic PMIPv6, FPMIPv6, and proposed scheme are summarized in Table 1.

5   Conclusion

In this paper, we study the details of PMIPv6 handover procedure based on the recent IETF draft document, and propose a low-latency handover scheme to reduce handover latency, especially *packet reception latency* and outgoing packet delivery latency for both intra- and inter-LMA movement. In addition, we compare handover latency of the proposed scheme, basic PMIPv6 by using timing diagram.

In order to increase the benefit of PMIPv6 and to enhance PMIPv6 for low latency and seamless handover, our proposed scheme defines adjacent MAGs for MAG to send MN's profile, especially HNP, in advance and uses bi-directional tunnel between current MAG and adjacent MAGs and buffering in the current MAG to reduce the handover latency and packet loss. In the future work, we plan to extend proposed scheme with the case where the route optimization, multi homing, and simultaneous movement scenarios are required. In addition, we will implement the proposed scheme on the embedded devices for multimedia streaming system and evaluate the performance.

References

1. Johnson, D., et al.: Mobility Support in IPv6. IETF RFC 3775 (2004)
2. Soliman, H., et al.: Hierarchical Mobile IPv6 Mobility Magement (HMIPv6). IETF RFC 4140 (2005)
3. Koodli, R.: Fast Handovers for Mobile IPv6. IETF RFC 4068 (2005)
4. Kempf, J.: Problem Statement for Network-Based Localized Mobility Management (NETLMM). IETF RFC 4830 (2007)
5. Kempf, J.: Goals for Network-Based Localized Mobility Management (NETLMM). IETF RFC 4831 (2007)
6. Gundavelli, S., et al.: Proxy Mobile IPv6. IETF draft-sgundave-mipv6-proxymip6-15 (May 2008)
7. Kim, P.-S., et al.: Proactive Correspondent Registration for Proxy Mobile IPv6 Route Optimization. International Journal of Computer Science and Network Security 7(11) (2007)
8. Laganier, J., et al.: Travelling without Moving: 802.11 Access Points backed by Secure NETLMM. ICCCN (2007)
9. Lee, J.-C., Kaspar, D.: PMIPv6 Fast Handover for PMIPv6 Based on 802.11 Networks. IETF draft-lee-netlmm-fmip-00 (July 2007)
10. Yokota, H., et al.: IETF draft-yokota-mipshop-pfmipv6-02 (February 2008)
11. Xia, F., et al.: IETF draft-xia-netlmm-fmip-mnagno-02 (November 2007)
12. Kim, G., Kim, C.: Low-latency Non-predictive Handover Scheme in Mobile IPv6 Environments. In: Niemegeers, I.G.M.M., de Groot, S.H. (eds.) PWC 2004. LNCS, vol. 3260. Springer, Heidelberg (2004)
13. Han, Y.-H., Jeong, D.: A Comprehensive Study on Handover Performance of Hierarchical Mobile IPv6. In: Sha, E., Han, S.-K., Xu, C.-Z., Kim, M.-H., Yang, L.T., Xiao, B. (eds.) EUC 2006. LNCS, vol. 4096. Springer, Heidelberg (2006)
14. Lei, J., Fu, X.: Evaluating the Benefits of Introducing PMIPv6 for Localized Mobility Management. TR IFI-TB-2007-02, University of Goettingen, Goettingen, Germany (2007)

Effects of Interaction between Transport and Application Layers on SIP Signaling Performance

Masataka Ohta

Faculty of Business Administration, Kanagawa University, Japan
m-ohta@kanagawa-u.ac.jp

Abstract. Session Initiation Protocol (SIP) is an application-layer protocol to handle sessions between two points. SIP is implemented on the
top of the transport protocols, such as user datagram protocol (UDP) or
transmission control protocol (TCP). SIP messages are transmitted by
UDP or TCP. We focus on an interaction between the application layer
(SIP) and the transport layer (UDP or TCP). The paper studies how
the interaction affects performance of SIP signaling. A significant performance difference was found to exist because of the interaction. In the
case of SIP over UDP, retransmissions of SIP messages decrease throughput. On the other hand, in the case of SIP over TCP, a large TCP buffer
causes a large call setup delay although TCP improves throughput.

1 Introduction

Most of Internet telephony services are realized using Session Initiation Protocol
(SIP)[2,3]. SIP is in the center of efforts toward the convergence of all telecommunications on to an IP-based network. The telephone services, which are currently
provided by the public switched telephone network (PSTN), will be one of the
major services in the IP-based network. However, there are hurdles that must
be overcome before Internet telephony can be adopted on a widespread basis. In
the PSTN, it is well known that an overload degrades the service quality, such
as throughput and call setup delay. It is expected that the Internet telephony
also encounters an overload. One of the purposes of the study is to investigate
what happens on the Internet telephony when an overload is applied. Some studies have reported that the performance of Internet telephony decreases during
a network overload [1]. When a SIP server is overloaded, the application(SIP)
and the transport layers interact each other. We investigate how the interaction
affects the performance of SIP signaling.

Since SIP is an application-layer protocol, SIP messages have to be transmitted by a transport protocol. The study is focused on the performance of SIP
signaling rather than that of media packets. RFC3261 states that supporting
user datagram protocol (UDP) [4] and transmission control protocol (TCP) [5]
is mandatory in SIP implementations. Thus, we consider UDP and TCP as the
transport protocols. RFC3261 has defined retransmission of SIP messages for

S. Balandin et al. (Eds.): NEW2AN 2008, LNCS 5174, pp. 122–133, 2008.
© Springer-Verlag Berlin Heidelberg 2008

unreliable transports (UDP). For reliable transports (TCP), SIP does not deal with the retransmission. Since the behaviour of SIP signaling is different from the transport protocols, we expect that the performance of SIP signaling depends on the transport protocols.

An overload can lead two types of congestion. One is a network congestion in which packets are lost in the IP layer. The other is a server congestion in which a load is concentrated at a particular server. As the result, the server will be overloaded. The paper focuses the server congestion. This type of congestion is observed in the PSTN. Telephone traffic sometimes is concentrated at a specific telephone exchange, then the exchange will be overloaded and suffered from the server congestion. In the IP-based network, it also is expected that traffic created by SIP users is concentrated at a particular SIP server causing the server congestion. We evaluate performance of SIP signaling under the server congestion, and investigate what happens when concentrated traffic is applied to a particular SIP server.

Most of existing studies have been focused on the network congestion. Camarillo et al. [8] studied SIP signaling performance by comparing performances under UDP, TCP, and stream control transmission protocol (SCTP). They have considered the head-of-line blocking effects of TCP caused by packet losses in the IP layer, which lead to a large delay in SIP message transmission. Lulling et al. [6] also studied effects of packet losses in the IP layer on a SIP message transmission focusing on TCP congestion controls. However, these studies did not evaluate end-to-end performance. They focused SIP message transmission between neighbour SIP proxies. We evaluate end-to-end performance focusing on the server congestion. Eyers and Schulzrinne [9] evaluated and compared call setup delays for H.323 over TCP and SIP over UDP. They did not consider the case of SIP over TCP. Fathi et al. [10] analysed call setup delay in a wireless environment. These studies examined the effect of packet losses on delay, but did not consider throughput and the server congestion.

We uses the network simulator ns-2 [11]. We implemented new types of agents, which act as user agents (UAs) and SIP proxy servers in the ns-2. We also developed a TCP agent, which accommodates connection management and window control. We need to develop a new TCP agent. With the existing TCP agents, an application process for TCP is assumed to be always ready for consuming everything in the receiving buffer. The agent, therefore, does not simulate blocking of receiving TCP buffer. To evaluate the interaction between the application and the transport, we have developed a more sophisticated buffer handling.

2   Session Initiation Protocol

Figure 1 shows a typical SIP message exchange used to establish a session. SIP is based on an HTTP-like request/response transaction model and uses several two-way handshakes. On receiving the request "INVITE," the provisional response "100Trying" is returned. This two-way handshake is between neighboring SIP nodes – a source UA and a SIP proxy, a SIP proxy and a SIP proxy, or a SIP proxy and a sink UA. On receiving the final response "200OK," the request "ACK" is

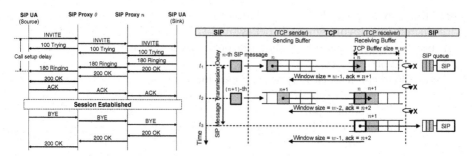

Fig. 1. A typical SIP message exchange

Fig. 2. Window control in TCP

returned to acknowledge the receipt of the final response. On receiving the request "BYE," the response "ACK" is returned. These two-way handshakes are between the end points – a source UA and a sink UA. Once the source UA receives the final response "200OK," the request "ACK" can be routed directly to the sink bypassing intermediate hops. The request "BYE" and the response "200OK" for "BYE" can also go directly. However, the paper assumes that each SIP message traverses all hops. The complete SIP signaling route can be recorded using the record-route header field in the SIP message [3]. A service provider should know call duration to calculate the service charge. To know the durations, the SIP proxies have to know the time at which the call is cleared in addition to the time at which the session is established. In other words, the SIP proxies have to receive all the SIP messages. In this situation, the request "ACK," "BYE," and response "200OK" have to traverse all hope.

In the study, we use the term "throughput" in the sense of the rate of successfully completed sessions, i.e., the number of calls that complete all of the message exchange shown in Figure 1 within a second. We also define "call setup delay" as the interval between the moment when a source UA sends the initial "INVITE" and the moment when the source UA receives "180Ringing." Figure 1 shows the call setup delay according to this definition.

3   SIP Transport Protocols

3.1   SIP over UDP

Since UDP sends data without confirmation, packet losses in a network can occur. For UDP, therefore, the application layer must detect and recover from packet loss. RFC3261 defines retransmission procedures to improve the reliability of transmitting SIP messages. As mentioned in Sec.2, SIP uses several two-way handshakes: INVITE-100Trying, 200OK-ACK and BYE-ACK. RFC3261 has defined two retransmission types – one is for INVITE transaction, which contains "INVITE-100Trying," and the other is for non-INVITE transaction, which contains "200OK-ACK" and "BYE-200OK."

In the INVITE transaction, the client transaction sends the message "IN-VITE," and on receiving "INVITE," the server transaction returns the response "100Trying". For unreliable transports (UDP), the client transaction retransmits the "INVITE" at an interval that starts at T1 seconds and doubles after every retransmission. The timer T1 is defined in RFC3261. The client transaction ceases retransmission when it receives the provisional response "100Trying" or when $64 \times$T1 has passed since the initial INVITE was sent. The default value for T1 is 500 ms [3], thus, the "INVITE" is retransmitted at intervals of 0.5, 1.0, 2.0, 4.0, 8.0 and 16.0 seconds. After 32 seconds without any response, the client transaction ceases retransmission.

The retransmission procedure of the non-INVITE transaction is basically the same. This time RFC3261 introduces timer T2. For UDP, the "200OK" or "BYE" is retransmitted at an interval starting at T1. The interval is doubled after every retransmission, capping off at T2. The default value of T2 is 4.0 seconds. Therefore, the "200OK" and "BYE" are retransmitted at intervals of 0.5, 1.0, 2.0, 4.0, 4.0, 4.0, 4.0, 4.0 and 4.0 seconds. After 32 ($64 \times$T1) seconds in total, the retransmission is ceased.

Although these retransmissions improve the reliability, the retransmissions increase a load applied to a SIP server and may affect the SIP signaling performance. The paper evaluates the effects of the retransmissions.

3.2   SIP over TCP

In case of TCP, SIP does not retransmit SIP messages. TCP flow control can protects from packet loss in the application (SIP) layer. However, the flow control makes the SIP message transmission delay large.

Figure 2 shows a scenario that leads to a large transmission delay. The scenario depends on the implementation, so we make several assumptions. Each TCP connections accommodate the receiving and sending buffers as shown in Figure 2. We let w [segments] be the TCP buffer size (Although TCP describes the size and, the sequence and acknowledgement number in bytes, we describe them in segments for simplicity). When SIP (sender side) passes n-th SIP message to a TCP sender at time t_1, the TCP sender stores it in the sending buffer to wait for transmitting by the IP layer. Then, the TCP sender transfers n-th SIP message to a TCP receiver. The TCP receiver receives the SIP message through the IP layer and stores it in the receiving buffer. If the SIP queue, which is in the application (SIP) layer, is not full, the message can be forwarded to the SIP queue immediately. Then, the TCP receiver clears the receiving buffer advertising the acknowledge number $n + 1$ [segment] and a window size of w to the TCP sender.

If the SIP queue is full, the message cannot be forwarded to the SIP queue. There can be two scenarios. One is that the received message is discarded at the SIP queue, just like UDP. The other is that the received message waits until the SIP queue becomes free. In the scenario, the n-th SIP message remains in the receiving buffer. The free space of the receiving buffer is reduced from w to $w - 1$. Accordingly, the TCP receiver advertises a reduced window size of $w - 1$ as shown in Figure 2. The TCP receiver also advertises the acknowledge number of $n + 1$, because the TCP receiver completes to receive n-th SIP message. In the figure, we assume that the SIP queue is full until time t_3. All SIP massages which arrive during the period between t_1 and t_3 are stored in the receiving buffer, and the advertising acknowledgement number increases. On the other hand, the advertised window size is reduced, because the received SIP messages occupy the receiving buffer and reduce the free space in the receiving buffer. As shown in the figure, when the $(n + 1)$th SIP message are transferred at time t_2, the message is stored in the receiving buffer. Then the TCP receiver advertises the acknowledgement number of $(n + 2)$ and the reduced window size of $w - 2$. We assume an application process assigned to each TCP receiver. The process is periodically (100 ms) invoked to attend transferring the message stored in the receiving buffer. When the application process finds an empty buffer in the SIP queue at time t_3, it transfers the n-th SIP message stored in the receiving buffer to the SIP queue. Then, the transmission of the n-th SIP message is completed and the TCP receiver clears the receiving buffer, advertising a window size of $w - 1$ segments to notify a change of the window size (window update). In this scenario, the SIP message transmission delay is $t_3 - t_1$ as shown in Figure 2. As explained here, although discarding a SIP message at the SIP queue is prevented, the SIP message transmission delay can be large when a SIP server is overloaded. If the TCP receiving buffer becomes full during the period between t_1 and t_3, the TCP receiver advertises the window size of 0. Then, the TCP sender does not transfer another messages. If the $(n + 1)$th SIP message arrives at the TCP sender when the window size is zero, the message is stored in the sending buffer. The $(n + 1)$th message waits for transmission until the window size becomes non-zero. If too many SIP messages arrive at the TCP sender before the window size becomes non-zero, the sending buffer will be full. In the case of non real-time protocol, such as FTP, an application will stops and waits to transfer when the buffer is full. Since SIP is a real-time protocol, we assume that SIP messages that encounters the sending buffer full are discarded. Thus, even if the TCP flow control protests from packet loss, the SIP message can be discarded. The TCP sender must send at least one octet of data even if the window size is zero, when no window update is received. The recommended retransmission interval is two minutes[5]. We assumes that the window update is received within two minutes.

3.3   Interaction between Transport and Application Layers

We summarise the interaction between the application (SIP) and transport layers caused by the server congestion. As explained above, there are three types of interactions.

SIP over UDP: When queue of a SIP proxy server is full, a SIP message is discarded. SIP detects this discard by a time out. SIP then retransmit the discarded message. In this case, UDP simply forwards the messages from/to the IP layer and the application layer. We express this interaction as UDP.

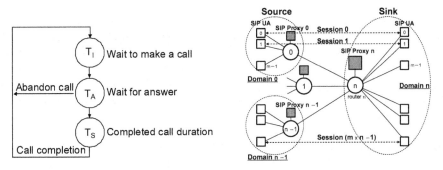

Fig. 3. User model for a source and a sink   **Fig. 4.** Network Model for SIP signaling
UA

SIP over TCP(discard): When queue of a SIP proxy server is full, a SIP
message is discarded just like the case of SIP over UDP. Since RFC3261 does
not define the retransmission for the reliable transport, SIP does not retransmit
the discarded message and the call is cleared. We express this interaction as
TCP(discard).

SIP over TCP(wait): When queue of a SIP proxy server in the application
layer is full, a received message is kept in the TCP receiving buffer in the transport layer to wait for the SIP queue becoming free. This prevents a message loss,
but enlarges a SIP message transmission delay. We express this interaction as
TCP(wait).

4   Network and Simulation Model

4.1   User Model for the UA

Figure 3 shows the user behavior. A Source UA makes another call in T_I seconds
after the UA terminates a call. The called user assumed to answer the call in
T_A seconds. The duration of session is T_S seconds. These values are assumed
to be exponentially distributed. Therefore, every source UA makes a call every
T_{call} $(= T_I + T_A + T_S)$ seconds. The user may abandon the call, if the user does
not hear a ring back tone or the called user does not answer the call. A user
is assumed to abandon a call after a time T_{abdn} seconds. T_{abdn} is assumed to
be normally distributed. The average value and standard deviation of T_{abdn} are
assumed to be 20 seconds and 20/3, respectively. The user may retry to make
a call after abandoning a call. However, we do not consider this to simplify the
situation. In the study, $T_I = 30$ sec, $T_A = 4$ sec, and $T_S = 30$ seconds in average.
Therefore, every source UA makes a call every 64 seconds.

4.2   Network Configuration

Figure 4 shows the SIP signaling network to be studied. As explained in Sec.1,
we focus on the server congestion. The network model is chosen so that SIP

signaling traffic generated by source UAs is concentrated to the SIP proxy n. So, the server for the SIP proxy n will be congested. This situation occurs when a huge number of users try to make a call to the same destination (domain).

SIP UAs are shown by white squares. Gray squares indicate SIP proxies and white circles indicate routers. Each SIP element is connected by links. Areas surrounded by the dotted circles indicate domains. Each source domain 0 to $n-1$ contains m SIP UAs and 1 SIP proxy. The sink domain shown in the right hand side contains $n \times m$ SIP UAs and 1 SIP proxy. $n \times m$ pairs of source and sink SIP UA are assumed to try establishing sessions. Since each sink UA belongs to the same domain, all SIP messages are routed to the SIP proxy n, which is in the sink domain. Consequently, the SIP proxy n is expected to be bottlenecked.

4.3   Queueing Model for the SIP Singaling Network

The simulations were carried out based on the queueing model shown in Figure 5. Each layer accommodates buffers and queues. Each SIP proxy has a queue through which it receives SIP messages from the transport layer. In the scenario, all SIP messages are routed to the SIP proxy n. So, the queue of SIP proxy n, shown as "Bottleneck queue" in Figure 5, is expected to be bottlenecked when an overload is applied. When TCP is used for the transport protocol, TCP

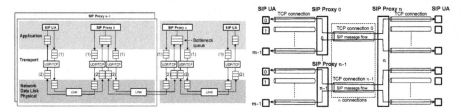

Fig. 5. Queueing Model for the SIP signaling network

Fig. 6. TCP connections in the SIP signaling network

connections are established between SIP elements. Each TCP connection has a buffer pair - sending ((1) in Fig.5) and receiving ((2) in Fig.5) buffer. Figure 6 shows TCP connections, where n TCP connections are established between SIP proxy 0 to $n-1$ in the source domains and the SIP proxy n in the sink domain. The SIP message flows created in the source domain i ($= 0$ to $n-1$) and the sink domain are bundled the same TCP connection-i. Namely, proxy-to-proxy communication involves multiple simultaneous SIP message flows. TCP connections are also established between SIP proxies and each SIP UA. These TCP connection bundles a single SIP message flow.

Since we focus on the server congestion, we do not consider the packet losses in the network, data link and physical layers. In the scenario, the SIP proxy n is expected to be bottlenecked and be suffered from the server congestion.

4.4   Traffic Model

As explained in Sec.4.1, every source UA makes a call every T_{call} ($= T_I + T_A + T_S$) seconds. As shown in Figure 1, 7 SIP messages (INVITE, 100, 180, 200, ACK, BYE, and 200) are involved in completing a call. Since all SIP messages are routed to the SIP proxy n, the offered load to the SIP proxy n ρ_n can be expressed as

$$\rho_n = \lambda/\mu_{sip} = \frac{7}{(T_I + T_A + T_S) \cdot \mu_{sip}} \cdot n \cdot m \tag{1}$$

where λ denotes the average SIP message arrival rate at the SIP proxy n, $n \cdot m$ is the total number of source UAs, and μ_{sip} is the average service (processing) rate of the SIP proxy. $\rho_n > 1$ means that the SIP proxy n is overloaded.

The service time is strongly dependent on an implementation, and is determined by many factors, such as SIP parsing, routing, a process scheduler of OS, and so on[12]. In the study, we simply assume the service time of the SIP proxy for each SIP messages. The service times are 24.096msec for INVITE, 2.048msec for 100Trying, 180Ringing and 200OK, and 4.096msec for BYE and 200OK. The service time for INVITE is larger than that of others, because that includes a delay for an address resolution and a processing time for a routing. These service times are assumed to constant. The average service time is $5.78 \times 10^{-3} (= 1/\mu_{sip})$ seconds, which means that the SIP proxy can handle 173 messages in a second. Since 7 SIP messages are involved in a call completion, the maximum throughput of the SIP proxy is $24.7 (= 173/7)$ [calls/sec]. Using assumed values for T_I, T_A, T_S and μ_{sip}, the offered load to the SIP proxy n is calculated as

$$\rho_n = 0.63 \times n \cdot m \times 10^{-3} \tag{2}$$

The offered load to the SIP proxy 0 to $n - 1$, denoted by ρ_i is

$$\rho_i = 0.63 \times m \times 10^{-3} \qquad (i = 0 \quad to \quad n - 1) \tag{3}$$

As explained in Sec.3, the average SIP message size is 658.3 bytes. The average transmission time of the link is 0.53 ms assuming that each link speed is 10 Mbps. Therefore, the usage rate of the link connecting the SIP proxy n and router n, denoted by ρ_{link}, is calculated as

$$\rho_{link} = \lambda \times 0.53 \times 10^{-3} = 0.058 \times n \cdot m \times 10^{-3} \tag{4}$$

Since $\rho_n \geq \rho_i$ and $\rho_n > \rho_{link}$, the SIP proxy n will be bottlenecked. We also assume that "INVITE" and "BYE" are 1024 bytes and the other SIP messages are 512 bytes (658.3 bytes on average). Therefore, we do not consider IP fragmentation.

5   Effects of Interaction between Transport and Application Layers

We assume that the buffer size for all of the SIP proxy is 40 [segments]. This means that 40 SIP messages can be queued in total. The buffer size of the queues in the network layer is assumed to be 50. For simplicity, we assume that sizes of SIP messages are smaller than MSS (Maximum Segment Size) so as not to causes a fragmentation. T1 and T2, which are the retransmission parameters for UDP, are also assumed to be 0.5 seconds and 4.0 seconds, respectively (these are the default values in RFC3261). All links in Figure 4 are assumed to be 10 Mbps(link speed) and 10 ms(propagation delay). Traffic parameters for the SIP signaling, such as the arrival rate of SIP message and service rate of SIP proxy, are shown in Sec.4.4.

Figure 7 shows throughput characteristics, which compare the effects of the interactions described in Sec.3.3, these are UDP, TCP(discard) and TCP(wait). In the simulation study, fixing n to 10, we change the value of m to change ρ_n. The throughput increases lineally until ρ_n approaches 1. No significant difference is observed for $\rho_n < 1$. The throughput shows the maximum value when ρ_n is around 1. The maximum value is almost the same as the maximum throughput of SIP proxy (24.7sec: see Sec.4.4). This is because that the bottleneck is the SIP proxy n in the scenario. Thus, the throughput of the SIP signaling network is limited with the value of the maximum throughput of the SIP proxy n. When ρ_n exceeds 1, namely the SIP proxy n is overloaded, the throughput decreases. In the case of UDP, the throughput decreases rapidly with ρ_n. The throughput is almost 0, when ρ_n is 3. In the case of TCP (discard), although the throughput is slightly improved, the throughput also decreases rapidly. TCP (wait) gives a better overload performance compared to the other cases. Figure 7 also shows the dependency of TCP buffer size. The figure shows the cases of TCP buffer size 10 and 20. In the figure, the number 20 in TCP(wait,20) means that TCP buffer size is 20, for instance. In the case of TCP (discard), the TCP buffer size does not influence on the throughput. On the other hand, in TCP (wait) increasing the buffer size from 10 to 20 slightly improves the throughput. In order to see what causes the throughput degradation, Figure 8 shows SIP message loss probability at the queue of SIP proxy n. As explained in Sec.3.3, TCP(wait) prevents the message loss in the application layer. So, the loss probability for TCP(wait) in Figure 8 is 0. In the case of UDP, SIP detects the message loss by a timeout. Followed by the timeout, the UAs and SIP proxies begin to retransmit SIP messages as explained in Sec.3.1. The retransmission amplifies the offered load, and the message loss probability increases as the offered load increases. The SIP

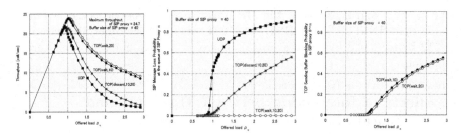

Fig. 7. Throughput Characteristics   **Fig. 8.** SIP message loss probability   **Fig. 9.** TCP sending buffer blocking probability

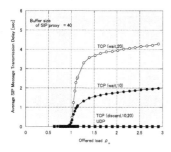

Fig. 10. Average SIP message transmission delay

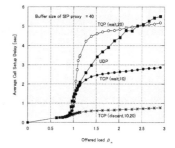

Fig. 11. Average call setup delay

message loss leads to retransmission, which consumes network resources, such as the CPU cycles of the SIP proxies, thereby decreasing the throughput. In case of TCP(discard), the message loss also increases as the offered load increases. Since SIP over TCP(discard) does not perform retransmissions, the loss probability is smaller than the case of UDP.

In case of TCP(wait), the TCP receiving buffer prevents discarding a SIP message at the SIP queue, as explained in Figure 2. When the TCP receiver finds that the queue of the SIP proxy n is full, it will not forward a received SIP massage to the SIP proxy. Instead, it preserves the received SIP message in the receiving buffer of the TCP connection reducing the window size. Then, the TCP sender reduces a transmission speed. This leads to increase occupancy of the TCP sending buffer. As the result, TCP sending buffer blocking occurs in SIP proxies $0 \sim n-1$. Figure 9 shows TCP sending buffer blocking probability in SIP proxy $0 \sim n-1$. The blocking probability for TCP(wait) increases with ρ_n. The blocking probability for TCP(wait) is almost the same as the SIP message loss probability for TCP(discard), and is smaller than the SIP message loss probability for UDP. The figure shows the cases that TCP buffer sizes are 10 and 20. The dependency of the TCP buffer size is not significant. Figure 10 shows the average SIP message transmission delay between the SIP proxies. The delay is the time observed in the transport layer from the instant that the SIP proxies $(0 \sim n-1)$ forward a message to the corresponding TCP sender to the instant that the TCP receiver forwards the message to the SIP proxy n. In the case of UDP and TCP(discard), the delay is almost 0, even if the overload is applied, because UDP and TCP(discard) give up delivering a message immediately when it finds that the queue of the SIP proxy n is full. In the case of TCP(wait), when rho_n exceeds 1, the delay increases rapidly. The delay is strongly dependent on the TCP buffer size. These delays come from the waiting time for passing a SIP message to the queue of the SIP proxy n, as explained in Figure 2. For a large TCP buffer size, the delay for TCP(wait) becomes very large. The SIP message transmission delay enlarges the call setup delay.

Call setup delay (defined in Sec.2) is an important performance measure from the user's perspective. Figure 11 shows the average call setup delay. When ρ_n exceeds 1, the call setup delays for UDP and TCP(wait) increase rapidly. In the

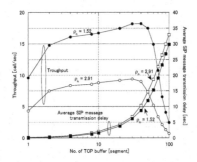

Fig. 12. Dependency of TCP buffer size for TCP(wait)

case of UDP, the loss of SIP messages in the application layer lead to the large call setup delay. Once a SIP message is lost, the message is retransmitted as explained in Sec.3.1. The retransmission intervals are doubled after each retransmission. The retransmission interval makes the call setup delay large. The call setup delay for TCP(wait) is strongly dependent on the TCP buffer size. For a large TCP buffer size, the delay can be very large. Thus, the TCP buffer size for TCP(wait) should not be too large. Comparing with Figure 10, we recognize that the SIP message transmission delay for TCP(wait) is dominant in the call setup delay for TCP(wait). In the case of TCP(discard), the call setup delay is rather small. In the case of TCP(discard), SIP does not perform retransmission, and TCP receiver does not wait until the queue of SIP proxy is not full. TCP(discard) gives a better performance concerning to the call setup delay.

TCP(wait) provides a better throughput than UDP and TCP(discard). However, TCP(wait) provides a large call setup delay for a large TCP buffer size as shown in Figure 11. Figure 12 shows a dependency of TCP buffer size on the throughput and SIP message transmission delay for TCP(wait). As shown the figure, there is the peak in the throughput. When the buffer size is too small or too large, the throughput decreases. The average SIP message transmission delay increases, as the buffer size increases. Especially, the delay grows rapidly when the buffer size exceeds 20. As the result, the call setup delay for a large TCP buffer size can be larger than that for UDP. Thus, we should carefully consider the TCP buffer size. Choosing an appropriate TCP buffer size, TCP(wait) can keep the call setup delay an acceptable value. If the call setup delay is the most important factor for a service provider, TCP(discard) is better choice. In the numerical example, the preferable TCP buffer size should be around 10.

6   Conclusions

We studied the effects of interaction between the transport and the application (SIP) layer. In the case of UDP, a SIP message is simply discarded when the message encounter full of buffer of a SIP proxy. For TCP, we considered two scenarios. One is TCP(discard) in which a SIP message is discarded just like the case of UDP. The other is TCP(wait) in which a SIP message is remained in TCP buffer and waits until buffer of a SIP proxy becomes free.

We have found several differences in the performance of SIP signaling which come from the interactions between the application and the transport layers. Considering these results, the paper suggests that TCP(wait) suits for the transport layer. However, we need careful consideration in TCP buffer size. TCP buffer size should not be too large so that the call setup delay is kept an appropriate value.

The paper does not considered the network congestion which leads packet losses in the network layer. The packet loss invokes a TCP congestion control which reduce a transmission speed. The congestion control may affects the SIP signaling performance. This issue is a future study.

References

1. Govind, M., Sundaragopalan, S., Binu, K.S., Saha, S.: Retransmission in SIP over UDP - Traffic Engineering Issues. In: Proc. of International Conference on Communication and Broadband Networking, Bangalore (May2003)
2. Schulzrinne, H., Rosenberg, J.: The Session Initiation Protocol: Internet-Centric Signaling. IEEE Communication Magazine 38(10), 134–141 (2000)
3. Rosenberg, J., et al.: "SIP: Session Initiation Protocol", RFC3261 (June 2002), http://www.ietf.org/rfc/rfc3261.txt
4. Postel, J.: User Datagram Protocol, RFC768, IETF (August 1980)
5. Postel, J.: "Transmission Control protocol", RFC 793 (September 1981)
6. Lulling, M., Vaughan, J.: A Simulation-based Comparative Evaluation of Transport Protocols for SIP. Computer Communications 29(4), 525–537 (2006)
7. Allman, M., Paxson, V., Stevens, W.: "TCP Congestion Control", RFC2581, IETF (April 1999)
8. Camarillo, G., Kantola, R., Schulzrinne, H.: Evaluation of Transport Protocols for the Session Initiation Protocol. IEEE Network 17(5), 40–46 (2003)
9. Eyers, T., Schulzrinne, H.: Predicting Internet Telephony Call Setup Delay. In: Proc. 1st IP-Telephony Workshop, pp. 107–126 (April 2000)
10. Fathi, H., Chakraborty, S., Prasad, R.: On SIP Session Setup Delay for VoIP Services Over Correlated Fading Channels. IEEE Trans. on Vehicular Technology 55(1), 286–295 (2006)
11. VINT Project: "Network simulator ns-2", http://www.isi.edu/nsnam/ns/
12. Wanke, S., Scharf, M., Kiesel, S., Wahl, S.: Measurement of the SIP Parsing Performance in the SIP Express Router. In: Pras, A., van Sinderen, M. (eds.) EUNICE 2007. LNCS, vol. 4606, pp. 103–110. Springer, Heidelberg (2007)

Modeling Long-Range Dependent VBR Traffic Using Synthetic Markov-Gaussian TES Models

I-Hui Li

Department of Information Management
Ling Tung University, Taichung 408, Taiwan
sanity@mail.ltu.edu.tw

Abstract. Recent measurement studies of network traffic and variable bit rate video indicate that the traffic exhibits long-range dependence (LRD). It becomes more and more important to model this kind of traffic. This paper presents a traces-generating framework based on TES (Transform-Expand-Samples) and simple synthetic Markov-Gaussian processes for modeling LRD traffic with variability over several time scales. All of the traffic studies showed that the measurement exhibits *approximate* second-order self-similarity. The network resource is limited and the *real* long-range dependent traffic has no room under the circumstances. The proposed framework can fit both the probability density function of the empirical traces and the autocorrelation function spanning over several time scales. Besides, we discuss the validity of approximate LRD modeling with the short-range-dependent approaches.

Keywords: long-range dependence, traffic modeling, self-similar, Gaussian-Markov processes.

1 Introduction

The world of networks becomes more and more complex. Plenty of information that can be transferred in the real life will be exchanged in the networks. Recent traffic measurement studies ([1][2][3][4][5]) show that the behavior of traffic is very different from traditional Poisson. It is not surprising because the employment of Poisson is based on the traffic characteristics of the POTS (plain old telephone system). Today's multimedia network is far from POTS, so the traffic characteristics are different from the traditional Poisson without surprise.

One characteristic of the observed traffic is that the autocorrelation presents the power law and non-summability. The finding shows that the slow-decaying autocorrelation spans over several time scales. It has motivated the assertion that the observed traffic exhibits second-order long-range dependence. This assertion is disputable. The question is whether the long-range dependence [6] can affect in the resource-limited networks. Moreover, it brings some debate within the traffic-modeling community about the philosophical property of the traffic. However, the observed traffic shows that the slow-decaying autocorrelation spans over several time scales. The traffic modeling mainly concerns about the performance evaluation with

S. Balandin et al. (Eds.): NEW2AN 2008, LNCS 5174, pp. 134–146, 2008.

given certain characteristics and about exploring the predictive performance when the characteristics change. The physical explanation of the choice of what kind of models is less important than above.

The finding of slow-decaying autocorrelation has inspired the research in the area of self-similar traffic models [7][8], and second-order self-similar traffic models such as Fractional Brownian Motions [9][10] and Fractional AutoRegression Integrated Moving Average. The tools for analyzing queuing behavior are very rare. It is also difficult to produce data traces with this kind of model because of its long-range dependence.

In this paper we suggest superposition of Markov-Gaussian processes to model the observed traffic. The Markov-Gaussian process is simple. It only models lag one autocorrelation. Each state depends upon the previous state before it and so on. By this relationship, any given state may be thought to be dependent directly or indirectly on all states preceding it and influential in all states successive to it, in sliding scale. The LRD must model all direct autocorrelation. So we employ superposition of Markov-Gaussian processes to model the LRD-like observed traffic.

The goal of our model is to reconstruct the slow-decaying autocorrelation that spans over several time scales. Besides, we discuss the question whether the LRD can be modeled by short-range dependent models such as Markov-Gaussian processes. However, we find that the observed LRD-like traffic with autocorrelation spanning over several time scales can be modeled by this approach.

2 Approach

2.1 Findings of Measurement Studies

For a fractional Gaussian noise the autocorrelation function is as following [11]:

$$r_k = 2^{-1}[|k+1|^{2H} - 2|k|^{2H} + |k-1|^{2H}] \tag{1}$$

Here H is the so-called Hurst parameter. For large lag k the autocorrelation is approximately

$$r_k = H(2H-1)k^{2H-1} \tag{2}$$

The recent measurement studies suggest that the traffic exhibits asymptotic second-order self-similarity as the fractional Gaussian noise and the characteristics can be described by the autocorrelation function r_k.

$$r_k = a(k)^{-\beta} \tag{3}$$

where a is a positive constant and $0 \le \beta \le 1$.

The autocorrelation function r_k of the measurement, such as LAN, VBR video, WWW, [2][3][12][13][14], is found to be a function like (3). The studies show that the slow-decaying is measured to span some number of time scales. So the findings of these measurement studies indicate that the autocorrelation $\hat{r}_k$ is a power-law-decaying function like (3) spanning some number of time scales.

$$\hat{r}_k = a(k)^{-\beta}, \ 1 \le k \le 10^n \tag{4}$$

where n denotes the number of time scales over which $\hat{r}_k$ of the measured traffic presents power-law-decaying. In the resource-limited networks we believe that the n is a finite number. It is sufficient to limit n in a certain range for fitting the data traces. We don't have to consider the real second-order self-similar processes (n = ∞) but only approximate self-similar processes (n is limited).

2.2 Markov-Gaussian Processes

The main proposition for the Markov-Gaussian model is that the system has only a one-step short memory. It has the following properties: Every state depends directly only on the preceding one.

$$E\langle X_t | X_{t-1}, X_{t-2}, X_{t-3},...\rangle = E\langle X_t | X_{t-1}\rangle \tag{5}$$

And the lag 1 correlation is

$$r_1 = E(X_t X_{t-1}) \tag{6}$$

To extend this sense of correlation, every state may be thought to be dependent directly or indirectly on all states preceding it. Furthermore, the lag k correlation becomes

$$r_k = (r_1)^k \tag{7}$$

For a discrete Markov-Gaussian process, it has the following relationship

$$X_t = r_1 X_{t-1} + n_t \tag{8}$$

where the n_t is the Gaussian noise.

2.3 Fitting of the Power-Law-Decaying Autocorrelation

This fitting is motivated by Madelbrot's approach [9]. We use a weighted sum of N standardized Markov-Gaussian processes

$$Y_t = \sum_{i=0}^{N-1} \sqrt{W_i} X_t^{(i)} \tag{9}$$

where $X_t^{(i)}$ denotes the i-th standardized Markov-Gaussian process.

Here the Markov-Gaussian processes are assumed to be pairwise uncorrelated. The self-correlation $r_k^{(ii)}$ is

$$r_k^{(ii)} = E[X_{t+k}^{(i)} X_t^{(i)}] = r_i^k \tag{10}$$

and the cross-correlation $r_k^{(ij)}$ is

$$r_k^{(ij)} = E[X_{t+k}^{(i)} X_t^{(j)}] = 0, \ i \ne j \tag{11}$$

Autocorrelation

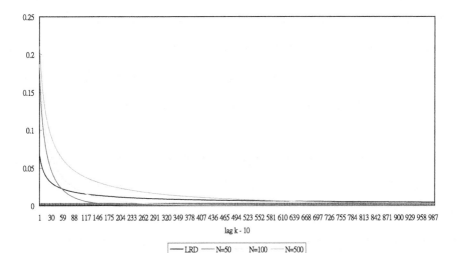

lag k - 10

—— LRD —— N=50   N=100   ······ N=500

Fig. 1. The autocorrelation of uniform-spaced fit

Then the autocorrelation function of Y_t is

$$r_k^{(Y)} = E(Y_t Y_{t+k}) / E(Y_t Y_t) = \sum_{i=0}^{N-1} W_i \, r_i^k / \sum_{i=0}^{N-1} W_i \tag{12}$$

If we set $r_i = e^{-u}$ and $N \to \infty$, the continuous analog of the above formula is

$$\tilde{r}_k^{(Y)} = \int_0^\infty e^{-uk} W(u) du \tag{13}$$

Note that $\tilde{r}_k^{(Y)}$ and $W(u)$ are a Laplace transform pair for parameter pair k and u. So we can get the weighting function $W(u) = L^{-1}(\tilde{r}_k^{(Y)})$.

If $\tilde{r}_k^{(Y)}$ is chosen as the power-law decaying autocorrelation $r_k'^{(Y)}$,

$$r_k'^{(Y)} = a(k)^{-\beta}, \tag{14}$$

then the weighting function can be derived as the following form.

$$W(u) = au^{\beta-1} \Big/ \Gamma(\beta) \tag{15}$$

Substituting (13) in (10), the autocorrelation of Y_t becomes

$$r_k^{(Y)} = \sum_{i=0}^{N-1} [a(-\log r_i)^{\beta-1} / \Gamma(\beta)] r_i^k / \sum_{i=0}^{N-1} W_i \tag{16}$$

and the error term is

$$err = \sum_{k=1}^{10^n} \left| \hat{r}_k - r_k^{(Y)} \right|^2 = \sum_{k=1}^{10^n} \left| \sum_{i=1}^{N-1} \left[ak^{-\beta} / N - ar_i^k (-\log r_i)^{\beta-1} / \Gamma(\beta) \right] \right|^2 \tag{17}$$

Fitting Error

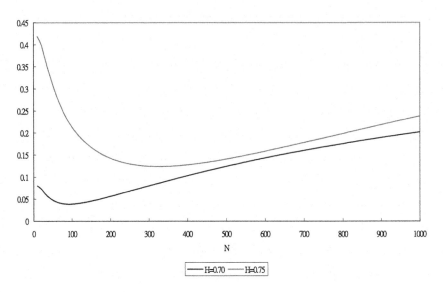

Fig. 2. The fitting error of uniform-spaced fit

The optimal N and error versus H

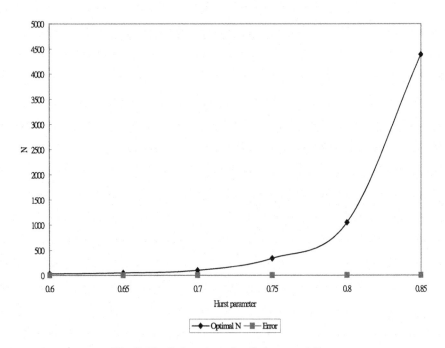

Fig. 3. The fitting error of uniform-spaced fit

Here the r_i's are chosen as incremented in uniform steps. If the number of time scale n and the long-range dependence parameter β are respectively 3 and 0.6 (H=0.7), Fig. 1 shows that the uniform-distribution r_i's in [0,1] can effectively fit the long-range dependent autocorrelation when N is large enough.

Fig. 2 shows the error term versus N. The curve has a clear valley around N=90 for H=0.7 and N=330 for H=0.75. Here we can find that the synthetic Markov-Gaussian approach can fit the autocorrelation function of the approximate self-similar processes.

Fig. 3 shows the optimal N versus Hurst parameter H. The optimal N value raises exponentially as H becomes large. However, the fitting error keeps very low even for large H. The measured LRD data from the recently studies indicate that the Hurst parameters are almost between 0.6 and 0.85. The measured data can be modeled by the proposed method.

3 Fitting to Starwars Variable Bit Rate Video

The two-hour long empirical sample of VBR video [1] is a typical multimedia long-range dependent data trace. The tail behavior of the marginal bandwidth is accurately described using heavy-tailed distributions (Pareto distributions) and the autocorrelation of the VBR video trace decays hyperbolically. So it is a good data trace to examine the proposed approach.

The proposed approach exploits N discrete Markov-Gaussian processes to synthesize the desired traces. Remember that

$$Y_t = \sum_{i=0}^{N-1} \sqrt{W_i} X_t^{(i)} \tag{18}$$

$$Y_t = \sum_{i=0}^{N-1} \sqrt{W_i} X_t^{(i)} \text{ and } X_t^{(i)} = r_1^{(i)} X_{t-1}^{(i)} + n_t^{(i)} \tag{19}$$

If $n_t^{(i)}$ is a normal distribution N(0, $1-(r_1^{(i)})^2$), then the mean μ_i and variance σ_i^2 of $X_t^{(i)}$ are respectively 0 and 1. Because of the pairwise-uncorrelated property of $X_t^{(i)}$ the mean and variance of Y_t are respectively 0 and $\sum_{i=0}^{N-1} W_i$. Define a new process Z_t:

$$Z_t = Y_t \Big/ sqrt(\sum_{i=0}^{N-1} W_i \sigma_i^2) = \sum_{i=0}^{N-1} \sqrt{W_i} X_t^{(i)} \Big/ sqrt(\sum_{i=0}^{N-1} W_i) \tag{20}$$

So the mean of Z_t is 0 and the variance of Z_t is 1.

Fig. 4 is the cumulative probability function $F_S(s)$ of the starwars VBR. The pdf of the traces is a Gamma/Pareto distribution [1] shown in Fig. 5. Let $F_Z(z)$ be the cumulative probability function of Z_t. Define a new self-similar synthetic process U_t

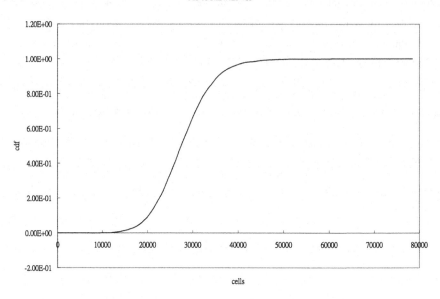

Fig. 4. The cumulative density function $F_S(s)$ of the starware VBR

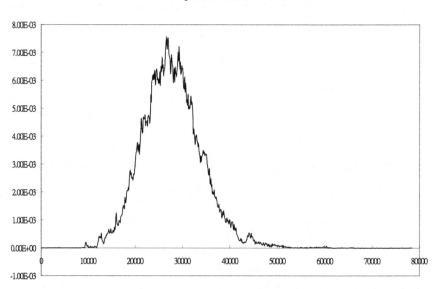

Fig. 5. The probability density function of the starware VBR

with the same cumulative probability as the starwars VBR. Then U_t can be generated by the process Z_t using the following transformation:

$$U_t = h(Z_t) = F_S^{-1}(F_Z(Z_t)) \quad \text{for } t = 1,2,3,... \tag{21}$$

With the TES (Transform-Expand Samples) method the generated trace is mapped into the desire trace by the marginal distribution of the empirical data. The long-range dependent parameter, Hurst parameter are not changed after the transformation. Mandelbrot [9] has shown that the estimated Hurst parameter is largely unaffected by the marginal distribution of the time series. Fig. 6 is the transformation of the generated traces to the starwars VBR traces.

Transformation

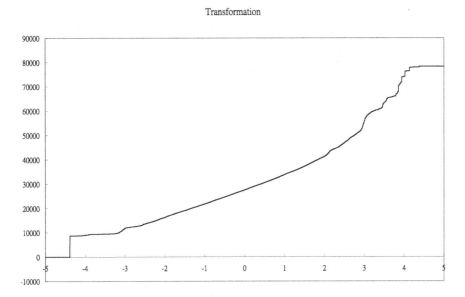

Fig. 6. Transformation function

4 Procedures of Generating Synthetic Self-similar Traces

The procedure of the proposed synthetic self-similar traffic generating method will be systematically described in this section. The procedure has a few steps: the first step is to analyze the data traces which we want to fit; secondly the optimal synthetic number and the transformation function must be decided based on the first step's analysis; at last the synthetic self-similar traces can be generated.

The procedure of generating synthetic self-similar traces:

(1) Exploring the long-range dependent parameter, Hurst parameter, H; The common graphical measurement schemes for the Hurst parameter are variance-time plots and pox plots. The techniques for estimating H based on the asymptotic properties

of the periodogram are known and provide confidence intervals for the estimated value of H. In particular, for Gaussian processes Whittle's approximate maximum likelihood estimator (MLE) has been studied extensively and has been shown to exhibit many desirable statistical properties. These estimating schemes can be referenced in [1][3][13].

(2) Deciding the number of time scales n over which we want to model the traffic; The traffic of the real world has limited room for long-range dependence and the self-similar spanning number of time scales is dependent on the limited resource of networks, for example, buffer size. Theoretically, any number of time scales n can be modeled in the proposed self-similar traffic-generating scheme.

(3) Analyzing and preparing the transformation function h(.) of the data traces; From the analysis of the cumulative distribution function of the measured data traces, the transformation function can be derived. With the transformation function the generated data traces can have the same probability density function as the measured data traces.

(4) Finding the optimal synthetic number N based on the calculated H and the fitting number of time scales n; When the Hurst parameter and the desired number of time scales are decided, the optimal synthetic number N can be found with the fitting error function.

(5) Generating the self-similar traces using the N synthetic Markov-Gaussian processes with the TES method; The parameters of the proposed self-similar traces-generating scheme are the Hurst parameter and the optimal synthetic number. With these parameters the produced data traces will have the asymptotic long-range dependence spanning over the desired number of time scale n. Then the produced data traces can be transformed to have the same probability density function as the measured data traces by the TES method without affecting the autocorrelation structure largely.

5 Results

Here the proposed approach is used to fit the starwars vbr traces. Fig. 7 shows the autocorrelation of the generated traces versus starwars vbr traces. It is very clear that the proposed approach has successfully fitted the long-range dependence of the starwars vbr traces. In this figure the short-range dependence of the generated traces seems to be not close to that of the starwars vbr traces. However, the long-range dependence is the most important factor to the queuing performance and to model the long-range dependence is much more important to the short-range dependence.

In Figs. 8-10, we show the tail distribution in the trace-driven simulation: with the starwars vbr and with the generated traces. These figures show some kinds of difference in very large tail but the generated tail is close to the empirical one to a reasonable range. The proposed method is able to model the probability density function and the correlation structure and it can predict the queuing behavior if the desired queue length is not too large.

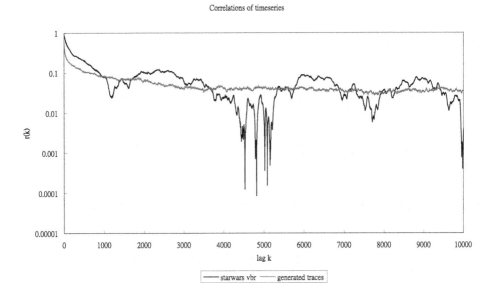

Fig. 7. Autocorrelation of generated traces

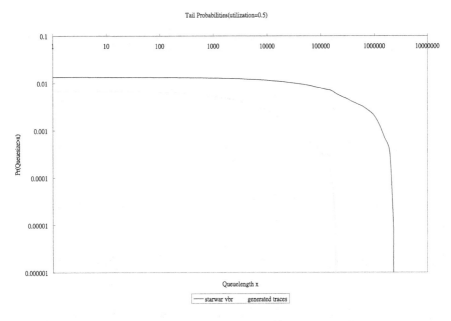

Fig. 8. Tail behavior of starwars vbr and generated traces at load=0.5

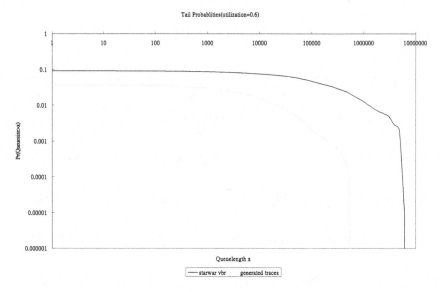

Fig. 9. Tail behavior of starwars vbr and generated traces at load=0.6

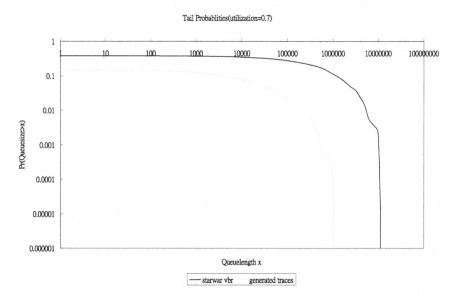

Fig. 10. Tail behavior of starwars vbr and generated traces at load=0.6

6 Conclusions

Recent studies of traffic measurement indicate that today's traffic of networks presents long-range dependence. Traditional short-range dependent models are claimed to be inappropriate to fit the self-similar traffic. Self-similar traffic is difficult to be modeled because of its long-range dependence. However, in a resource-limited network, there is no room for real long-range dependent traffic. All studies of traffic measurement can only show that the traffic exhibits the approximate second-order self-similarity.

Here an approach based on synthetic Markov-Gausssian processes is proposed. A Markov-Gaussian process is very simple. It needs only a one-step memory, lag-1 correlation. If the Hurst parameter and the number of time scales over which we want to model the traffic are decided, the self-similar traces can be generated using the synthetic Markov-Gaussian processes. The proposed approach can fit not only the correlation structure but also the probability density function of the traffic.

Starwars vbr traffic is one of the famous long-range dependent traces. Here we use the starwars vbr to examine the proposed approach. The results show that the correlation structure of the generated traces is close to the tendence of the starwars vbr and the queuing tail behavior of the generated traces can predict the performance of the data traces if the tail value is not too large. In a resource-limited network environment, the traffic can only present approximate self-similarity.

It is interesting to note that even being able to model long-range dependent correlation structure and the probability distribution function does not appear sufficient to allow for accurate prediction for very large tails. To model the long-range dependence with short-range dependent processes is not sufficient to predict very long-term queuing behavior. Anderson and Nielson [15] have the same findings in their studies.

References

1. Garrett, M.W., Willinger, W.: Analysis, Modeling and Generation of Self-Similar VBR Video Traffic. ACM Computer Communication Review 24, 269–280 (1994)
2. Paxson, V., Floyd, S.: Wide Area Traffic: The Failure of Poisson Modeling. IEEE/ACM Transactions on Networking 3(3), 226–244 (1995)
3. Leland, W.E., Taqqu, M.S., Willinger, W., Wilson, D.V.: On the Self-Similar Nature of Ethernet Traffic. IEEE/ACM Transactions on Networking 2(1), 1–15 (1994)
4. Chandramouli, Y., Neidhardt, A.: Application level traffic measurements for capacity engineering. In: The 2002 ACM SIGMETRICS international Conference on Measurement and Modeling of Computer Systems, June 15 - 19, pp. 260–261 (2002)
5. Wang, Z., Liu, J.: Traffic Measurement Mechanisms for High Precision Internet Applications. In: Eighth ACIS International Conference on Software Engineering, Artificial Intelligence, Networking, and Parallel/Distributed Computing (SNPD 2007), pp. 66–69 (2007)
6. Rezaul, K.M., Grout, R.V.: Identifying Long-range Dependent Network Traffic through Autocorrelation Functions. In: 32nd IEEE Conference on Local Computer Networks (LCN 2007), pp. 692–697 (2007)

7. Min, G., Ould-Khaoua, M., Kouvatsos, D.D., Awan, I.U.: A Queuing Model of Dimension-Ordered Routing under Self-Similar Traffic Loads. In: 18th International Parallel and Distributed Processing Symposium-Workshop 14 (IPDPS 2004), p. 251a (2004)
8. Julio, C., Pacheco, R., Roman, D.T.: Accuracy of Time-Domain Algorithms for Self-Similarity: An Empirical Study. In: 15th International Conference on Computing (CIC 2006), pp. 379–384 (2006)
9. Mandelbrot, B.B.: A Fast Fractional Gaussian Noise Generator. Water Resources Research 7(3), 543–553 (1971)
10. Mandelbrot, B.B., Vanness, J.W.: Fractional Brownian Motions, Fractional Noises and Applications. SIAM Review 10(4), 422–437 (1968)
11. Norros, I.: On the Use of Fractional Brownian Motion in the Theory of Connectionless Networks. IEEE JSAC 13(6), 953–962 (1995)
12. Sahinoglu, Z., Tekinay, S.: On Multimedia Networks: Self-Similar Traffic and Network Performance. IEEE Communications Magazine, 48–52 (January 1999)
13. Erramilli, A.: Experimental Queueing Analysis with Long-Range Dependent Packet Traffic. IEEE/ACM Transactions on Networking 4(2), 209–223 (1996)
14. Crovella, M.E.: Self-Similar in WWW Traffic: Evidence and Possible Causes. IEEE/ACM Transactions on Networking 5(6), 835–845 (1997)
15. Anderson, A.T., Nielsen, B.F.: A Markovian Approach for Modeling Packet Traffic with Long-Range Dependence. IEEE JSAC 16(5), 719–732 (1998)

Performance of Multi-service System with Retrials due to Blocking and Called-Party-Busy

S.N. Stepanov[1], O.A. Kokina[1], and V.B. Iversen[2]

[1] Intellect Telekom
109044, Moscow, Melnikova 29, Russia
{kokina,stepanov}@i-tc.ru
[2] COM·DTU, Technical University of Denmark,
DK-2800 Kgs. Lyngby, Denmark
vbi@com.dtu.dk

Abstract. In this paper we construct a model of a multi-service system with an arbitrary number of bandwidth flow demands, taking into account retrials due to both blocking along the route and to called-party-busy. An approximate algorithm for estimation of key performance measures is proposed, and the problem of dimensioning the system is considered.

Keywords: teletraffic, multi-service line, retrials, dimensioning.

1 Introduction

The necessity of studying the influence of retrials in next generation networks is obvious for the several reasons. The main reason among these is that by taking account of the subscriber behavior we can in more realistic way reconstruct the input traffic. This aspect is very important, especially when considering intelligent properties of new networks, where following the basic features of such systems a subscriber (or some device producing requests for bandwidth) will have a lot of possibilities for playing an active role in the dialogue with the servicing system. This realistic approach in reconstruction of the input process causes many difficulties in the theoretical study of models with retrials because in these models time intervals between successive call arrivals are dependent random variables.

The investigation of this phenomena can be done in frame of queueing models with retrials (reviews of the early papers in this field can be found in [1]–[2]). Unfortunately, models of multi-service systems with possibility of retrials are not well-studied because of a complicated nature of the random processes that are used for their description. The number of retrial flows taken into account adds the same number of components in the corresponding random process. Each component takes an infinite number of values. All of this complicates the finding of the performance measures. The aim of this paper is to construct an approximate algorithm that can be used for estimation of key performance measures of multi-service systems with arbitrary number of input flows and possibility of retrials due to insufficient amount of bandwidth or called-party-busy.

S. Balandin et al. (Eds.): NEW2AN 2008, LNCS 5174, pp. 147–155, 2008.

2   Model Description

Let us consider a single broadband transmission link with a transmission speed v measured in basic bandwidth units. A number of n of different traffic flows uses the link in a sharing mode at call level. A call of k'th flow uses b_k bandwidth units for the time of connection that is exponentially distributed with parameter equal to α_k.

We assume that calls from the k'th traffic flow can be blocked for two reasons: (1) insufficient bandwidth on first link for call servicing or (2) in case of sufficient bandwidth for call servicing the absence of sufficient amount of bandwidth along the route to the called subscriber including the situation of called-party-busy. The first reason of blocking depends on statistics of multi-service system bandwidth sharing. The second reason of blocking is modelled for calls of k'th flow by probability p_k which we call blocking probability of called device.

In case of blocking by any reason a subscriber forming k'th flow of traffic repeats request for bandwidth with probability H_k after random time having exponential distribution with parameter equals μ_k. The scheme of model functioning is shown in Fig. 1 .

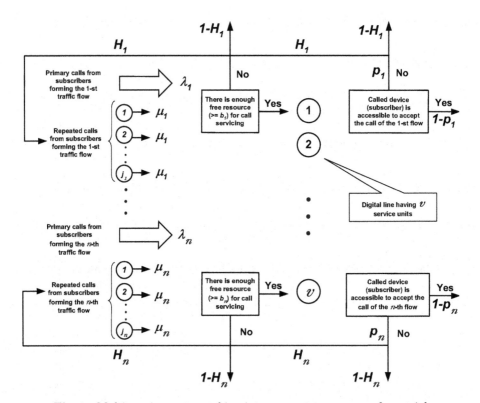

Fig. 1. Multi-service system taking into account two reasons for retrials

For k'th flow, let j_k denote number of subscribers in repeating mode and i_k denote number of calls being serviced. The state of the model is described by a vector

$$s = (j_1, \ldots, j_n, i_1, \ldots, i_n),$$

where j_k, $k = 1, 2, \ldots, n$, varies in interval $[0, \infty)$, and values $i_1, \ldots, i_n$ satisfies the inequality $i_1 b_1 + \ldots + i_n b_n \leq v$. The state space of the model considered consists of the vectors s satisfying the formulated conditions. The model's dynamic behavior is described by a random Markov process of the type $r(t) = \{j_1(t), \ldots, j_n(t), i_1(t), \ldots, i_n(t)\}$, defined on the state space S. Let us denote by B_k the set of states $(j_1, \ldots, j_n, i_1, \ldots, i_n) \in S$, where requests for bandwidth for k'th traffic flow are refused. It means that components of the model state $(j_1, \ldots, j_n, i_1, \ldots, i_n)$ satisfy the inequality $i_1 b_1 + \ldots + i_n b_n + b_k > v$.

3   Performance Measures and State Equations

Let us denote by $P(j_1, \ldots, j_n, i_1, \ldots, i_n)$ the stationary probabilities of $r(t)$. To guarantee the existence of a stationary regime we suppose that

$$\max_{1 \leq k \leq n} H_k < 1, \quad \max_{1 \leq k \leq n} p_k < 1, \quad \min_{1 \leq k \leq n} \mu_k > 0. \tag{1}$$

Key performance measures of the model considered look as follows:

$\pi_{p,k}$: portion of blocked primary requests for bandwidth for calls of k'th traffic flow:

$$\pi_{p,k} = \sum_{(j_1, \ldots, j_n, i_1, \ldots, i_n) \in B_k} P(j_1, \ldots, j_n, i_1, \ldots, i_n).$$

$\pi_{r,k}$: portion of blocked repeated requests for bandwidth for calls of k'th traffic flow:

$$\pi_{r,k} = \frac{\sum\limits_{(j_1, \ldots, j_n, i_1, \ldots, i_n) \in B_k} P(j_1, \ldots, j_n, i_1, \ldots, i_n) j_k}{\sum\limits_{(j_1, \ldots, j_n, i_1, \ldots, i_n) \in S} P(j_1, \ldots, j_n, i_1, \ldots, i_n) j_k}.$$

J_k: mean number of repeating subscribers for k'th traffic flow:

$$J_k = \sum_{(j_1, \ldots, j_n, i_1, \ldots, i_n) \in S} P(j_1, \ldots, j_n, i_1, \ldots, i_n) j_k.$$

I_k: mean number of occupied bandwidth units for servicing requests of k'th traffic flow:

$$I_k = \sum_{(j_1, \ldots, j_n, i_1, \ldots, i_n) \in S} P(j_1, \ldots, j_n, i_1, \ldots, i_n) i_k b_k.$$

Λ_k: total intensity of primary and repeated requests for bandwidth for k'th traffic flow:

$$\Lambda_k = \lambda_k + J_k \mu_k.$$

$\Lambda_{b,k}$: total intensity of blocked primary and repeated requests for bandwidth for k'th traffic flow:

$$\Lambda_{b,k} = \lambda_k(\pi_{p,k} + (1 - \pi_{p,k})p_k) + J_k\mu_k(\pi_{r,k} + (1 - \pi_{r,k})p_k).$$

$\pi_{t,k}$: portion of blocked primary and repeated requests for bandwidth of k'th traffic flow

$$\pi_{t,k} = \frac{\Lambda_{b,k}}{\Lambda_k}.$$

M_k: mean number of retrials per primary call attempt for k'th traffic flow

$$M_k = \frac{J_k\mu_k}{\lambda_k}.$$

To find the performance measures of interest it is necessary to construct and solve the system of state equations. To do this we should restrict the maximum number of repeated subscribers for each traffic flow. Let us suppose that for k'th traffic flow the number of repeated subscribers is less or equal than N_k, $k = 1, 2, \ldots, n$. For state $(j_1, \ldots, j_n, i_1, \ldots, i_n)$ the total number of occupied service units is denoted by $i = i_1 b_1 + \ldots + i_n b_n$. The system of state equations that relates $P(j_1, \ldots, j_n, i_1, \ldots, i_n)$ looks as follows:

$$P(j_1, \ldots, j_n, i_1, \ldots, i_n)\left\{\sum_{k=1}^{n} \lambda_k \left(1 - p_k(1 - H_k)\right) I(i + b_k \leq v, j_k < N_k)\right. \quad (2)$$

$$+ \sum_{k=1}^{n} \lambda_k(1 - p_k) I(i + b_k \leq v, j_k = N_k) + \sum_{k=1}^{n} \lambda_k H_k I(i + b_k > v, j_k < N_k)$$

$$+ \sum_{k=1}^{n} j_k\mu_k(1 - p_k H_k)I(i + b_k \leq v) + \sum_{k=1}^{n} j_k\mu_k (1 - H_k) I(i + b_k > v)$$

$$\left. + \sum_{k=1}^{n} i_k\alpha_k\right\}$$

$$= \sum_{k=1}^{n} P(j_1, \ldots, j_n, i_1, \ldots, i_k - 1, \ldots, i_n) \lambda_k(1 - p_k) I(i_k > 0) +$$

$$+ \sum_{k=1}^{n} P(j_1, \ldots, j_k - 1, \ldots, j_n, i_1, \ldots, i_n) \lambda_k H_k I(j_k > 0) \times$$

$$\times (p_k I(i + b_k \leq v) + I(i + b_k > v)) +$$

$$+ \sum_{k=1}^{n} P(j_1, \ldots, j_k + 1, \ldots, j_n, i_1, \ldots, i_k - 1, \ldots, i_n) \times$$

$$\times (j_k + 1)\mu_k (1 - p_k) I(i_k > 0, j_k < N_k) +$$

$$+ \sum_{k=1}^{n} P(j_1, \ldots, j_k + 1, \ldots, j_n, i_1, \ldots, i_n)(j_k + 1)\mu_k(1 - H_k)\, I(j_k < N_k) \times$$

$$\times (p_k\, I(i + b_k \leq v) + I(i + b_k > v)) +$$

$$+ \sum_{k=1}^{n} P(j_1, \ldots, j_n, i_1, \ldots, i_k + 1, \ldots, i_n)(i_k + 1)\,\alpha_k\, I(i + b_k \leq v).$$

Here the function $I(\cdot)$ equals one if the condition formulated in brackets is true and equals zero other ways. The normalization condition is valid for $P(j_1, \ldots, j_n, i_1, \ldots, i_n)$

$$\sum_{(j_1, \ldots, j_n, i_1, \ldots, i_n) \in S} P(j_1, \ldots, j_n, i_1, \ldots, i_n) = 1.$$

The system of state equations can be solved numerically by Gauss-Seidel iteration algorithm a for number of traffic flows 2–3 when number of service units is restricted by a few dozens. Let us use this opportunity for numerical study of key performance measures dependence on the subscriber behavior in case of blocking.

Let us consider the multi-service system with the following values of input parameters $v = 100$, $n = 2$, $\mu_1 = \mu_2 = 10$, $b_1 = 1$, $b_2 = 20$, $\lambda_1 = 100$, $\lambda_2 = 25$, $\alpha_1 = 1$, $\alpha_2 = 1$, $p_1 = 0, 5$, $p_2 = 0$. Fig 2 shows the dependence of I_1 and I_2 on increasing of probability of retrials for subscribers forming the first traffic flow. The value of $H_2 = 0$. The results show that – with all other parameters fixed – an increasing probability of retrial for subscribers of the first traffic flow, the

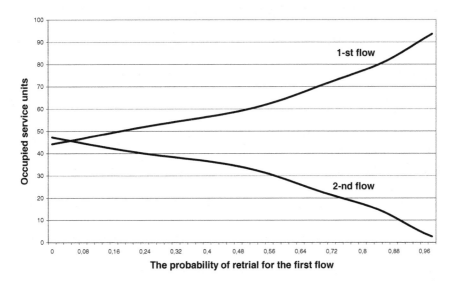

Fig. 2. The dependence of I_1 and I_2 on increasing of probability of retrial for subscribers forming the first traffic flow

subscribers forming the second flow are practically forced out. This results once again emphasize the importance of taking into account the subscriber behavior in case of blocking when dimensioning multi-service systems.

4   Approximate Algorithm

The idea of the approximate algorithm is based on the assumption that when the time interval between successive retrials made by one subscriber increases then the process of repeated attempts can be considered as Poissonian. Because it is known that values of performance measures are weakly dependent on the length of interval between successive retrials it allows to use Poissonian assumption for approximate evaluation of models with taking into account the subscriber behavior in the form of repeated attempts. This approach is quite often used for approximate calculation of stationary performance measures of models with retrials (for details, see[1]).

This approach has a number of good features. The most important among them is that the proposed method makes the problem of estimating the system's performance measures equivalent to estimating of performance measures of the same model but with probability of retrials equals to zero. Let us call this model as Poissonian. As a result all effective algorithms of analyzing the Poissonoian model can be used for estimating the performance measures of the corresponding model with retrials. For the considered model this problem can be effectively solved by the recursive algorithms [3]–[6]. The Poissonian model is insensitive to the service time distribution. So the obtained results can be used also for the generalized version of the considered model taking into account the retrials and a general distribution of service time.

Let us by x_k denote the unknown intensity of combined (primary and repeated) flow of requests for bandwidth produced by subscribers forming k'th traffic flow. To find unknown values of x_k, $k = 1, 2, \ldots, n$ we will use the conservation laws that relate intensities of offered and served traffic flows for the model considered. Let us suppose that $N_k = \infty$, $k = 1, 2, \ldots, n$ and multiply (2) consequently by $j_1, \ldots, j_n, i_1, \ldots, i_n$. After summing up the obtained expressions we have the relations:

$$\Lambda_k = \Lambda_{b,k} + I_k, \quad \Lambda_k - \lambda_k = \Lambda_{b,k} H_k, \quad k = 1, 2, \ldots, n. \tag{3}$$

Let us suppose that (3) is valid for the model with Poissonian assumption concerning flows of retrials. It gives the system of implicit equations:

$$x_1 = \frac{\lambda_1}{1 - (p_1 + \pi_1(x_1, \ldots, x_n)(1 - p_1))H_1},$$

$$\ldots \qquad \ldots \qquad \ldots \tag{4}$$

$$x_n = \frac{\lambda_n}{1 - (p_n + \pi_n(x_1, \ldots, x_n)(1 - p_n))H_n}.$$

Here $\pi_k(x_1, \ldots, x_n)$ is the portion of lost requests for k'th traffic flow, depending on unknown values of $x_1, x_2, \ldots, x_n$. The system (4) is solved by successive substitutions. To do this we should know the algorithm of estimation the values of $\pi_k(x_1, \ldots, x_n)$.

Let $P(i)$ be the portion of time for Poissonian model to be in the state when exactly i service units are occupied. The values of $P(i)$, $i = 0, 1, \ldots, v$ can be found by means of the recurrence formula ([3]–[5]):

$$P(i) = \begin{cases} 1, & i = 0, \\ \frac{1}{i} \sum\limits_{k=1}^{n} x_k(1 - p_k)b_k P(i - b_k), & i = 1, 2, \ldots, v. \end{cases} \tag{5}$$

Here $P(i - b_k)$ equals to zero if $i - b_k < 0$. The values $\pi_k(x_1, \ldots, x_n)$ are found from expressions

$$\pi_k(x_1, \ldots, x_n) = \sum_{i=v-b_k+1}^{v} P(i).$$

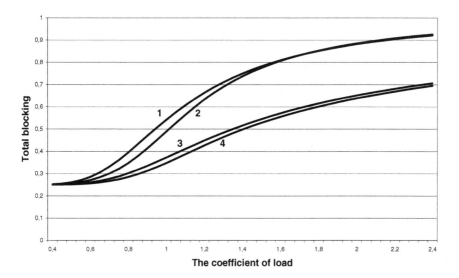

Fig. 3. The results of estimation of the portion of blocked primary and repeated requests for bandwidth for the first traffic flow (exact - curve 3 and approximate - curve 4) and the second traffic flow (exact = curve 1, and approximate = curve 2) with increasing of the load per one service unit a, $\lambda_1 = \frac{av}{nb_1}$, $\lambda_2 = \frac{av}{nb_2}$.

5   Numerical Results

Let us numerically illustrate the accuracy of approximate algorithm by considering the model of multi-service system with the following values of input parameters $v = 50$, $n = 2$, $\mu_1 = \mu_2 = 10$, $b_1 = 2$, $b_2 = 5$, $H_1 = 0,7$, $H_2 = 0,7$, $\alpha_1 = 1$, $\alpha_2 = 1$, $p_1 = 0,25$, $p_2 = 0,25$. Fig. 3 shows the results of estimation of the portion of blocked primary and repeated requests for bandwidth for the first traffic flow (exact = curve 3, and approximate = curve 4) and the second traffic flow (exact - curve 1 and approximate - curve 2) with increasing load per service unit a, $\lambda_1 = \frac{av}{nb_1}$, $\lambda_2 = \frac{av}{nb_2}$.

Analysis of the numerical results presented allows us to conclude that the approximate procedure has quite a good accuracy in estimating key performance measures of multi-service system with arbitrary number of input flows and possibility of retrials because of absence the sufficient amount of bandwidth or called-party-busy. The error of estimation is decreasing to zero for small and large values of intensities of ariving requests.

Let us consider the problem of dimensioning multi-service systems taking retrials into account. Let us estimate the quality of servicing by maximum value of total blocking $\pi_{\max} = \max\limits_{1 \leq k \leq n} \pi_k$. Traditional solution of this problem consists of consequent increasing of bandwidth v and verifying the inequality $\pi_{\max} < \pi$. The problem with implementation of traditional scheme is that the intensity of input traffic does not depend on the number of service units. If part of the input traffic is caused by retrials the usage of traditional scheme considerably increases the final answer if we compare obtained solution with solution based on the model with taking into account retrials.

Let us illustrate this by numerical example by considering the model of the multi-service system with parameters: $n = 2$, $b_1 = 1$, $b_2 = 5$, $H_1 = 0,9$, $H_2 = 0,9$, $\mu_1 = 10$, $\mu_2 = 10$, $\alpha_2 = 1$, $\alpha_1 = 1$, $p_1 = 0$, $p_2 = 0$, $\pi = 0,05$. Fig. 4 shows the solution of dimensioning problem with and without taking the retrials into account. Intensities of input traffic are as follows: $\Lambda_1 = 17,211$, $\Lambda_2 = 11,145$. The difference between two solutions is 42 service units.

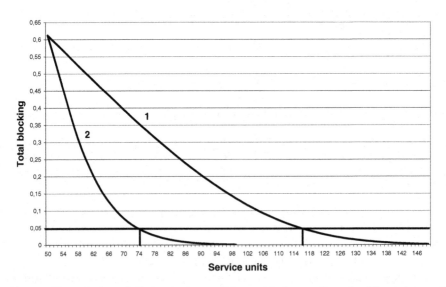

Fig. 4. The solution of dimensioning problem with (curve 2) and without (curve 1) taking into account the retrials

6   Conclusion

The model of multi-service system with arbitrary number of input flows of demands and possibility of retrials because of insufficient amount of bandwidth or called-party-busy is considered. Approximate algorithm for estimation of key performance measures is suggested. The accuracy of approximate approach is studied numerically. The problem related to the dimensioning of the system is considered.

References

1. Stepanov, S.N.: Chislennye metody rascheta sistem s povtornymi vyzovami (Numerical methods for calculation of the retrial systems). Nauka, Moscow (1983)
2. Falin, G.I., Templeton, J.G.C.: Retrial Queues. Chapman & Hall, New York (1997)
3. Fortet, R., Grandjean, C.: in a loss system when some calls want several devices simultaneously. Electrical Communications 39(4), 513–526 (1964); ITC–4, Fourth International Teletraffic Congress, London, UK, July 15-21(1964)
4. Kaufman, J.S.: Blocking in a shared resource environment. IEEE Transactions on Communications 29(10), 1474–1481 (1981)
5. Roberts, J.W.: A service system with heterogeneous user requirements – applications to multi–service telecommunication systems. In: Pujolle, G. (ed.) Performance of Data Communication Systems and their Applications, pp. 423–431. North–Holland Publ.Co (1981)
6. Iversen, V.B., Stepanov, S.N.: The unified approach for teletraffic models to convert recursions for global state probabilities into stable form. ITC19. In: Proceedings 19th International Teletraffic Congress, Beijing, China, August 29- September 2, pp. 1559–1570 (2005)

The Impact of Self-similarity on Traffic Shaping in Wireless LAN

Adam Domański[2], Joanna Domańska[1], and Tadeusz Czachórski[1,2]

[1] Institute of Theoretical and Applied Informatics
Polish Academy of Sciences
Bałtycka 5, 44-100 Gliwice, Poland
{joanna,tadek}@iitis.gliwice.pl
[2] Institute of Informatics
Silesian Technical University
Akademicka 16, 44-100 Gliwice, Poland
adamd@polsl.pl

Abstract. IEEE 802.11 MAC protocol ist the de facto standard for wireless local area networks (WLANs). In today's Internet, the emerging widespread use of real-time voice, audio, and video applications makes QoS (Quality of Service) a key problem. At the MAC layer, 802.11e defines extensions to enhance the QoS performance of 802.11 WLAN. This MAC layer solution leaves such issues of QoS as QoS guarantee and admission control to the traffic control systems at the higher layers. This article tries to show that implementation of some mechanisms of traffic shaping causes some improvement of the level of QoS in WLANs. First we analyse the influence of the traffic shaping in WLANs stations by the mechanism of token bucket filter. Next the analysis of the behaviour of Access Point with AQM mechanism was carried out. The conducted research has shown that it is possible to achieve certain level of QoS thanks to the implementation of traffic shaping mechanisms. We make our experiments comparatively for the traffic sources with self-similarity and without it. Our results confirm the necessity of taking into account the self-similar character of wireless traffic.

1 Introduction

Current market trends show that wireless communication is the most developing transmission technology, becoming the base for many audio and video applications. Evolution of this technology is a two-way process, dealing with mobile telephony and wireless local area networks (WLANs) [1].

First, the speed of data transfer rate in cellular networks is gradually growing. Nowadays second generation (2G) of mobile telephony systems is not suitable for efficient digital data transfer thus 2G-systems were extended by 2.5G technology containing packed-switched capabilities: data transmission mechanism High-Speed Circuit-Switched Data (HSCSD), packed switched technologies General Packet Radio Service (GPRS) and Enhanced Data rates for GSM Evolution

S. Balandin et al. (Eds.): NEW2AN 2008, LNCS 5174, pp. 156–168, 2008.
© Springer-Verlag Berlin Heidelberg 2008

(EDGE). Another step is third generation (3G) Universal Mobile Telecommunications System (UMTS) and mobile telephony communications protocol High-Speed Downlink Packet Access (HSDPA).

Second, there is greater and greater employment of the mobile wireless local area networks (WLANs). Achieved bitrates (above 1 Mbit/s) are considerably higher than that in mobile telephony standards GPRS and EDGE. The cost of local installations is relatively low so this technology has gained tremendous attention in recent years. WLANs are mostly implemented as Wi-Fi (IEEE 802.11), Bluetooth (IEEE 802.15.1) or WiMAX (IEEE 802.16).

At present the most popular standard of wireless networks is IEEE 802.11 [2]. This standard is widely available because of its low cost and relatively high capacity. Meanwhile, Quality of Service - sensitive applications such as multimedia streaming service become increasingly popular, and these applications require certain real time constraints such as bandwidth, delay, and jitter to be satisfied. In the section 2 the improving of IEEE 802.11 to manage some aspects of QoS is described. A MAC layer extension for QoS, IEEE 802.11e, has been recently ratified as a standard. This MAC layer solution, however, addresses only the issue of prioritized access to the wireless medium and leaves such issue as QoS guarantee and admission control to the traffic control systems at the higher layers [7]. This article investigates the impact of TBF-mechanisms implemented in WLAN stations on some aspects of QoS in Wireless LAN. The authors also try to show that the implementation of AQM-mechanisms in Access Point has also the great impact on the level of QoS in WLAN. Section 4 describes the simulation model of the 802.11 with above mentioned extensions and presents the obtained results.

Many empirical and theoretical researches on Ethernet [8], Wide Area Network [10,9], WWW traffic [11] and VBR video traffic [12] have shown the self-similar characteristic of the network traffic. These results cannot be directly applied to the wireless LANs due to the difference in their MAC protocol implementation. Section 3 gives some definition of self-similarity and describes recent empirical studies of WLAN traffic characteristic. The conclusion of this studies causes the using in our simulation model the self-similar traffic sources (see section 4).

The summary and conclusions drawn from the research conducted in this work are included in the section 5.

2   Quality of Service in Wireless LAN

The 802.11 standard family was developed by the Institute of Electrical and Electronics Engineers (IEEE) [3], in order to deal with the modern wireless connectivity needs. The IEEE 802.11 standard includes delailed specifications for both the medium access control (MAC) and the physical layer (PHY). The original 802.11 MAC protocol includes two modes of operation characterized by coordination functions:

- DCF - compulsory distributed co-ordination function - is an asynchronous data transmission function, which best suits delay intensive data (e.g. email, ftp),
- PCF - optional point co-ordination function - the polling-based function is utilized in delay sensitive data transmission (e.g. real time audio or video)

DCF defines two access mechanism to employ packet transmission:

- The default scheme is called the basic access mechanism, in which station transmit data packet after deffering the medium is busy.
- An optional way of transmitting data packets, namely, the request to send/ clear to send RTS/CTS reservation scheme. This scheme uses small RTS/CTS packets to reserve the medium before large packets are transmitted in order to reduce the duration of a colliision.

The IEEE 802.11 standard is the most widely deployed wireless local area network infrastucture. However, it cannot provide QoS support for the increasing number of multimedia applications. Thus, a lot of research works have been carried out to enhance the QoS support in 802.11 networks [13]. So far, most research works for the QoS support in IEEE 802.11 have focused on controlling the priority to access the wireless medium by modifying the MAC protocol [5,6] The IEEE 802.11 Task Group E (802.11e) has defined enhancements to the original 802.11 MAC to provide QoS. The 802.11e standard introduces the Hybrid Coordination Function (HCF), which combines functions from DCF and PCF with enhanced QoS-specific mechanisms and frame types. HCF has two modes of operation - Enhanced Distribution Coordinate Access (EDCA) and HCF Controlled Channel Access (HCCA). EDCA and HCCA are contention-based and polling-based mechanisms for channel access. EDCA contention access is an extention to DCF and provides prioritized access to the wireless medium. HCCA uses a Hybrid Coordinator to centrally manage the wireless medium access to provide parameterized QoS. Like PCF, HCCA uses a polled-based mechanism to access the medium. This MAC layer solution, IEEE 802.11e, however, addresses only the issue of prioritized access to the wireless medium and leaves such issue as QoS guarantee and admission control to the traffic control systems at the higher layers [7].

This article try to show that it is possible to achieve certain level of QoS thanks to the implementation of traffic shaping mechanisms.

3   Characteristics of the Traffic in Wireless LAN

Clasically, the traffic intensity, seen as a stochastic process, was represented in queueing models by short term dependencies [14]. However, the analysis of measurements shows that the traffic has also long-terms dependencies and has self-similar character. It is observed on various protocol layers and in different network structures [8,9,10,11,12].

The term "*self-similar*" was introduced by Mandelbrot [15] in description of proceses in the field of hydrology and geophysics. It means that a change of time scale does not influence statistical properties of the process. A stochastic process X_t is self-similar with Hurst parameter $H(0.5 \leq H \leq 1)$ if for a positive factor g the process $g^{-H}X_{gt}$ has the same distribution as the original process X_t, [16]. Mathematically, the difference between short-range dependent and long-range dependent (self-similar) processes is as follows [17]:

For a short-range dependent process:
- $\sum_{r=0}^{\infty} \mathrm{Cov}(X_t, X_{t+\tau})$ is convergent,
- spectrum at $\omega = 0$ is finite,
- for large m, $\mathrm{Var}\,(X_k^{(m)})$ is asymptotically of the form $\mathrm{Var}\,(X)/m$,
- the aggregated process $X_k^{(m)}$ tends to the second order pure noise as $m \to \infty$;

For a long-range dependent process:
- $\sum_{r=0}^{\infty} \mathrm{Cov}(X_t, X_{t+\tau})$ is divergent,
- spectrum at $\omega = 0$ is singular,
- for large m, $\mathrm{Var}\,(X_k^{(m)})$ is asymptotically of the form $\mathrm{Var}\,(X)m^{-\beta}$,
- the aggregated process $X_k^{(m)}$ does not tend to the second order pure noise as $m \to \infty$,

where the spectrum of the process is the Fourier transformation of the autocorrelation function and the aggregated process $X_k^{(m)}$ is the average of X_t on the interval m:

$$X_k^{(m)} = \frac{1}{m}(X_{km-m+1} + ... + X_{km}) \qquad k \geq 1.$$

There are several methods used to check if a process is self-similar. The easiest one is a visual test: one can observe the behaviour of the process through the scales of time. The other one is the estimation of aggregated index of dispersion IDC or aggregated coefficient of variation CV. The aggregated index of dispersion is equal to the variance of the number of arrivals within the interval m divided by the average number of arrivals during the same interval:

$$IDC(m) = \frac{\mathrm{Var}\,(mX_k^{(m)})}{E(mX_k^{(m)})}$$

and CV is

$$CV(m) = \frac{\sqrt{\mathrm{Var}\,(mX_k^{(m)})}}{E(mX_k^{(m)})}$$

For a self-similar processes, IDC increases on several time scales and CV is much more than 1 for small time intervals. Estimation of Hurst parameter is the most frequently used method to check if a process is self-similar: for non-self-similar processes $H = 0.5$; for $0.5 < H \leq 1$ process is self-similar; the closer H is to 1, the greater the degree of persistance of long-range dependence. The parameter can be estimated by various methods, among others by the analysis of variance-time plot [16]. The variation of aggregated self-similar process is equal to:

$$\text{Var}\left(X_k^{(m)}\right) \approx \text{Var}\left(X\right)m^{-\beta}, \quad \text{or} \quad \log\text{Var}\left(X_k^{(m)}\right) \approx \log\text{Var}\left(X\right) - \beta\log m$$

so the log-log plot of $\text{Var}\left(X_k^{(m)}\right)$ versus m is a straight line with slope $-\beta$, $0 < \beta < 1$, and $H = 1 - \beta/2$.

Self-similarity of a process means that the change of time scale does not influence the process: the original process and the scaled one are statistically the same. It results in long-range dependence and makes possible the occurrence of very long periods of high (or low) traffic intensity. These features have a great impact on a network performance. They enlarge the mean queue lengths at buffers and increase the probability of packet losses, reducing this way the quality of services provided by a network [18]. According to Stallings [18], "Self-similarity is such an important concept that, in a way, it is surprising that only recently has it been applied to data communications traffic analysis". As mentioned above, many empirical and theoretical researches have shown the self similar characteristics of the network traffic. These results cannot be directly applied to the wireless LANs due to the difference in their MAC protocol implementation. Recent empirical study on WLANs demonstrates that the wireless traffic exhibits self-similarity on a wide range of time scales [19,20,21,22] As a consequence of this fact there is a need of the proper selection of traffic sources when modelling behaviour of this type of networks.

4   The Model of the 802.11 Network - Numerical Results

In models created for the needs of this article we use the traffic source based on the MMPP (Markov modulated Poisson process) source introduced by S. Robert [23,24] to represent the self-similar traffic.

The time of the model is discrete and divided into unit length slots. Only one frame can arrive during each time-slot. In the case of memoryless, geometrical source, the frame comes into system with fixed probability α. In the case of self-similar traffic, packet arrivals are determined by a n-state discrete time Markov chain called modulator. It was assumed that modulator has $n = 5$ states $(i = 0, 1, \ldots 4)$ and packets arrive only when the modulator is in state $i = 0$. The elements of the modulator transition probability matrix depend only on two parameters: q and a – therefore only two parameters should be fitted to match the mean value and Hurst parameter of the process. If p_{ij} denotes the modulator transition probability from state i to state j, then it was assumed that $p_{0j} = 1/a^j$, $p_{j0} = (q/a)^j$, $p_{jj} = 1 - (q/a)^j$ where $j = 1, \ldots, 4$, $p_{00} = 1 - 1/a - \ldots - 1/a^4$, and remaining probabilities are equal to zero. The passages from the state 0 to one of other states determine the process behaviour on one time scale, hence

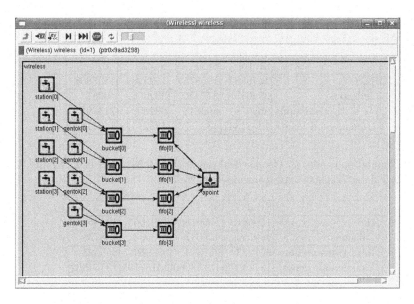

Fig. 1. 802.11 model, traffic controlled by TBF alghoritm - OMNET++

the number of these states corresponds to the number of time-scales where the process may by considered as self-similar.

This model enables us to represent, with the use of few parameters, a network traffic which is self-similar over several time-scales. For such a source model, one must fit only two parameters: expectation and the Hurst parameter (plus the number of states in Markov chain n; it defines the number of time-scales on which the process has self-similar character).

This section contains the analysis of the 802.11 MAC network with several workstations and one Access Point (AP). We also analyse the 802.11 network with modifications like AQM implemented in AP and TBF packet scheduling algorithms implemented in workstations. Our analysis were carried out in the OMNeT++ simulation environment [25]. OMNeT++ is a public-source, component-based, modular and open-architecture simulation environment. OMNeT++ is free for academic and non-profit use; commercial users must obtain a license. The example screenshot of our model graphical representation in OMNeT++ is presented on figure 1.

To emphasize the importance of using self-similar traffic sources the comparative research has been carried out for the self-similar and poisson (non self-similar) source. Input traffic intensity was chosen as $\alpha = 0.5$ or $\alpha = 0.081$, and due to the modulator characteristics, the Hurst parameter of self-similar traffic was fixed to $H = 0.8$. For both considered in comparisons cases, i.e. for geometric interarrival time distribution (which corresponds to Poisson traffic in case of continuous time models) and self-similar traffic, the considered traffic intensities are the same. A detailed discussion of the choice of the source parameters is presented in [26].

Table 1. Simulation results for self-similar sources $\alpha = 0.081$ $\mu = 0,05$, Empty slots: 8209875, All slots: 72000000, All transmissions: 2737886

	St. 0	St. 1	St. 2	St. 3
RTS	1076582	1058753	1092284	1135808
Collisions	393013	377450	382086	400671
Nb of transmissions	665076	663818	692600	716392
Transmission length	13321347	13247352	13827542	14317817

Table 2. Simulation results for Geometrical sources $\alpha = 0.081$, $\mu = 0,05$, Empty slots: 9593991, All slots: 72000000, All transmissions: 2627087

	St. 0	St. 1	St. 2	St. 3
RTS	1295812	1296189	1294961	1294563
Collisions	449356	449647	449083	448736
Nb of transmissions	657223	657314	656959	655591
Transmission length	13114073	13133069	13137195	13151662

Our model of 802.11 network uses compulsory distributed co-ordination function (DCF) with RTS/CTS reservation scheme. The model assumes that Access Point is also a Medium Access Controller. Our first goal is to capture the aspect of collision distribution on the level of RTS/CTS signals and the usage of the transmission channel for this mechanism and to compare the results for self-similar and poisson traffic. Tables 1 and 2 show the influence of input traffic characteristics for the number of RTS, the number of collisions and the number of transmissions for all workstations.

In figures 2 and 4 the RTS distribution with low load conditions is presented. Figures 3 and 5 present the RTS distribution for self-similar and geometrical traffic in the case when network is heavily loaded. When the network is oveloaded the number of collision are greater than the number of successfully sended reservation frames. Because of that in the next model the traffic generated in the workstations is shaped by the Token bucket filter mechanism (fig. 1).

Figures 6 and 7 show the impact of TBF (Token Bucket Filter) mechanism for the RTS distribution.

Table 3 presents more detailed results.

Table 3. Comparison of transmission for geometrical and selfsimilar sources with TBF

Model	factor of network utylization	factor of network collisions
geo $\alpha = 0,5$	0,327996486	1,114468403
geo $\alpha = 0,081$	0,729666653	0,02495586
self $\alpha = 0,5$	0,690259625	0,486739722
self $\alpha = 0,081$	0,759917472	0,0215725
geo with TBF $\alpha = 0,5$	0,452072075	1,108089167
self with TBF $\alpha = 0,5$	0,728667884	0,399881278

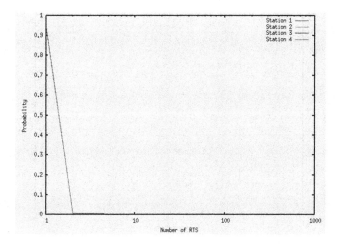

Fig. 2. RTS distribution (Self-similar sources) $\alpha = 0.081$, $\mu = 0.05$

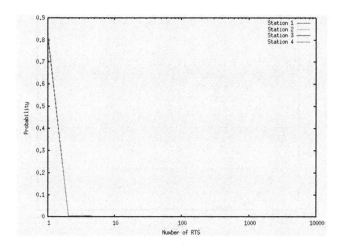

Fig. 3. RTS distribution $\alpha = 0.5$ $\mu = 0.05$ (Self-similar sources)

For $\alpha = 0,081$ both geometrical and selfsimilar sources cause low utilisation of network - the factor for the channel usage reaches up to 70 percent. Also the collision factor remains low. There is no reason for shaping the outgoing traffic with the TBF mechanism for network with such a redundancy of resources. Differently, for $\alpha = 0,5$ the usage of network falls to 32 percent for geometrical source. The usage of TBF mechanisms for the networks with the surplus of

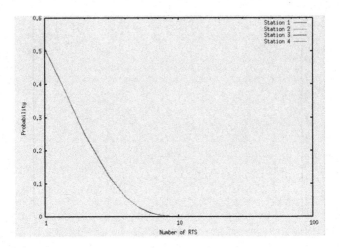

Fig. 4. RTS distribution $\alpha = 0.081$, $\mu = 0.05$ (Geometrical sources)

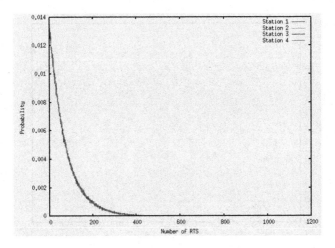

Fig. 5. RTS distribution $\alpha = 0.5$, $\mu = 0.05$ (Geometrical sources)

resources has also decreased the efficiency. For the case of overloaded network the application of TBF mechanism has improved the transmission parameters. This results confirm the sense of using traffic shaping mechanisms to achieve some level of QoS (controlled load) in 802.11 network.

Our last model has treated AP as a intermediary between LAN and WLAN networks. The same transmission parameters as those above mentioned have

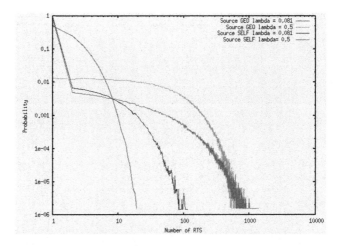

Fig. 6. RTS distribution - GEO ans SELF sources

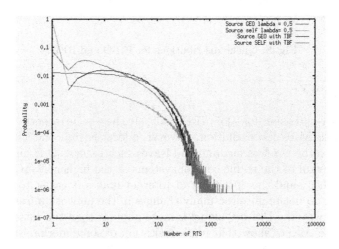

Fig. 7. RTS distribution - GEO ans SELF sources with TBF

been used. The received results were close to those obtained earlier. However it has been observed that high packet loss appeared in the buffer storing packets that enter the wireless network. This problem has been solved by applying Active Queue Management. The application of the simplest RED algorithm [28] has resulted in the significant decrease of losses (fig. 8).

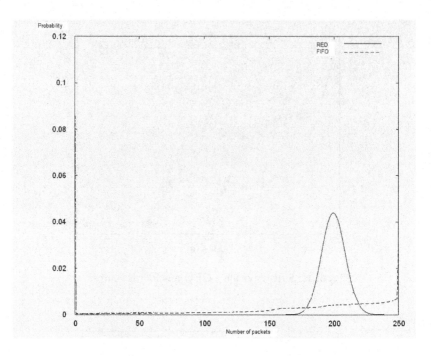

Fig. 8. Queue distributions for FIFO and RED

5   Summary

A MAC layer extension for QoS, IEEE 802.11e, has been recently ratifed as a
standard. This MAC layer solution, however, addresses only the issue of prioritized access to the wireless medium and leaves such issue as QoS guarantee and
admission control to the traffic control systems at the higher layers.

On the other hand the specificity of Internet makes it easier to apply algorithms which do not require too many changes in the network infrastructure (it
is the reason why the Diffserv model is more popular than the Intserv).

This article tries to show that implementation of some mechanisms of traffic
shaping causes some improvement of the level of QoS in WLANs. Moreover, it
has been shown that the implementation of the QoS mechanisms for network
with considerable redundancy did not make any sense.

Modifying the rules of the access point behaviour by adding active queue
management functionality at the link between LAN and WLAN network can
significantly improve parameters of transmission. It is certainly important to
realise that in this way it is impossible to obtain full QoS. However, when applying some techniques, well-known in the wired networks and implemented in
advanced routers for a long time (advanced methods of scheduling packets in
queues [28]), it is possible to achieve the satisfying level of QoS. The results
obtained for the overloaded network show the improvement of throughput while
applying the TBF mechanism. The implementation of the AQM mechanism has

allowed for the significant decrease in loss of packets in the incoming queue to the wireless network.

In the paper it has also been shown that it is necessary to take into account the self-similar feature while modelling the wireless network. Last research has shown that the traffic in WLAN networks reveals the self-similarity feature. The obtained results for the geometrical and self-similar sources have differed significantly. Therefore one should use self-similar traffic sources while modelling of WLAN.

Acknowledgements

This research was financed by Polish Ministry of Science and Higher Education project no. N517 025 31/2997.

References

1. Stallings, W.: Wireless Communications and Networks. Prentice-Hall, Englewood Cliffs (2002)
2. O'Hara, B., Petrick, A.: IEEE 802.11 handbook, A designer's companion. IEEE Press, Los Alamitos (1999)
3. IEEE Standard for Wireless LAN Medium Access Control (MAC) and Physical Layer (PHY) specifications. ISO/IEC 8802-11:1999(E) (August 1999)
4. Chatzimisios, P., Boucouvalas, A.C., Vitsas, V.: Performance analysis of the IEEE 802.11 MAC protocol for wireless LANs. International Journal Of Comunications Systems (18) (2005)
5. Vaidya, N.H., Bahl, P., Gupta, S.: Distributed fair schedulig in a wireless LANs. In: Sixth Annual International Conference on Mobile Computing and Networking, Boston, USA (June 2000)
6. Benveniste, M., Chesson, G., Hoeben, M., Singla, A., Teunissen, H., Wentink, M.: EDCF proposed draft text. IEEE working document 802.11-01/131r1 (2001)
7. Dukju, K., Seungjae, H., Hojung, Ch., Rhan, H.: A traffic control system for IEEE 802.11 networks based on available bandwidth estimation. In: Wireless Communications and Mobile Computing 2008, vol. 8, pp. 407–419. John Wiley & Sons, Chichester (2008)
8. Willinger, W., Leland, W.E., Taqqu, M.S.: On the self-similar nature of ethernet traffic. IEEE/ACM Transactions on Networking (1994)
9. Domańska, J., Domański, A., Czachórski, T.: Self-similarity of network traffic and the performance of QoS mechanisms. Archiwum Informatyki Teoretycznej i Stosowanej 16(3) (2004)
10. Paxson, V., Floyd, S.: Wide Area Traffic: A Failure of Poisson Modeling. IEEE/ACM Transactions on Networking (June 1995)
11. Crovella, M., Bestavros, A.: Self-similarity in World Wide Web Traffic: Evidence and Possible Causes. IEEE/ACM Transactions on Networking (December 1997)
12. Garret, M., Willinger, W.: Analysis, modeling and generation of self-similar VBR video traffic. ACM SIGCOMM, London (September 1994)
13. Qiang, N., Turletti, T.: QoS Support for IEEE 802.11 WLAN. In: Wireless LANs and Bluetooth, pp. 432–438. Nova Science Publishers, New York (2005)

14. Kleinrock, L.: Queueing Systems, vol. II. Wiley, New York (1976)
15. Mandelbrot, B., Ness, J.V.: Fractional Brownian Motions, Fractional Noises and Applications. SIAM Review 10 (October 1968)
16. Beran, J.: Statistics for Long-Memory Processes. Chapman and Hall, Boca Raton (1994)
17. Cox, D.R.: "Long-range dependance: A review". Statistics: An Appraisal (1984)
18. Stallings, W.: High-Speed Networks: TCP/IP and ATM Design Principles. Prentice-Hall, Englewood Cliffs (1998)
19. Liang, Q.: Ad Hoc Wireless Network Traffic - Self-Similarity and Forecasting. IEEE Communications Letters 6(7) (2002)
20. Oliveira, C., Kim, J.B., Suda, T.: Long-Range Dependence in IEEE 802.11b Wireless LAN Traffic: An Empirical Study. In: IEEE Annual Workshop on Computer Communications (2003)
21. Lee, I.W.C., Fapojuwo, A.O.: Characteristic of Wireless LAN Traffic. Wireless Networks and Emerging Technologies (2004)
22. Qin, Y., Yuming, M., Taijun, W., Fan, W.: Hurst parameter estimation and characteristics analysis of aggregate wireless LAN traffic. Communications, Circuits and Systems 1, 339–345 (2005)
23. Robert, S.: Modélisation Markovienne du Trafic dans Réseaux de Communication. PhD thesis, Ecole Polytechnique Fédérale de Lausanne, Nr 1479 (1996)
24. Robert, S., Boudec, J.Y.L.: New models for pseudo self-similar traffic. Performance Evaluation 30(1-2) (1997)
25. OMNeT++ Community Site, http://www.omnetpp.org
26. Domańska, J.: Procesy Markowa w modelowaniu natężenia ruchu w sieciach komputerowych. PhD thesis, IITiS PAN, Gliwice (2005)
27. Zhai, H., Kwon, Y., Fang, Y.: Performance analysis of the IEEE 802.11 MAC protocol in wireless LANs. Wireless Communications And Mobile Computing (4) (2004)
28. Domańska, J., Domański, A., Czachórski, T.: The Drop-From-Front Strategy in AQM. In: Koucheryavy, Y., Harju, J., Sayenko, A. (eds.) NEW2AN 2007. LNCS, vol. 4712, pp. 61–72. Springer, Heidelberg (2007)

Service Concentration Node in Internet Multimedia Subsystem (IMS)

Didem Gozupek[1] and Aziz Sever[2]

[1] Dept. of Computer Engineering, Boğaziçi University, Istanbul, Turkey
[2] Argela Technologies, Istanbul, Turkey
didem.gozupek@boun.edu.tr, aziz.sever@argela.com.tr

Abstract. Convergence of the cellular world to an all-IP domain has stimulated the development of Internet Multimedia Subsystem (IMS) service architecture, which is touted by service providers and vendors as the next generation solution for breaking the barriers not only between cellular and IP worlds, but also between wireline and wireless networks. Although many GSM operators and legacy network service providers are adapting rapidly to the IMS world, the pace of this adaptation process is still quite low in certain parts of the world. In this paper, we propose a novel IMS concept referred to as Service Concentration Node (SCN), which enables the operators to provide their customers with IMS services that are offered in other domains with which the operator has a signed agreement, and thereby facilitates the utilization of a vast amount of IMS services in various operator networks. The formulated technical framework as well as its accompanying business model is quite promising in enriching the IMS services experience in next generation networks.

1 Introduction

We have witnessed a tremendous growth in various technologies such as WiFi, VoIP, enterprise IP, IP-TV, and WiMax, where IP is the common link shared between all of them. IMS architecture is envisaged to allow the true realization of convergence, by opening up doors for the operators to provide their subscriber base with the delivery of media rich services independent from the location of the user and the access technologies. The initial standard was originally defined by 3GPP [1][2][3][4]. ETSI TISPAN and the ITU are also working on the IMS architecture [5].

IMS service architecture is comprised of a layered system consisting of interoperable modular components, which enable the service providers to implement and manage new services in a rapid and efficient way. Doing so, it also eliminates the reliance on a single vendor for components. This service layer is built on top of the IMS Core Layer, which consists of Call Session Control Functionalities (CSCF's), namely P-CSCF (Proxy-CSCF), Serving-CSCF (S-CSCF), and Interrogating-CSCF (I-CSCF), as well as Breakout Gateway Control Functionality (BGCF) and Home Subscriber Server (HSS). In the current IMS architecture, the services of a user reside in the user's home network domain, communicating with the S-CSCF in

S. Balandin et al. (Eds.): NEW2AN 2008, LNCS 5174, pp. 169–176, 2008.
© Springer-Verlag Berlin Heidelberg 2008

the user's home domain. Our proposed scheme enables the end user to utilize the IMS services offered in another IMS domain through the Service Concentration Node (SCN), which communicates with the user's S-CSCF in its home domain. Typically, the operator of the user's home network has a signed agreement with the IMS service providers communicating through the SCN. This way, the operators that currently lack the IMS services infrastructure are able to provide their customer base with enriched multimedia IMS services, such as Sponsored Call and Video RBT applications [6].

The rest of the paper is organized as follows: Section 2 provides a brief overview of the IMS architecture. Section 3 consists of the overall architecture and signaling flow of our proposed scheme, whereas Section 4 gives the detailed design of the SCN. Finally, Section 5 concludes the paper.

2   IMS Overview

3GPP defines IMS as a new subsystem consisting of a new mobile network infrastructure that enables the convergence of data, speech, and mobile network technology over an IP-based infrastructure. As depicted in Figure 1, the architecture can be considered to be consisting of 3 layers; namely IMS Services Layer, IMS Core Layer, and IMS Convergence Layer.

The top layer is the IMS Services Layer, where applications and service components reside. A multitude of services such as voice and video telephony, instant messaging, presence services, push to talk, etc. are delivered through the IMS interface to the IMS Core Layer, where session control is executed. The SIP Application Server (SIP-AS) is an IMS Application Server on which these IMS services are built. The application logic is written using the API's of SIP- AS, typically defined as JSR 116 [7]. There are key server components that help serve the multimedia content delivery, such as the Media Server, Streaming Server, Instant Messaging and Presence Server. The SIP-AS's communicate with the S-CSCF, which resides in the IMS Core Layer.

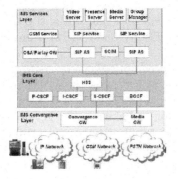

Fig. 1. IMS Architecture

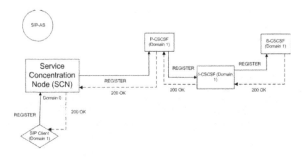

Fig. 2. Registration procedure in the proposed scheme

The middle layer is the IMS Core Layer, which consist of CSCF's, HSS, and BGCF. CSCF's are the functional areas within the IMS that provides all of the session and call control, helping set up, manage sessions and forward messages between IMS networks. It is the central routing engine and policy enforcement point for the network, and uses the SIP protocol for call control [8]. The CSCF is also broken out into three specialized functions known as the S-CSCF, I-CSCF, and P-CSCF. S-CSCF routes the SIP signaling to and from the subscribers via SIP-AS's, in accordance with the service profile information held for each subscriber in the HSS. On the other hand, I-CSCF provides location services when a message or service must traverse multiple IMS domains, whereas P-CSCF handles all of the requests to and from the user and forwards them appropriately.

HSS is a centralized database that consists of all the pertinent user information, such as home network location, security information, and user profile information. The importance of HSS lies in the fact that it removes the burden of having to replicate this information for different services and different types of access networks. The S-CSCF and SIP-AS's communicate with the HSS using Diameter protocol. S-CSCF extracts the service information of the user from HSS in the form of Initial Filter Criteria, which defines the trigger points for services and pointers to SIP-AS's where the specific service application logic exists.

The bottom layer is the IMS Convergence Layer, which bridges the IMS and legacy networks and is comprised of three key components defined by IMS: Signaling Gateway, which translates between SIP and SS7 signaling, and Media Gateway, which translates between IP and the legacy transport network.

3 Service Concentration Node (SCN) Overview

Figure 2 illustrates the registration procedure in our proposed scheme. First of all, the operator of IMS domain-1 has an agreement with the operator of IMS domain-0, where the SCN and the AS's of the offered IMS services reside. Domain-1 operator provides its customers with an SCN-enabled SIP client, which basically sends towards SCN the REGISTER request having the domain-1 URI. SCN keeps a list of the served domain names. Upon receipt of this REGISTER message, SCN checks whether this message is one of the served domains.

If yes, then it saves the location; i.e., URI-IP address mapping, of this user. Then it loads the Initial Filter Criteria (IFC) of this user. This way, the trigger points, namely where the IP addresses of the AS's of the services that this user has subscribed to reside, are loaded. Afterwards, without performing any authentication or location update procedure, SCN forwards this request to the P-CSCF of the home domain, which is domain-1. The REGISTER message is then passed towards the S-CSCF of the home domain, which performs the authentication procedure and then sends back an 200 OK message, after a possible 401 UNAUTHORIZED message. Note here that the authentication is done in the home domain, whereas the location saving and Initial Filter Criteria loading are implemented in SCN.

4   Service Concentration Node (SCN) Overview

The call flow initiated from an SCN-enabled subscriber is illustrated in Figure 3. Upon receiving the INVITE message, the SCN realizes that this user has a SIP application to be offered first. Therefore, the INVITE message is forwarded to the pertinent SIP-AS, the IP address of which was retrieved during loading of the Initial Filter Criteria. The SIP-AS implements its service logic; for instance, if this is a Sponsored Call application, it initiates another call to the Media Server, and after the media is played to the user, it tears up its new call with the Media Server, and then forwards the original INVITE message to the SCN [6]. If the SIP-AS is willing to stay on the path during the entire call flow, then it inserts its IP address to the Record-Route header of the INVITE message before forwarding it back to the SCN. Afterwards, SCN checks whether the destination domain of this INVITE message is one of the served domains. If yes, then it forwards this INVITE message to the pertinent domain. The normal SIP call flow then proceeds with 180 Ringing, 200 OK, and ACK messages.

On the other hand, the call flow initiated towards an SCN-enabled client is depicted in Figure 4. When the INVITE message is received, if the user has a terminating side service, like Video RBT [6], then the INVITE message is forwarded to the pertinent SIP-AS. Similar to the previous case, the SIP-AS

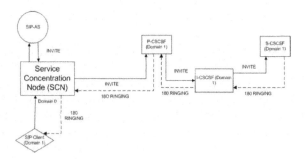

Fig. 3. Call flow initiated from an SCN-enabled client

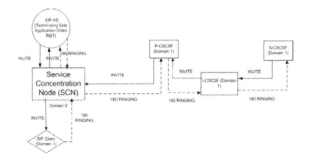

Fig. 4. Call flow terminated at the SCN-enabled client

may prefer to stay on the path during the entire flow (which is a must for Video-RBT application) by inserting a Record-Route header containing its own IP address to the INVITE message. Afterwards, the INVITE message is directly delivered to the location of the terminating user, which was saved during the registration process.

5   Detailed Design

Service Concentration Node (SCN) is implemented on top of the SIP-to-Mobile Gateway [9], where communication between software entities is accomplished via Messaging and Distribution Framework (MAD) [10]. The functionalities of the software entities in Figures 5 and 6 are as follows: SIP SIGNALING entity is responsible for SIP signaling, whereas SIP PROXY is the main call control module. SIP REGISTRAR performs the registration related functionalities and SUBSCRIBER MANAGER is responsible for maintaining the subscriber related information, such as saving the subscriber's location in the database. SIP SIGNALING is written using C++, while all the rest are written in Java.

When a REGISTRATION message corresponding to domain-1 is received by SIP SIGNALING, it sends an equivalent MAD message; i.e., RegistrationReq Proxy. SIP PROXY realizes that this message corresponds to one of the served sip

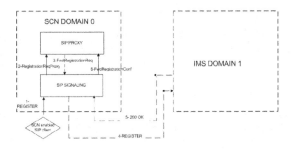

Fig. 5. Registration procedure phase 1

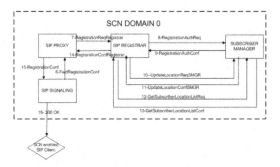

Fig. 6. Registration procedure phase 2

domains; therefore, it first loads the Initial Filter Criteria of this message, then basically forwards this message to the pertinent IMS domain. After 200 OK message is received from the IMS domain-1, SIP SIGNALING entity sends FwdRegistrationConf message to SIP PROXY. SIP PROXY is now aware of the fact that it has to save the location information of this user without performing any authentication procedure, because the user has already been authenticated in its home domain. Hence, it executes the sequence of messages depicted in Figure 6.

After receiving the FwdRegistrationConf message, SIP PROXY sends a RegistrationReq message to SIP REGISTRAR, which sends appropriate messages to SUBSCRIBER MANAGER such that the location; i.e., URI-IP address mapping, of the user is saved in the database. SIP REGISTRAR first sends a RegistrationAuthReq message to SUBSCRIBER MANAGER, with a *doAuthenticate* flag set to false. This way, SUBSCRIBER MANAGER realizes that it will not perform any authentication for this user, and immediately sends a RegistrationAuthConf message with result true. On the other hand, in the original SIP-to-Mobile Gateway, a Location Update in the SS7 network is performed right after this step, which equips the SIP-to-Mobile Gateway with VLR functionality. However, in the SCN, no location update in the SS7 network is performed, since the actual residence of the SCN-enabled user is its home network domain, which is domain-1 in this case, rather than the SCN domain-0.

After receiving the RegistrationAuthConf message, SIP REGISTRAR sends UpdateLocationReqSMGR message to the Subscriber Manager, which includes the Contact URI's of this user in its REGISTER message. Subscriber Manager saves the location information; i.e., uri-IP address mapping, in the database and returns back an UpdateLocationConfSMGR message to the SIP REGISTRAR. Note that this location update corresponds to the location of the user in the subscriber database of SCN, rather than the location in the SS7 network. Afterwards, SIP REGISTRAR retrieves the saved location information via Get SubscriberLocationListReq message. Upon successful reception of the GetSubscriberLocationListConf message, SIP REGISTRAR sends RegistrationConf Registrar message to SIP PROXY, which in turn sends RegistrationConfRegistrar message to SIP PROXY. Finally, SIP PROXY sends RegistrationConf

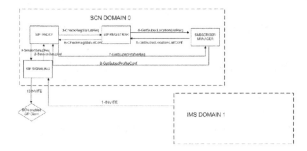

Fig. 7. Detailed call flow to an SCN-enabled client

message to SIP SIGNALING, which then sends an 200 OK message corresponding to the initially received REGISTER message.

Figure 7 depicts the detailed call flow from an IMS domain- 1 to an SCN capable client, whose home domain is also domain-1 but is an SCN user. Firstly, SIP SIGNALING software entity sends a SessionSetupInd message to SIP PROXY in response to receiving the INVITE message. SIP PROXY then checks whether the called address is registered by sending a CheckRegStatusReq message to SIP REGISTRAR, which then finds out that it is indeed registered, since the location of it was saved during registration. SIP REGISTRAR retrieves the registration status of the called address by sending a GetSubscLocationListReq message to SUBSCRIBER MANAGER. Finally, SIP PROXY forwards this message by sending SessionSetupReq message to SIP SIGNALING, which then sends the previously received INVITE message to its destination.

6   Conclusion

In this paper, we proposed a novel IMS concept, referred to as Service Concentration Node (SCN). Contrary to the existing IMS architecture, SCN concept enables the users to utilize IMS services offered in domains different than their home domain. This way, operators that currently lack the infrastructure to offer rich IMS services to their subscriber base are able to do so by signing an agreement with another operator that can offer these services. Note that SCN is also able to communicate with other IMS nodes such as Service Capability Interaction Management (SCIM) module. We have also outlined the overall as well as the detailed design of our proposed SCN scheme for both registration and call flow processes.

References

1. 3GPP TS 23.228-720: IP Multimedia (IM) Subsystem-Cx and Dx Interfaces
2. 3GPP TS 24.228: Signalling flows for the IP multimedia call control based on SIP and SDP

3. 3GPP TS 23.002: Network Architecture
4. 3GPP TS 23.218: IP Multimedia (IM)Session Handling; IP Multimedia (IM) Call Model
5. ETSI TISPAN TR 180 001 NGN-Release 1
6. IMS Forum, White Paper, Rich Multimedia Applications on IMS Framework
7. JSR(Java Specification Request) 116: SIP Servlet API, Java Community Process Program
8. IETF RFC 3261 SIP: Session Initiation Protocol
9. SIP2Mobile Gateway, US Patent No: 20070243891
10. Method and System for Communicating Between Application Software, US Patent No: 20060136931

Observing the Impact of QoS Negotiation on the Signaling Load of the IMS

Juan Miguel Espinosa Carlín

RWTH Aachen University
Communication and Distributed System
Ahornstr. 55, 52074 Aachen, Germany
espinosa@i4.informatik.rwth-aachen.de
http://www.nets.rwth-aachen.de

Abstract. Defined by the 3GPP, the IP Multimedia Subsystem (IMS) is becoming the de facto overlay architecture for enabling service delivery in converged environments. In such scenarios, the QoS requirements of the delivered services vary from user to user, and from service to service. To cope with this necessity, the IMS allows subscribers to personalize the QoS settings of the session that they want to establish. In order to know the impact that the requested QoS parameters have on the available network resources, it is of paramount importance to know how the signaling load in the core network is influenced by them. With this goal in mind, this paper shows how the IMS SIP signaling load is affected when users change their QoS parameters when negotiating a session. The observations were done on a testbed based on the Open IMS Core implementation of the Fraunhofer FOKUS Institute at Berlin and on the IMS Client developed at the University of Cape Town.

1 Introduction

Nowadays, users want to be able to access their services in a uniform way without regard of the type of subscription they have (e.g. fixed, mobile, data), the terminal that they are using for connecting to the network, or the network connectivity existing in their environments (e.g. WLAN, GPRS, UMTS) [1]. One of the main constraints imposed by this paradigm is the need to enable a unified service control architecture that allows the delivery of rich multimedia services that satisfy the expectations of the users. With this goal in mind, the 3GPP introduced the IP Multimedia Subsystem (IMS) as a service control architecture aimed to realize network and service convergence.

One of the key features of the IMS is that it offers users the possibility of personalizing the QoS settings of the session that they want to establish. Although the core IMS session negotiation mechanisms are already defined, further research is needed in order to know the impact that the requested QoS parameters have on the available network resources. With this goal in mind, this paper presents the results of a set of observations done regarding the impact of different QoS parameters on the SIP signaling of an IMS session. The testbed used

S. Balandin et al. (Eds.): NEW2AN 2008, LNCS 5174, pp. 177–186, 2008.

for observation is based on the UCT IMS Client [2] and on the Open IMS Core
Project [3].

The rest of this article is structured as follows. Section 2 gives an overview
of the IMS architecture. Then, Sect. 3 describes the hardware and software
environment in which the observations were done. Section 4 presents the results
of the observations for different QoS parameters, and discusses their impact on
the SIP signaling load. Finally, the conclusions and pointers towards future work
are given in Sect. 5.

2   The IP Multimedia Subsystem

The IMS is based on protocols developed by the Internet Engineering Task Force
(IETF). The basic ones are the Session Initiation Protocol (SIP) for session control and signaling, the Diameter protocol for doing Authentication, Authoriza-
tion, and Accounting (AAA), and the Common Open Policy Service (COPS)
protocol, used for enabling policy administration, configuration and enforcement. For an IMS core interconnecting only IP-based networks, the simplified
architecture of the system is depicted in Fig. 1.

As shown, the transport layer groups all the IP-based networks that allow
users to access their services via the main IP backbone. For each one of these
access networks, a specific technology is defined, and the terminal used to interact
with the service must be compatible with these definitions (e.g. GSM, UMTS,
WiFi, WiMAX).

Next, the IMS layer includes all the functions that implement the logic for
service delivery and for enabling AAA in the system. The main nodes in this
level are:

- One or more SIP-based servers, collectively called Call Session Control Functions (CSCF's), which are the essential nodes in charge of processing all the
  SIP signaling in the IMS. Depending on its role, a CSCF can be categorized
  as a Proxy-CSCF (P-CSCF) when it is the first point of contact between the
  IMS terminal and the network, as an Interrogating-CSCF (I-CSCF) when it
  is a proxy located at the edge of an administrative domain, or as a Serving-
  CSCF (S-CSCF) when it performs session control and provides routing services. All the interfaces between the CSCF's are base on SIP.
- One or more Home Subscriber Servers (HSS's), which are the repositories for
  the related subscriber information required to handle multimedia sessions.
  The interfaces between the HSS and the CSCF's are based on the Diamater
  protocol.

Finally, the service layer contains the AS's that implement the services that
will be delivered to the end users. For all the AS's, the interfaces to the S-CSCF
are based on SIP, while the interfaces to the HSS's are based on the Diameter
protocol. Each AS belongs to one of the following types:

- SIP Application Server (SIP AS): An AS that hosts and executes native IMS
  services based on SIP. Depending on the kind of services that it delivers, it

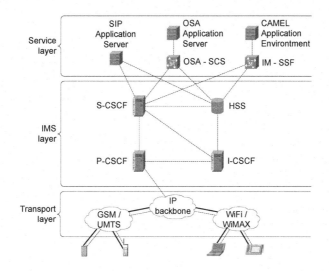

Fig. 1. Simplified IMS Architecture

can act as a SIP proxy, as a SIP redirect server, as a SIP user Agent, or as a SIP Back-to-Back User Agent.

– Open Service Access - Service Capability Server (OSA-SCS): A gateway that provides an interface to access services deployed in an OSA environment.
– IP Multimedia Service Switching Function (IM-SSF): A function that enables the reuse of Customized Applications for Mobile networks Enhanced Logic (CAMEL) services.

3   Evaluation Environment

This section gives an overview of the hardware and software that were used to realize the IMS clients and core network, and depicts the deployed scenario in which the observations were done.

3.1   Open IMS Core Project

The Open IMS Core Project [3] is an implementation of the three kinds of CSCF's and of a lightweight HSS. The four components are all based on Open Source software. The Open IMS CSCF's (Proxy, Interrogating, and Serving) were developed as extensions to the SIP Express Router (SER) [4], while the FOKUS Home Subscriber Server (FHoSS) is a lightweight implementation of an HSS.

The testbed was configured with one IMS domain (`open-ims.test`), with the default parameters for listening SIP requests. Additionally, a DNS server was set up to properly resolve the donfigured domain name. For each one of the HSS, the two IMS subscribers configured by default (Alice and Bob) were used.

3.2 UCT IMS Client

The UCT IMS Client [2] was developed by the Communications Research Group at the University of Cape Town, South Africa. The client was used with the default configuration for both Alice and Bob. For the case of Alice, the IMS configuration parameters are the following:

- Public User Identity: `sip:alice@open-ims.test`
- Private User Identity: `alice@open-ims.test`
- Proxy CSCF: `sip:pcscf.open-ims.test:4060`
- Realm: `open-ims.test`
- QoS: `None`, `Mandatory`, and `Optional`
- QoS Type: `Segmented`
- Access Network: `IEEE-802.11b`

3.3 Testing Scenario

The architecture of the deployed networking scenario is shown in Fig. 2.

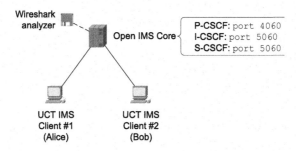

Fig. 2. Networking Scenario Deployed for Testing

In order to know how the network load is affected by different QoS parameters, it is necessary to have a close look to the SIP flows that exchanged between the parties involved in the signaling. The tool used for this purpose was Wireshark, a known protocol analyzer that features live capture and offline analysis of network streams.

4 Observation Results

The operations done by each entity in the IMS for the aim of setting up a session are described in 3GPP TS 24.229 [5]. In order to clearly understand how different QoS parameters affect the SIP signaling, the six scenarios shown in Table 1 where reproduced in the testbed.

Table 1. Tested QoS Session Requirements

Scenario	Alice	Bob
1	None	None
2	Optional	None
3	Optional	Optional
4	Mandatory	None
5	Mandatory	Optional
6	Mandatory	Mandatory

4.1 Session Setup without QoS Negotiation

The following assumptions were done for all the applied tests:

- Alice and Bob are registered in the network.
- The client configuration of both Alice and Bob is the same.
- The supported video and audio codecs are the same for both IMS clients.

The SIP sequence traced in the testbed for the first scenario is shown in Fig. 3. The procedure starts with Alice's client sending an INVITE request (1) directed to Bob. After receiving the request, the network replies Alice with a 100 Trying response (2), confirming that the set up was started. After being processed inside the network, the INVITE request (3) is forwarded to Bob. This INVITE sends back its correspondent 100 Trying (4). At this point, is important to mention that because no QoS negotiation is going to take place, Bob's client only sends a 101 Dialog Establishment response (5 and 6) to inform about the set up status.

The next step consists on Bob's client alerting him about the incoming request. This action is indicated to Alice by sending her a 180 Ringing response (7) with an SDP [6] body for indicating Alice about the characteristics accepted by Bob's

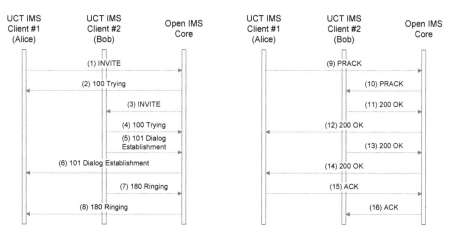

Fig. 3. Traced SIP signaling for setting up a non-QoS session

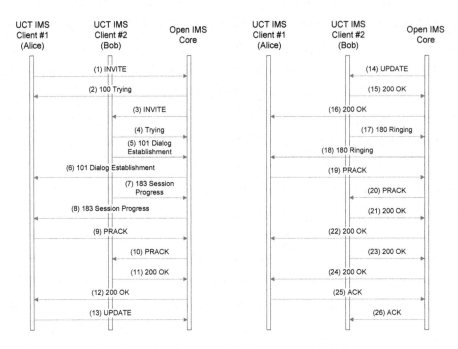

Fig. 4. Traced SIP Signaling for Setting up a QoS Session

client. This notification is forwarded by the network to Alice (8). After receiving this message, Alice's client send back a PRACK request (9) with a SDP description to indicate Bob's client on the agreed codecs used for transmitting the session media. This PRACK request (10) is routed back to Bob, which sends the final 200 OK response (11) regarding the media parameters. This response is routed back to Alice following the usual SIP signaling path (12). After accepting the call, Bob's device sends a 200 OK response (13) that is then forwarded to Alice (14). Finally, Alice's client generates the last ACK to confirm that the generation of media-place traffic with the agreed codecs can be started (15 and 16).

4.2   Session Setup with Mandatory QoS Negotiation

The traced SIP sequence for scenarios 4, 5, and 6, is shown in Fig. 4. It is important to mention that, although Bob's requirement was changed from None to Optional, and then to Mandatory, the found signaling was exactly the same.

As in the previous case, the procedure starts with Alice's client sending a INVITE request (1) directed to Bob. This request includes a SDP body describing the type of session that Alice's client wants to create. After receiving the request, the network replies Alice with a 100 Trying response (2), confirming that the session set up was started. Then, the INVITE request (3) is further forwarded to Bob. This INVITE is answered with its correspondent 100 Trying response (4). After, Bob's client sends a 101 Dialog Establishment response (5) to inform about the set up status, which is further forwarded to Alice's client (6).

Because in these scenarios, at least one of the parties demands the negotiation of a Mandatory QoS as described in RFC 3312 [7], Bob's client sends back an additional 183 Session Progress provisional response (7). This message also includes a SDP [6] body, which contains the media streams and codes that Bob is able to accept for this session. At this point, Bob's client can start its resource reservation, because it knows the parameters needed for it. As in the case of the 101 Dialog Establishment, the 183 Session Progress response (8) is sent back to Alice via the same route. The provisional response is received by Alice and, assuming that both clients agreed on the codes used, Alice sends a PRACK request (9) including the definitive SDP configuration for the session. In parallel with the generation of this request, Alice's client starts the mechanisms for resource reservation. The PRACK request (10) is sent to Bob traversing the required proxies in the network. When the request is received, Bob's client generates the corresponding 200 OK response (11) for confirming the media streams and codecs that will be used for the session. At this time, Bob's terminal may still be involved in its resource reservation process, and most likely it will not be complete. Additionally, the 200 OK response (11) indicates Alice's client that Bob's client wants to receive an indication when Alice is ready with her resource reservation. The 200 OK response (12) is then forwarded to Alice.

Once that Alice's terminal has finished with its resource reservation, it sends an UPDATE request (13) to inform Bob's client that the procedures on its side are finished. This is indicated through a specific SDP body added to the request. The request is routed to Bob via the same proxies as before (14). Upon reception, Bob's client generates the corresponding 200 OK and, assuming that it has already finished with the resource reservation procedures on its side, it sends the response back to Alice (15 and 16).

The SIP flow continues with Bob's client alerting him about the incoming request. This action is indicated to Alice by sending her a 180 Ringing response (17) with a SDP body for indicating Alice about the characteristics accepted by Bob's client. This notification is forwarded back to Alice (18). After receiving this message, Alice's client will likely generate a locally stored tone to indicate Alice that Bob's terminal is ringing. Additionally, Alice's client sends back a PRACK request (19), which is routed back to Bob (20). When the request is received, Bob's device sends back a 200 OK response (21). This message is routed back to Alice following the usual SIP signaling path (22). After accepting the call, Bob's device sends another 200 OK response (23) which is further forwarded to Alice (24). Finally, Alice's client generates the last ACK response (25 and 26) to confirm that the generation of media-place traffic with the agreed codecs can be started. At this point the QoS session setup is complete, and both Alice and Bob can generate their respective audio and video media streams.

4.3   Session Setup with Optional QoS Negotiation

The trace found for scenarios 2 and 3 is shown in Fig. 5. As in the case of the previous observation, the change in Bob's configuration from None to Optional has no impact on the amount of exchanged signaling.

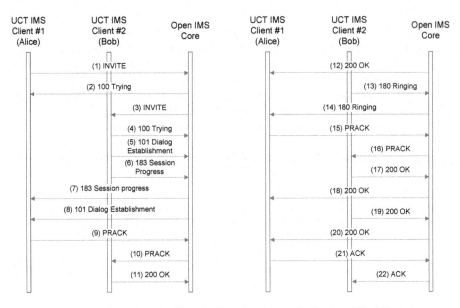

Fig. 5. Traced SIP Signaling for Setting Up a Session with Optional QoS Requirements

This trace is almost the same as the one for the case in which a Mandatory QoS is desired. The only difference is that, because the QoS negotiation is set to Optional, neither the UPDATE request for notifying that the QoS reservation is finished, not the corresponding OK response, are present in the flow; the rest of the flow is the same as the one traced for the previous observation.

4.4   Impact on the Signaling Load

The number and type of SIP requests traced for each one of the tested scenarios are resumed in Fig. 6.

As expected, scenario 1 is the one with the shortest sequence, with only 16 SIP request and responses. Because in this scenario no QoS negotiation takes place, the session establishment is straightforward. For the case of scenarios 2 and 3, the number of SIP messages is 22 (an increase of 37.5% with respect to scenario 1), while for the case of scenarios 4, 5, and 6, the number of messages increases to 26 (an increase of 62.5% also with respect to scenario 1). Although in both cases the increment is caused by the overhead of the QoS reservation mechanisms used at the time of negotiation, in scenarios 2 and 3 the QoS is Optional, so neither UPDATE requests nor their corresponding OK responses are set by the clients: these 4 messages (i.e. two UPDATE requests and two OK responses) make the difference with scenarios 4, 5, and 6.

The amount of data sent during the negotiation phase of each scenario, is reported in Fig. 7. As shown, the scenarios in which some QoS negotiation is needed are the ones that consume more bandwidth. From around 16 KB for scenario 1, the amount of transferred data grows up to 21 KB for scenarios 2

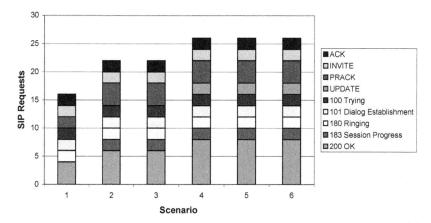

Fig. 6. Number and Type of SIP Requests for each Scenario

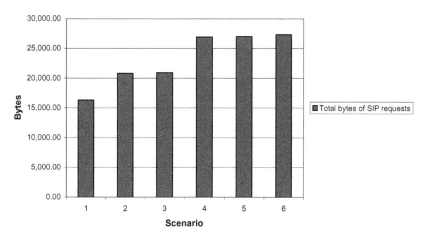

Fig. 7. Transferred Bytes for each Scenario

and 3 (an increment of around 30% with respect to scenario 1), and up to 27 KB (an increment of around 70%) for scenarios 4, 5, and 6.

5 Conclusions and Future Work

This paper reported the impact that QoS parameters have on the SIP signaling load when two subscribers negotiate an IMS session. For this aim, an IMS testbed based on well known software components was set up. The observations were performed for the setup procedures with no QoS requirements, with mandatory QoS requirements, and with optional QoS requirements. The results allow us to conclude that the amount of observed SIP signaling increases between 37.5% and 62.5% with the definition of Optional and Mandatory QoS parameters.

Moreover, the amount of data sent by the clients and the core network increases between 30% and 70% in the cases in which some QoS parameters are present.

Future work includes further observations for the roaming scenarios involving the set up of sessions in complex roaming scenarios, and the invocation of additional IMS services at the time of register (e.g. Presence).

References

1. Schonhowd, R.: Telecom And IT Strategists Must Pursue Converged Service Delivery To 2010 And Beyond. Forrester Research Report (February 2007) (Last retrieved on the 25.03.2008),
   http://www.forrester.com/Research/Document/Excerpt/0,7211,41543,00.html
2. University of Cape Town: UCT IMS Client (2007) (Last retrieved on the 15.11.2007),
   http://uctimsclient.berlios.de
3. Fraunhofer FOKUS: OpenIMSCore.org — The Open IMS Core Project (2007) (Last retrieved on the 20.04.2008), http://www.openimscore.org
4. iptel.org: SIP Express Router (2006) (Last retrieved on the 15.11.2007),
   http://www.iptel.org/ser
5. 3rd Generation Partnership Project: Internet Protocol (IP) multimedia call control protocol based on Session Initiation Protocol (SIP) and Session Description Protocol (SDP); Stage 3. 3GPP TS 24.229 (September 2007)
6. Handley, M., Jacobson, V.: SDP: Session Description Protocol. RFC 2327 (Proposed Standard) (April 1998) Updated by RFC 3266
7. Camarillo, G., Marshall, W., Rosenberg, J.: Integration of Resource Management and Session Initiation Protocol (SIP). RFC 3312 (Proposed Standard) (October 2002) Updated by RFCs 4032, 5027

A Novel Approach to Optimize Information Dissemination in IMS Presence System

Rongheng Lin, Hua Zou,Yao Zhao, and Fangchun Yang

State Key Lab of Networking and Switching Technology
Beijing University of Posts and Telecommunications
Beijing, 100876, China
Ronghenglin@gmail.com, {zouhua,zhaoyao,fcyang}@bupt.edu.cn

Abstract. Presence is an important service in IMS. The user in IMS would use presence as a necessary component. But the more user uses, the harder presence information transmits. Previous researches show that the traffic increases rapidly with the presentity number grows. In this paper, we propose an approach to optimize information dissemination in IMS Presence System. Comparing with previous researches, a new architecture is set up. In this architecture, presence information transmission is modeled as a growing tree. A k sequence method is proposed to construct the tree. And some key algorithms are introduced for the tree adjustment. With a mathematical analysis, the new method is more suitable for presence information dissemination. And in the experiment, the result shows that the method can do a better job with smaller traffic and jitter.

Keywords: Presence information, Dissemination, Tree Network, Traffic.

1 Introduction

Presence is regarded as one of the important services in IMS. Presence service records all peers' status and traces all the changes, even a tiny difference. The service provides two models for the presence information dissemination, which are known as pull and push. Presence helps transform the communication paradigm from a device-centric to a user-centric model. Callers benefit through presence by learning about the availability of the person whom they are trying to reach and how he wishes to be contacted—before they reach that person.

The presence service in IMS is a client-server based system. Obviously, presence server needs to maintain all information of the registered users. Moreover, not only the static information, but also the dynamic changes are maintained. So when system becomes bigger and more complex, user status changes may cause a lot of information dissemination. The performance of the server becomes a bottleneck. Meanwhile, frequent status changes would cost lots of network bandwidth. Bandwidth and the traffic are also bottlenecks.

On the other hand, people change the architecture to a distributed system (p2p), which is suitable for reducing the pressure of the server. But the system becomes

S. Balandin et al. (Eds.): NEW2AN 2008, LNCS 5174, pp. 187–198, 2008.
© Springer-Verlag Berlin Heidelberg 2008

harder to manage and maintain. Furthermore, in IMS system, some devices access from the air interface. Because the P2P structure will take up lots of radio resources, it is hard for radio based devices to act as a peer in P2P system.

To solve the problem, we propose an approach to optimize information dissemination in IMS presence system. With our method, a new style of architecture is set up. With the algorithm of schedule nodes, presence information could be delivered by schedule, and with direction.

The rest of paper is organized as follows: Chapter 2 gives some related work about the presence information dissemination and explains why they aren't very sufficient for our purpose. Chapter 3 proposes the main idea of our approach. In chapter 4, we analyze the traffic, delay and jitter difference between the new and the original method. In chapter 5, a simulator is proposed to validate the new method.

2   Related Works

There are many presence information dissemination researches. We can divide them into two categories: a client-server based research and a distributed system based research.

Most of client-server based researches focus on the traffic optimization. And current research on protocol is almost based on SIP. In their options, the messages amount must be reduced. RLS is an extension to sip specific event notification mechanism for subscribing a list of resources. RLS [8] is mainly useful when clients are on a low bandwidth link. But RLS mostly can't reduce the size and the number of messages. So a conditional subscription technology [10] is composed. The idea of the technology is to deliver messages that the state changes since the previous notification. But the method is only suitable for pull model, not for push model. Partial publication mechanism [7, 9] using a similar idea can reduce the document size to a 1/4 size on average. Yet another method that uses common notification for multiple watchers can also reduce the amount of notification messages [12]. In addition, in order to reduce message size, a compression method (SIGCOMP) [11] is introduced. SIGCOMP can reduce message size to 1/3 [3, 4, 6].

Though all methods can reduce part of the traffic, the server performance is still a bottleneck; it is because so many interactions still exist. With the cost analysis, the cost is still rapidly growing with the larger number of presentities. Also, the traffic is growing up rapidly [5].

Skype is an example of distributed system that delivers presence information. Though Skype is known for its high quality voice, experiment shows its poor performance of presence information dissemination [2]. As the P2P system is an autonomous system, there is not a peer that knows all the other peer states. Besides, peers also can't even know if the presence information is spread to all the peers that need it. So it is hard for peers to spread the information. Therefore, the P2P system is not very suitable for presence information which is pushed.

As concerned above, there are several problems in current research. Firstly, traffic is still a problem. It is not effectively enough for a server to handle with tons of requests. The more requests handled, the more hardware needed. The server may be lack of CPU, memory and traffic. The second one is the network structure. Telecom service always has a time requirement. The client-server structure will encounter some problems in this situation. When the client's number becomes huge, the server needs to handle with times of requests than before. It is hard for the server to guarantee the same quality of service. Lots of requests would be placed in a message queue so that some clients would not meet their time requirements. In order to overcome this problem, more hardware, even more clusters are added in the system. So the system becomes more complex and harder to maintain. Instead of a client-server system, the P2P system is not a good solution, either. As the difficulty for managing the P2P system, it is hard to use in a telecommunication system. On the other hand, the P2P system is not very suitable for presence information spread.

Following the above considerations, we propose a new method, which may be more than just Client-Server or P2P. The method is designed for a large scale system. As the network needs to be managed, a centre point is reserved. And in order to reduce the pressure of the server, a distributed nodes structure is designed. Those distributed nodes are intelligent, as they maintain some relationships and context.

3   The Novel Approach for Information Dissemination

The main idea is to use a manageable distributed architecture to deliver the presence information. In a large scale system, there is thousands of presence information. It is necessary to classify them [1]. So the first step is to classify the information. And then, the distributed architecture can be designed on those clusters.

3.1   Presence Information Classification

It is known that the number of the presence information must be huge. There must be some policies to classify the information into different kinds. There are some relationships among presentities. A user would only like to receive the information of friends who are on his/her buddy list. And there are a crowd of people who need to receive a same set of presence information that describes their common friends' status. So the division policy must follow those relationships. If not, the division would be meaningless for presentities. A social network method would be used in this scenario. In a social network, this crowd of people would become a community. Presence information would be shared in this community. In opposite, the presence information out of community is useless. So this paper would use the community to relay the presence information.

The community recognition will be an offline task. A graph method or some data mining method would be applied to generate the community. In this paper, we would just use these algorithms in social network analysis to generate the community.

3.2   The Structure of the Delivery System

As it is described in part 2, neither a client-server based system, nor a pure P2P system can deliver the presence information very well. In this part, the author will present a mixed structure presence system that can be easily used to deliver the presence information with small cost of the presence server.

Because the IMS system needs to be manageable, the presence server is still the centre node in the new system.

The new architecture is as follows:

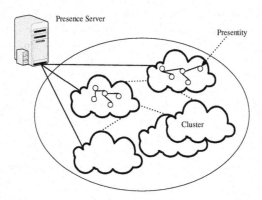

Fig. 1. The new architecture

The node here is called presentity. The presentities are organized in a cluster way. As mentioned before, the similar nodes are clustered in a community. The presence information dissemination would be described as the message transmission within and among the clusters.

Presentity would become more intelligent so that it can receive and forward presence information. Presence server sends the presence information to the corresponding cluster. Then the information is delivered node by node in the cluster. A tree model can be used for the inter-cluster transmission. The transmission path would be described as a tree path. So the presence server would become the root node of a tree. The presentity would become a tree node, which needs to receive and forward the presence information. Moreover, presentities retain their children information.

The tree is set up when the first time presence information is delivered. And then the tree topology may change when the presentity status changes. How to adjust the tree will be presented later.

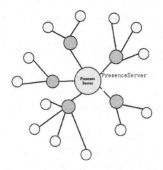

Fig. 2. The Tree for Dissemination

Definition 1. T= (root, nodes, edges, relation). Root represents the presence server, and the nodes are the collection of the presentities in the network. There are two kinds of node, the non-leaf node and the leaf node. The non-leaf node is an intelligent one, as it needs to forward information and maintain the tree structure. The leaf node is a simple one which just receives the presence information. Edges represent the communication path between nodes. Relation shows the relationship between nodes (eg: parent, child). The tree shows a transmission path of presence information.

Definition 2. The tree's height represents the number of max hops. If we assign the edge with weight that represents the transmission time, the height would be the longest message transmission time.

Definition 3. Degree. The degree shows the max number of children that a node would have. The community tree is not always a completed tree.

How does the tree work? The process includes a tree setup stage and a tree flooding& adjustment stage.

The tree setup stage includes these steps:

a)   The presence server would compute the communities. A community would be represented by a tree.

b)   Preset some parameters of a community, such as the height of the tree and the degree of the tree.

c)   Set up a k sequence. The k sequence is formed based on the active presence status in the current cluster. The method to set up the k sequence is called k sequence method which would be provided later.

d)   The k sequence indicates the tree structure. A community tree is set up according to the k sequence.

The presentities can be divided into two categories in IMS system: a radio based user and a wired user. The radio based users are those who use radio interface to access the network. Since the radio resource is limited, it is unsuitable for a radio user to act as a non-leaf node. So the radio based user would act as a leaf node.

As the above concerned, we propose a method called K-sequence to implement the tree model. The method would sort the active users in a cluster at first. The sort algorithm is quite simple by just putting radio based users after all wired users to form a sequence.

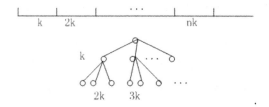

Fig. 3. The K Sequence and the Tree

And then, the sequence is logically divided into k based slots. A tree would be set up logically according to the sequence.

The reason why the tree setup procedure is a logic one is that there is not a real tree. There is only a k sequence that is a representation of a tree. And the node maintains information of its parent and children in the logical tree.

After we form the sequence, the root node would pass down the presence information according to the structure. First, the root node passes the presence information to the first level nodes (first k nodes are chosen from the k sequence). The first level nodes would also choose their k children and flood the information to them. The procedure goes on until the sequence meets its end.

Meanwhile, the middle (non-leaf) node will indicate its parent with its backup nodes information. And the backup nodes are always the node's children. So parent nodes are aware of grandchildren's information that might be useful for the sub tree re-construction.

In order to specify the above procedure, we consider marking the node with node number.

Assume the node number is just the same as the number in the sequence. Assume N is the total length of the sequence. Node i's parent is P_i, and its children is C_{ij}. The relationship would be represented by these formulas.

$$P_i = i \,/\, k \tag{1}$$

$$C_{ij} = i \times k + j \tag{2}$$

Both of formulas (1) (2) indicate the relationship between a parent and its children. Formula 1 shows that the parent id can be computed by the child id. It is quite obvious that K stands for the degree of the tree. Formula 2 shows how to compute the child id from the parent. J stands for the sequence number of the child in its parent.

When the tree is ready, it can be used to transmit the presence information. Usually, the presence information transmits through the preset tree structure. But when some presentities are "logout" or "offline", the tree structure needs to be refracted. The tree structure refracting process is called tree adjustment. So, the tree adjustment results in information dynamic dissemination path.

In fact, there are two concepts in the message dissemination that use the tree model, which are the tree flooding and the tree adjustment.

The "tree flooding model" shows how the message flows in a cluster. Presence information goes according to the tree node structure. The flooding here means a node sends messages to all of its children nodes when it forwards messages.

"Tree adjustment model" shows how the tree would be adjusted when some nodes are down. It is obvious that a node which is in a "logout" state would not receive or forward any message. So the tree would change its structure when the presence information changes.

The detailed procedure of the tree adjustment is:

1) Every middle(non-leaf) node would maintain the information of its children and parent. In most situations, a node uses them to transmit the presence information.

2) When a node is down, its parent would find the warning quickly. Some backup nodes are used to replace the nodes. (the backup nodes are always the children of the broken node)

3) If the backup nodes are also down, a sub tree construction procedure would begin which would construct a sub tree to replace the broken sub tree.

4) There is also a procedure that shows how to add a node. When a node is added, it would be assigned as a leaf. So it can receive the message from its new parent.

As shown in the Table 1 description, the tree adjustment can be done when message transmits. A visual procedure can be viewed from the Figure 4. The node can be replaced by its children. As we replace the node, we are in fact operating the k sequence.

Table 1. The Algorithm of Tree Adjustment for Middle Node

```
Function  transmitAndAjust(nodeNumber, info){
     Update the node with updateInfo.
     Loop( From 1 to children's number){
       child=getCurrentChild()
         //prepare to send the information to child
         Check the child to send is "logout"?
       if (logout){
           get the backup nodes
           check if the nodes are logout?
           Current= the one isn't 'logout'
       If (all backnodes logout )
         recontructSubTree(child,info,true)
             else {
               Current= the one isn't 'logout'
         Current's node number change to its parent
           recontructSubTree(current,info,false)}}
             else  transmitAndjust(child,info)
       }}
Function  reconstructSubTree(node, info,flag){
  //the flag indicate if need to replace the first one
    sequence= recomputeSequence(node,info,flag)
    buildTreeAndTransmit(node,sequence)}
```

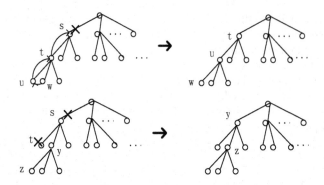

Fig. 4. The adjustment of a tree

4 Math Analysis

We first analyze the traffic of this method, comparing with the original client server method.

The traffic of presence server can be divided into two parts: the subscription message and the notification message. As the new method focuses on the push operation, the subscription traffic is the same as before.

The notification traffic would reduce greatly in the new method. With the original client server method, the server needs to send all active presentities the related update information. But now, only the degree times' messages are sent out in a community.

Assume that the original traffic is Tr_{old}, and the new method is Tr_{new}. As we divide the subscribers into clusters, so if there are N cluster, then the Tr_{old} can be represented by formula (3).

$$Tr_{old} = \sum_{i=1}^{N} Tr_{old}(i) \tag{3}$$

Each $Tr_{new}(i)$ would have some relationship with $Tr_{old}(i)$. In the original way, the server would send every update to all of the subscribers. But the new method would only send k notification messages to a community. If the number of people is c_i, and the active user number is c'_i. Then we can have a formula that the Tr_{new} is

$$Tr_{new} = \sum_{i=1}^{n} (\frac{k_i}{c'_i} \times Tr_{old}(i)) \tag{4}$$

$$c'_i \leq c_i \text{ If } (c'_i = 0), \text{ then } k_i/c'_i = 0$$

$$c_i \leq \sum_{j=1}^{h_i} k_i^{j} \tag{5}$$

Formula (4) indicates the relationship between the $Tr_{old}(i)$ and the Tr_{new}. The old method needs to send out c' messages while the new method only needs to send ki.

Formula (5) shows the constraints between Ci and ski, along with the Ci'. Because the dissemination is based on the tree structure, the node number would not be bigger than the tree node number. So c_i would not be bigger than the node number of a completed k tree with hi height. Obviously c' is not bigger than c_i. When a community has not active users, the transmission is zero for sure.

Not only the traffic is the factor needed to be concerned, but also the delay of the transmission is another factor. The delay of transmission contains two parts (formula (6) (7)), whose are the queue waiting time and the transmission time. Queue waiting time is generated by the server.

$$delay\_old = Lo + tr\_delay \qquad (6)$$

$$delay\_new = Ln + k \times tr\_delay \qquad (7)$$

Formula (6) describes the situation when the old method is applied. When a server is online, it needs to send a lot of notifications in a short time. So the server always puts the notifications in queue. But if the notification message number is huge, the queue time Lo would become bigger and bigger. Moreover in this situation, as the server's traffic is so huge, nodes near the server would also bear the pressure. Those nodes become a bottleneck of the system. As a feedback, the transmission time (tr_delay) would also go up. And the total time would increase.

In the new situation, a server only needs to send k(the degree) times for a community, so the queue size and the queue waiting time would be reduced greatly. Hence the Ln would be much smaller than the Lo. Though there are more hops that the message transmission needs, the time is under control. The max hop would not be bigger than the tree height. And the tree height is also not bigger than 4. Besides, as the dissemination is distributed, there is no node that needs to bear a big pressure. So there is not a network bottleneck. The transmission time can be guaranteed.

Another factor that must be concerned in a server design is jitter. Jitter describes the server working environment and affects the server performance and the configuration. A high jitter would cause the server's instability and maybe some other problems.

As for the old method, the jitter of the server is caused by the frequent status changes of the presentities. In our new system, the jitter would be reduced.

5 Experiment

5.1 The Simulator Data Set

As the presence information itself contains some kind of relationship, so it would be hard to just use a random function to generate presentites. It is known that the Orkut is an online community, so it has some social network feature. Moreover, the Orkut account is also the Google account. There is also a web presence in Orkut. So we generate our dataset based on the Orkut dataset. We use a grabbing program to pull the "userid" from the www.orkut.com. For privacy consideration, we don't grab any user information.

The dataset contains a 3 million friend relationship. We generate our simulator dataset based on it.

5.2 The Simulator Architecture

For the experiment environment, we construct a simulator. The simulator includes the server and client generator.

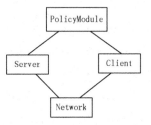

Fig. 5. Simulator Modules

As the figure describes, there are four modules in the simulator. The policy module is to preset the policy and other parameters of the server and client. The network module is the environment for server and client execution. Server and client modules stand for the real presence server and presentity.

We first preset the policy module with the initial parameters. The initial parameters include the working model (original method or new method).The network module would generate the server and client based on the parameters. When the program stops, we can get final data from the network module.

5.3 Experiment Analysis

We generate 10000 presentities. And there are 20 communities; each of them has 500 members.

In the first experiment, we observe the active presentity number from 1000 to 10000. And the presentities change their state below a possibility of 20%. The result shows that the original method would be like a line growing with certain slope. (Figure 6) That line would increase rapidly as the active presentities are added. And

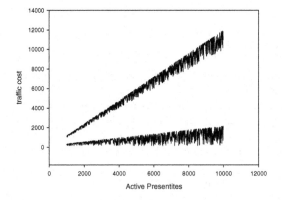

Fig. 6. The Traffic Comparison

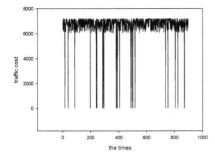

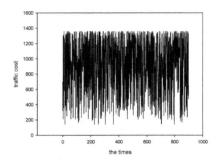

Fig. 7. The Jitter of the Original Method   **Fig. 8.** The Jitter of the New Method

the new method would only form a nearly horizon line (Figure 6). The traffic in the graph includes both subscription and notification messages.

Saw teeth in the line are almost affected by the subscription message and the random presence update information.

In order to show the jitter, we fix the presentities number at 6000. At one moment, 20% of the presentities change their status. That means there is random presence information at a given time. This experiment is for the jitter of the server performance. The result shows that the former method would encounter a big traffic jitter in the situation, which ranges from 0 to 7400 (Figure 7). However, the new method shows us a little jitter, which is just a range from 100 to 1300 (Figure 8).

6   Conclusions and Future Work

In this paper, we propose a new architecture for presence information dissemination, along with the key algorithms in the architecture. In our future work, we will apply the approach to a more complex and bigger network environment. The experiment would be more complex and realistic. Also we will focus on the relationship between the chosen parameters and the performance. Maybe the community merge mechanism would be introduced to achieve a better result. Moreover, the context aware technology would also be used in this scenario to make the dissemination more convenience.

References

1. Carzaniga, A., Rosenblum, D.S., Wolf, A.L.: Design and evaluation of a wide-area event notification service. ACM Transactions on Computer Systems 19, 332–383 (2001)
2. Lisha, G., Junzhou, L.: Performance analysis of a P2P-based VoIP software, p. 11. Guadelope, French Southern Territories (2006)
3. Urrutia-Valdes, C., Mukhopadhyay, A., El-Sayed, M.: Presence and availability with IMS: Applications architecture, traffic analysis, and capacity impacts. Bell Labs Technical Journal 10, 101–107 (2006)
4. Vishal, S.K., Hening, S.H.: Presence traffic optimization techniques,
   http://Sdev.voip.co.uk/~theo/misc/presence-traffic-optimization-techniques.pdf

5. Alam, M.T., WuCost, Z.D.: Cost Analysis of the IMS Presence Service. In: 1st Australian Confrence on Wireless Broadband and Ultra Wideband Communication, AusWireless 2006, Sydney, March 13-16 (2006)
6. Pous, M., Pesch, D., Sesmun, A.: Performance evaluation of a SIP based presence and instant messaging service for UMTS. In: 4th International Conference on 3G Mobile Communication Technologies, 2003. 3G 2003 (Conf. Publ. No. 494), June 25-27, pp. 254–258 (2003)
7. Lonnfors, M., Leppanen, E., Khartabil, H., Urpalainen, J.: Presence Information Dataformat (PIDF) Extension for Partial Presence, draft-ietf-simple-partial-pidfformat-07 (work in progress) (July 2006)
8. Roach, A., Campbell., B., Rosenberg, J.: A Session Initiation Protocol (SIP) Event Notification Extension for Resource Lists RFC 4662, (August 2006)
9. Niemi, A., Lonnfors, M., Leppanen, E.: Publication of Partial Presence Information, draft-ietf-simple-partial-publish-05 (work in progress) (July 2006)
10. Niemi, A.: An Extension to Session Initiation Protocol (SIP) Events for Issuing Conditional Subscriptions, draft-niemi-sip-subnot-etags-01 (work in progress) (June 2006)
11. Martin, M.G., Bormann, C., Ott, J., Price, R., Roach, A.: The SessionInitiation Protocol (SIP) and Session Description Protocol (SDP) Static Dictionary for Signaling Compression (SigComp), RFC 3485 (February 2003)
12. Khartabil, H., Leppanen, E., Lonnfors, M., Costa-Requena, J.: Functional Description of Event Notification Filtering, RFC 4660 (September 2006)

Separation of Responsibilities between Application Servers and Media Servers in NGNs: A Practical Approach

Alessandro Amirante, Tobia Castaldi, Lorenzo Miniero,
and Simon Pietro Romano

University of Napoli Federico II
Via Claudio 21, 80125 Napoli, Italy
{alessandro.amirante,tobia.castaldi,lorenzo.miniero,spromano}@unina.it

Abstract. In this paper we deal with the separation of concerns among
the entities involved in multimedia sessions in Next Generation Networks
(NGNs). In particular, the focus is on the explicit separation of responsibilities between Application Servers, which are in charge of the services
application logic, and Media Servers, whose task is instead the low-level
manipulation and delivery of media streams. The paper describes the
standardization proposals coming from both the 3GPP and the IETF
from a very practical point of view, and presents the reader with the
currently only available implementation of the MEDIACTRL architecture. Furthermore, the paper is also devoted to describing some typical
multimedia scenarios from the presented perspective, by providing an in-
depth analysis of the related sequence diagrams and protocol messages.

Keywords: IP Multimedia Subsystem, Media Servers, Application
Servers, Multimedia Applications.

1 Introduction

Telecommunication Services have seen in recent years a wide growth in terms of
heterogeneity of both the exploitable devices and the services they might wish to
access. This has led to the investigation of ways to properly benefit from such an
evolution without suffering from the issues this heterogeneity inevitably raises.
All these investigations fall under the general conceptual definition of the so-
called Next Generation Networks (NGNs) which envisage an "all-IP" approach
to transport all services, together with the related information. In particular,
several efforts have been spent for a few years in the attempt to obtain some
kind of service convergence in order to introduce an overall independence with
respect to network infrastructures within a standardization framework.

It is the case of the IP Multimedia Subsystem (IMS), which is being fostered
by the 3GPP (3rd Generation Partnership Project) as the proper solution to
obtain what is commonly known as fixed-mobile convergence, in order to allow
an easy integration among WiFi/cellular networks, the Public Switched Telephone Network (PSTN) and the Internet. The main objective of IMS has since

S. Balandin et al. (Eds.): NEW2AN 2008, LNCS 5174, pp. 199–211, 2008.
© Springer-Verlag Berlin Heidelberg 2008

its birth resided in trying to reduce both capital and operational expenditures (i. e. CAPEX and OPEX) for service providers, while at the same time providing operational flexibility and simplicity. The dispute around the most suitable signalling protocol among the IMS components has almost immediately been solved in favor of SIP (Session Initiation Protocol), a standard protocol defined within the IETF (Internet Engineering Task Force).

The IMS is mainly conceived as a framework to ease the deployment of multimedia-oriented services, including Voice over IP (VoIP), online gaming, video-conferencing, which are all to be provided on a single integrated infrastructure. Such an emphasis on the strong role played by media in the IMS architecture has led many efforts in the standardization towards the investigation of proper ways to cope with a fair share of responsibilities among the involved components. In particular, separate functionality is required from the components that are supposed to implement the application logic behind services and the components providing support for multimedia capability to the services themselves. In the IMS architecture, the former role is taken by Application Servers (AS), while the latter is taken by a logical component called Media Resource Function (MRF). To specify the interfaces between the two the standardization bodies have devoted many efforts in defining an architectural framework together with all the required protocols of interaction.

Such a direction of study is exactly the purpose of this paper, in which we describe the current state of the standardization efforts from a practical point of view, by also presenting to the reader some first-hand experiments based on real implementations of the specifications. In particular, the paper is organized in 6 sections. In Section 2 we will present a more detailed overview of both the context and the motivations that drove us to focus on such a work. The envisaged architecture is presented in detail in Section 3, which focuses on both the introduced roles. After these mostly introductory sections, we will delve into the details of the practical approach of our contribution, by first describing our implementation efforts in Section 4 and then introducing real-world scenarios in Section 5. Final remarks are provided in Section 6.

2   Context and Motivation

As introduced above, the more users ask for value added applications, the more an evolution of the obsolete network infrastructure and architecture is needed. To achieve the goal of granting a transparent (with respect to both devices and access networks) and better fruition of both contents and services, major efforts are being directed in separation of concerns among network components and heterogeneity of access. The separation of responsibilities is not a new proposal when thinking about multimedia capabilities. Several approaches have been presented in the past to cope with such an issue, some of them even within a standardization context. The first approach that comes to mind is the H.248 protocol, also known as MeGaCo as it was called when standardized by the IETF. This protocol, based on XML payloads, allowed applications to invoke media services from a remote

Media Server by means of a low-level API. Despite a widespread deployment of implementations supporting this specification, its low-level approach has recently moved the interested parties into researching an alternative way of dealing with media services. The preferred path of research almost suddenly became a SIP-based approach, SIP being the de-facto standard for IP-based multimedia applications and services. This led to several proposals, some already standardized and some still being specified. Such an abundance of proposals obviously resulted in a potential issue for implementors, considering a single standard reference protocol to be used in conjunction with SIP was still missing. As to the aforementioned matter, we touch on two almost independent activities currently being carried out that, although born in distinct contexts, are directed towards a common result: *IP Multimedia Subsystem* and *Media Server Control.*

The *IP Multimedia Subsystem* is a next generation network architecture that is currently being defined by the *Third Generation Partnership Project* (3GPP) with the goal of filling the gap between the cellular and Internet world. IMS envisages an access-agnostic architecture that supports a wide range of IP-based services and is based upon the *Session Initiation Protocol* (SIP), for all what concerns the signaling between the parties involved in a communication. Accordingly to the principle of separation of responsibilities, the IMS specification identifies several components, each in charge of a different task. We herein focus on those elements that come into play when considering multimedia services, i.e. the *Application Server* (AS) and the *Media Resource Function* (MRF). The former is the component in which resides the business logic, whilst the latter is in charge of manipulating the media flows. More specifically, the Media Resource Function is a macro-component made of the two sub-elements called, respectively, *Media Resource Function Controller* (MRFC), which is a signaling plane node, and *Media Resource Function Processor* (MRFP), which is a media plane node that implements all media-related functions.

The approach taken by the *Internet Engineering Task Force* is conceptually the same taken by the 3GPP, that is separating the application logic from the media processing. After investigating the needed requirements, the MEDIACTRL Working Group was recently opened, which specifies the Media Server (MS) as a centralized component that an Application Server (AS) can interact with by means of a dedicated protocol in order to implement multimedia applications. Such standardization efforts, and our related work in this field, will be the subject of the following sections.

3   Application Servers and Media Servers: The MEDIACTRL Approach

The MEDIACTRL WG, as already introduced, aims at specifying an architectural framework to properly cope with the separation of concerns between AS and MS in a standardized way. As such, the MEDIACTRL architecture envisages several topologies of interaction between AS and MS, the most general one being an m:n topology. Nevertheless, although some discussion on such topologies has

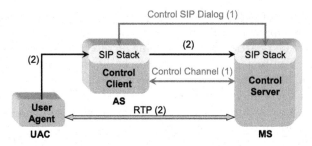

Fig. 1. Architecture: protocols interaction between AS, MS and UACs

already started, the focus is still on a 1 : 1 interaction (see Fig. 1), which will also be the topology this paper will refer to.

The current specification of the framework [1] envisages a modular approach when looking at the functionality to provide. This means that, inside the same MS, different inner components take care of orthogonal processing that may be required. To achieve this, the framework currently specifies a general invocation mechanism with opaque payloads, whereas such payloads can be directed to the proper component according to a header in the request itself. This way, new components providing additional media capability can be added at any time without affecting the specification of the core framework. Such components are called Control Packages; so far four different packages have been proposed: (i) a package providing basic Interactive Voice Response functionality [2]; (ii) a package for joining connection and implementing conferencing [3]; (iii) a package for dealing with VoiceXML [4]; (iv) a package we have written ourselves to involve floor control [5].

When looking at the protocol itself, the interaction between an AS and a MS relies on a so-called Control Channel. This channel is where MEDIACTRL-related messages flow between the AS (the client side) and the MS (the server side). An AS would instruct a MS into a specific media operation by placing a request on this channel, and the MS would reply to such request, as well as notify events, through the same channel. Of course, such a channel can only be set up as the result of a transport-level connection. This implies that either the AS or the MS must previously know the transport address of the other party in order to set up the connection. To allow this, a COMEDIA-based approach has been defined: the AS and MS make use of a SIP dialog to negotiate and set up the Control Channel.

Once this channel has been opened, a way to have the AS and MS authenticate each other is of course needed. This is needed in order to make sure that the client opening the control channel with the server is actually the same that initiated the SIP dialog in the first place. Such an authentication is accomplished by means of a dedicated Control Channel method called SYNCH: the newly connected AS has to send a properly constructed SYNCH message to the MS right after the connection has been opened, otherwise the connection is torn down.

More details upon the MEDIACTRL specification and the roles envisaged in the interaction will be presented in the following subsections.

3.1 Application Server: The Brain

The AS component plays a role which is of paramount importance inside ME-DIACTRL. It is in charge of appropriately controlling a MS in order to provide advanced services to end-users, and as such it is where all the application logic related to the services resides. To make a very simple example, it can be seen as the *brain* in the architecture, the entity making decisions and controlling all the actions accordingly.

The establishment of a Control Channel has already been introduced in the previous section. This establishment is a prerequisite to any further interaction between the AS and the MS.

For what concerns end-users, instead, being the AS a frontend to them, it is also in charge of terminating the signaling, in this case SIP. Considering the media functionality is actually provided by the MS and not by the AS itself, a way to have the AS transparently attach the users' media connections to the MS is needed: these media connections would then need to be properly manipulated to implement the service itself. In order to achieve this result, the specification currently envisages the use of a 3rd Party Call Control (3PCC) mechanism: the AS terminates the SIP signaling with the end-users and forwards their requests to the MS in order to have all media connections negotiated accordingly. These media connections are subsequently referenced by both the AS and the MS when needed (e.g. when the AS wants the MS to play an announcement on a specific user's connection).

The use of 3PCC, while powerful and relatively straightforward, can potentially lead to several issues when having to deal with complex use cases and/or scenarios. Nevertheless, these issues are not in the scope of this paper, and consequently discussions upon them are left to complementary documents.

3.2 Media Server: The Arm

Just as the AS is responsible for all the business logic, the MS is conceived to take care of every facet of the media processing and delivery. Its operations are realized according to the directives coming from the controlling AS. To recall the previously presented example, while the AS is the brain, the MS is the *arm*.

Such a distinction in roles makes it clear that the MS is supposed to be directly responsible for all the media on a low level. This means that, besides acting as the termination point for media connections with end users (whereas signalling is terminated by the AS), the MS is also responsible for manipulating these media connections according to incoming directives, policies and previous negotiations. Examples of operations a MS is supposed to be able to achieve include:

- mixing, transcoding and adaptation of media streams;
- low-level manipulation of media streams (e.g. gain levels, video layouts, and so on);
- playing and recording of media streams from and to both local and network environments;

- storing and retrieving of external references of any kind (e.g. media streams, VoiceXML and SRGS directives, and so on);
- tone and DTMF detection;
- Text-To-Speech and Speech Recognition.

Whereas a limited set of such operations is implicitly accomplished by the MS as a consequence of the initial SIP negotiations that make it aware of end users (e.g. user A only supports GSM audio, while user B also supports H.263 video as long as the bit rate is limited to 10kbps), the most relevant tasks for the MS come from the requests made by authorized ASs. For instance, within the application logic of a conferencing scenario an AS may first attach an end user to the MS in order to have them negotiate the available media between each other, and subsequently instruct the MS into playing an announcement (e.g. "Digit the PIN") followed by a DTMF collection (to have the user digit the conference pin number), and then into joining the user into the conference mix itself.

All these requests, together with the related responses and event notifications, flow through the already discussed control channel just as the previously described SYNCH transaction does. Some examples of this kind of interaction will be presented in subsequent sections, where a couple of typical use case scenarios will be described with respect to both sequence diagrams and protocol contents.

4 MEDIACTRL: An Open Source Implementation

It is quite clear according to the previous sections that the standardization work on the MEDIACTRL architecture is still ongoing and far from being a complete specification. However, work on it is progressing fast, and several documents have already been written which provide much of the required functionality. Thus, following the well known (even though almost forgotten) IETF motto "rough consensus, running code", we chose to implement the currently specified architecture. This was done in order to help researchers and developers understand the architecture and possibly dig out the flaws that may be hidden inside its specification. It came out to be the first (and at the time of writing the only) existing available implementation of the MEDIACTRL architecture. The prototype, which has been developed on both the server and client sides, currently includes the core framework, the new MEDIACTRL protocol and some of the control packages that can be employed in order to implement custom media manipulation. All the relevant details upon the implementation choices will be presented in the following subsections.

It is worth noting that our prototype implementation also paved the way for a call flow document we are carrying on in the MEDIACTRL Working Group. In fact, having a real world implementation of the protocol easily allowed us to reproduce popular use case scenarios involving media by means of the MEDIAC-TRL architecture, and consequently first-hand protocol interactions between AS and MS. Both the prototype and the document have been successfully presented at many MEDIACTRL meeting sessions, and helped provide feedback for the extant specification.

4.1   Application Server: SIP Servlet

Starting from the client side, we had to take into account what was required from a MEDIACTRL-enabled Application Server besides the business logic itself. Considering that the specification clearly points out that such an AS must support SIP, both for setting up the control channel with a MS by means of the previously introduced COMEDIA negotiation and for attaching User Agents to the MS through a 3PCC mechanism, the most obvious choice came out to be SIP Servlets. In fact, the SIP Servlet API defines a high-level servlet extension for SIP-based servers, thus enabling SIP applications to be deployed and managed based on the servlet model.

To implement the AS prototype through SIP Servlets, we developed it as a WeSIP[1] application. WeSIP is a SIP and HTTP Converged Application Server built on top of the popular open source SIP Proxy OpenSER. The software has been successfully tested with such a configuration; however, being a SIP servlet component, it was conceived to run "as it is" on any other standard container, and in fact it was successfully tested with the Sailfin[2] container as well.

The AS is in charge of appropriately controlling a Media Server in order to provide advanced services to end-users. Fig. 2 gives a simplified view of the protocol behavior of an AS interacting with a Media Server. For the details about such interaction, the interested reader can refer to [6].

The code implementing the Application Server has been entirely written in Java. It is structured in three packages. Two packages are dedicated to the implementation of the MEDIACTRL protocol messages, whereas a third one contains all the classes needed in order to carry out the Application Server functions. This last package includes two SIP servlets devoted, respectively, to control channel establishment and 3rd Party Call Control (3PCC). The business logic (i.e. the specific application orchestrated by the AS) is implemented in a separate thread which looks after all transactions occurring with the Media Server.

For the sake of conciseness, we will skip the details of the above described implementation. The interested reader can refer to the official project web site [7] for all information about both code and documentation.

4.2   Media Server: C++ Application

Coming to the MS itself, instead, the requirements were quite different than the one we identified for the client side. In fact, as explained in the previous section, a MEDIACTRL-enabled MS envisages low level media processing and manipulation besides custom media delivery by means of RTP.

---

[1] See http://www.wesip.eu
[2] See https://sailfin.dev.java.net/

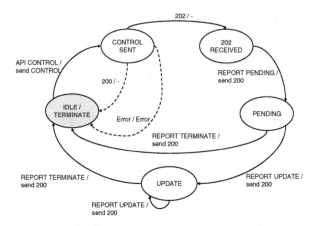

Fig. 2. The Application Server perspective in MEDIACTRL

This obviously suggested us that such an implementation would have strict real-time requirements, which led us into choosing C/C++ as the programming language to use to implement it.

However, the preliminary step to any actual implementation of the framework was indeed the investigation of the possible protocol states. This led us into specifying a state diagram for the MS as depicted in Fig. 3, which paved the ground for the implementation work itself.

While most of the code was written from scratch, we also decided to avoid the *not invented here* syndrome and thus reuse as many existing open source components as possible in the implementation.

The most important protocols to deal with were of course SIP and SDP. We needed a powerful and flexible SIP stack, capable to provide us with means to properly handle the 3PCC mechanism, as well as means to easily extend the standard SIP protocol with respect to both headers and bodies, in order to cope with the additional functionality envisaged in the MEDIACTRL specification (e.g. the COMEDIA negotiation and the ctrl-package SDP attribute). Thus, we chose the C++ *reSIProcate* library [8] as our SIP stack, which provided us with powerful means to deal with SIP behavior. This library is indeed well known in the open source community, and has many active IETF participants amongst its contributors. Besides, it has also been successfully used in many commercial projects as well, including two widespread SIP softphones, namely X-Lite and Eyebeam.

When coming to RTP, instead, we decided to make use of the open source C library *oRTP* [9], which we had already successfully used in a companion project of ours, the Confiance VideoMixer, which involved RTP manipulation for several different user agents at the same time.

Considering all the currently specified packages make a heavy use of XML for the framework messages bodies, we also had a strong need for a reliable XML parser. Our choice fell on the widespread *Expat* library [10], a very well known lightweight open source C component. The actual parsing of the XML contents, anyway, was left to an additional component called *Boost::regex*.

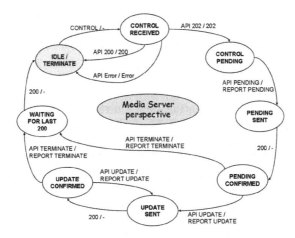

Fig. 3. Media Server SCFW State Diagram

Nevertheless, the list of protocols and meta-languages to handle did not end here. In fact some packages currently require the support for retrieval of external references. It is the example, for instance, of the Basic IVR package, which allows in its specification to externally refer resources to be used within dialogs, as remote media streams to play out in announcements, stored XML files containing the actual directives for the package, SRGS grammar bodies and so on. In order to appropriately satisfy this requirement, we made use of another well known open source component called *libcurl* [11]. This library allows for an easy retrieval of external files of any kind by supporting a wide range of protocols, including HTTP, FTP and many others.

Of course, all these libraries only allowed us to handle the protocols needed by the MEDIACTRL framework. Nevertheless, the framework would not be complete without an actual media support with respect to codecs, in order to properly implement the needed transcoding functionality. This led us to make use of additional open source components, the most important being the very well known *ffmpeg* [12] piece of software. This software in fact provides an easy API, called *libavcodec*, to deal with decoding and encoding of media streams, and proved very useful especially when implementing our video support for the framework.

All the aforementioned libraries were then integrated in the code we implemented from scratch ourselves, which will be briefly described in the following lines with respect to the requirements we met.

Particular attention was required by the way media connections are conceived in the specifications. In fact, such media connections can be addressed at different levels of granularity, according to whether the specific media label is included or not in the connection identifier. Hence, we specified wrapper classes and related callbacks in order to properly drive the flow of media accordingly.

For what concerns the MS itself, it was conceived to be modular with respect to both codecs (audio and video) and control packages, for which we designed a dedicated API. This allowed us to separate the design of the core framework itself from the realization of proof of concept control packages and codecs used to test the MS functionality. We realized a limited set of codecs that add basic media capability to the MS. Our current implementation offers good support to audio; at the time of this writing, work on video has also started and is progressing in parallel with the standardization work inside the WG (which has only recently started investigating such a direction). Regarding the control packages, we focused on two of the previously introduced ones, specifically Basic IVR and Conferencing. This allowed us to reproduce many real-world scenarios by having the AS properly orchestrate requests to the MS, as it will become clearer in the following section.

For the sake of brevity, we won't delve into the details of the implementation in these pages. The whole documentation regarding code, classes and the way they interact with each other, is available for any interested reader on the project home page [7].

5   Use Case Scenarios

The presented implementation efforts allowed for the testing of proper real-world use case scenarios involving media processing and delivery. Many of these scenarios have been included in the already mentioned call flow document, where snapshots of the full protocol interaction between AS and MS for each of them are provided.

Having implemented two orthogonal packages with respect to the provided functionality, we were able to design several heterogeneous scenarios, ranging from simple echo tests to phone calls, conferences and voice-mail applications. Each scenario could be reproduced by properly orchestrating different requests and by correlating the resulting output, just like a state machine. As a reference scenario to describe in this paper we chose the *Echo Test* one, which basically consists of a UAC directly or indirectly "talking" to itself. Other scenarios would be achieved taking quite similar approaches, by integrating different transactions in different fashions within the application logic. Once again, we invite the interested readers to refer to the complete draft [6] and to the prototype project homepage [7] for more details about all the scenarios.

The upcoming scenario description will focus on the specifically scenario-related interaction between the AS and the MS and the results in the UAC experience. This implies that the example assumes that a Control Channel has already been correctly established and SYNCHed between the reference AS and MS as described in the previous sections. Similarly, the 3PCC session among the AS, the MS and the interested UAC is also assumed to have already happened.

Once all the preliminary steps have taken place, the actual scenario can be reproduced and analyzed. We will herein provide the description of the simplest transaction that can be achieved, i.e. a Direct Echo Test approach, where the UAC directly talks to itself. Such a choice derives from a need of conciseness, since more complicated transactions could not be explained in few lines.

5.1   Direct Echo Test

In the Direct Echo Test approach, the UAC is directly connected to itself. This means that each frame the MS receives from the UAC is sent back to it in real-time. The popular open source PBX Asterisk makes use of such an approach in its echo test application.

In the framework this can be achieved by means of the conference control package, which is in charge of the task of joining connections and conferences. Specifically, the package method the AS has to make use of is called <join>, and a sequence diagram of a potential transaction is depicted in Fig. 4.

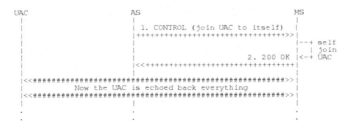

Fig. 4. Self Connection: Framework Transaction

All the transaction steps have been numbered to ease the understanding and the following of the subsequent explanation lines:

- The AS requests the joining of the connection to itself by sending a CON-TROL request (1), specifically meant for the conferencing control package (msc-conf-audio/1.0), to the MS: since the connection must be attached to itself, the id1 and id2 attributes are set to the same value, i.e. the connectionid;
- The MS, having checked the validity of the request, enforces the join of the connection to itself; this means that all the frames sent by the UAC are echoed back to it; to report the success of the operation, the MS sends a 200 OK (2) in reply to the MS, thus ending the transaction.

The complete transaction, that is the full bodies of the exchanged messages, is provided in the following lines:

1. AS → MS (SCFW CONTROL)

```
SCFW 74b0dc511949 CONTROL
Control-Package: msc-conf-audio/1.0
Content-Type: text/xml
Content-Length: 87

<?xml version="1.0"?>
<join id1="1536067209~913cd14c" \
    id2="1536067209~913cd14c">
</join>
```

2. AS ← MS (SCFW 200 OK)

```
SCFW 74b0dc511949 200
Content-Type: text/xml
Content-Length: 70

<?xml version="1.0"?>
<response status="200" reason="Join successful"/>
```

Such a transaction is the simplest form of transaction that can occur through the Control Channel. In fact, the reply to the CONTROL message is immediately provided to the AS in a 200 message. This is not always true, since asynchronous events related to the original request may occur and consequently influence the AS state behavior.

6   Conclusions and Future Work

In the introductory sections of this paper we have focused on the current lack of an agreed standard in the field of remote Media Servers. Such a lack logically represents the presence of many open issues for implementors and vendors. To try and overcome these issues a working group called MEDIACTRL was formed within the IETF, and many efforts have been spent towards the specification of a complete and reliable architecture capable to provide a solution to the several requirements that Next Generation Networks envisage. We then presented our prototype implementation of such an architecture, together with details upon our design choices and some high level perspective of a sample scenario amongst the most popular ones.

From the evaluation of our implementation and the reproduced scenarios, we came to the conclusion that the standardization efforts spent on the MEDI-ACTRL architecture is definitely on the right path. The modularity of control packages in particular looks like a good choice, as the many already achievable scenarios seem to confirm. We claim that this work can prove to be an invaluable tool for first hand experiments concerning the ongoing specifications.

Of course, there is a lot of work that still needs to be done. This not only includes continuous updates along with the specification themselves, but also additional features that are currently lacking in the prototype and may prove more than useful for more in depth evaluation of the framework, as well as a richer experience for end users.

Acknowledgments

This work has been carried out with the financial support of the European projects NetQoS and Content. Such projects are partially funded by the EU as part of the IST Programme, within the Sixth Framework Programme.

References

1. Boulton, C., Melanchuk, T., McGlashan, S.: Media Control Channel Framework. draft-ietf-mediactrl-sip-control-framework-01 (February 2008)
2. Boulton, C., Melanchuk, T., McGlashan, S.: A Basic Interactive Voice Response (IVR) Control Package for the Media Control Channel Framework. draft-boulton-ivr-control-package-06 (February 2008)
3. Boulton, C., McGlashan, S., Shiratzky, A.: A Conference Control Package for the Media Control Channel Framework. draft-boulton-conference-control-package-04 (February 2008)

4. Boulton, C., Melanchuk, T., McGlashan, S.: A VoiceXML Control Package for the Media Control Channel Framework. draft-boulton-ivr-control-package-06 (February 2008)
5. Miniero, L., Amirante, A., Castaldi, T., Romano, S.P.: A Binary Floor Control Protocol (BFCP) Control Package for the Media Control Channel Framework. draft-miniero-bfcp-control-package-00.txt (February 2008)
6. Miniero, L., Amirante, A., Castaldi, T., Romano, S.P.: Media Control Channel Framework (SCFW) Call Flow Examples. draft-miniero-mediactrl-escs-01.txt (November 2007)
7. MEDIACTRL - IETF Media Server Control Prototype, http://mediactrl.sourceforge.net
8. reSIProcate, http://www.resiprocate.org
9. oRTP., http://www.linphone.org
10. The Expat XML Parser., http://expat.sourceforge.net
11. cURL and libcurl, http://curl.haxx.se
12. FFmpeg, http://ffmpeg.mplayerhq.hu

Systematic QoS Class Mapping Framework over Multiple Heterogeneous Networks

Misun Ryu, Youngmin Kim, and Hongshik Park

Information and Communications University
119, Munji-ro, Yuseong-gu, Daejeon, 305-714, Korea
{rms0,injesus01,hspark}@icu.ac.kr

Abstract. A network defines its own Quality-Of-Service (QoS) class and has QoS support mechanisms. So, effectively to support end-to-end QoS in heterogeneous networks, a certain unified control is needed, however, it causes scalability problem as management complexity and implementation difficulty. There is a strong need to provide simple interoperability with QoS support so we present a QoS Class Mapping (QCM) framework: building blocks should be defined such as parameter mapping and class mapping. And we improve the framework, called as QCM-ASM, to support not only flawless class mapping but also fine-granular QoS in any circumstance. At last, another framework with adaptive QoS Class Selection (AQCS) mechanism, named as AQCM-ASM framework, is proposed. AQCS mechanism can prevent resource starvation of lower priority class and provide an effective resource distribution. As an experimental result, we demonstrate a performance of the proposed frameworks. The performance results show characteristics of each framework.

Keywords: Heterogeneous networks, QoS, interoperability.

1 Introduction

The trend of recent and current developments in network is toward Next Generation Networks (NGN) [1], a packet-based network providing the differentiated QoS over heterogeneous networks. It is also willing to accommodate not only traditional applications but also future emerging applications having a variety of QoS requirements. The future emerging applications will have finer QoS requirements such as virtual reality (VR) applications [3] providing an advanced human-computer interface and allowing a user to interact with other users in a simulated 3D environment. Namely, applications with a particular and finer requirement will be appeared and should be supported in NGN.

There are many studies on ways for the delivery of end-to-end QoS to applications over heterogeneous networks. And it can be categorized into two approaches. One is a central approach by a unified control and the other is a distributed approach by a QoS interworking. The unified control mechanism across several QoS domains has an ability to efficiently manage network resources like [4][5]. [4] proposed some methods that QoS requirement distributes per node/network along the application's routing path. [5] investigated the requirements, architecture and protocols across different network

S. Balandin et al. (Eds.): NEW2AN 2008, LNCS 5174, pp. 212–221, 2008.

environment. It was said that when signaling runs across several QoS domains, it requires that QoS control information should be universal. Centrally to manage QoS control information for heterogeneous networks causes scalability problems. In other words, it is not easy to be implemented, managed and maintained.

QoS interworking between different QoS domains can also simply support end-to-end QoS. Appropriate QoS class mapping can provide not absolute but differentiated QoS. In general previous studies on QoS mapping focused on parameter mapping and class mapping between different networks. With regard to mapping between parameters, the mapping between ATM and IP [8][9][12], between DiffServ and IntServ [11] and between ATM and FR [12] were discussed. These studies have different formulations according to network combination, respectively. Consequently, they would be too complicated to implement on an interworking function. There are also many researches related with class mapping in the different networks. [13][14] showed QoS class mapping among IP, DiffServ PHBs and UMTS. [15] presented QoS classes mapping between DiffServ and 802.11e, between 802.11e and 802.1D and between 802.1D and DiffServ.

In this paper, our goal is to propose a scalable and easy implementing framework to support end-to-end QoS over the heterogeneous network through flawless QoS class mapping. So, we assume that there is no central management function such as globally used signaling protocol and every network operates its own QoS mechanism.

This paper is organized as follows. Following this instruction, we describe a fundamental QCM (QoS Class Mapping) framework in the section 2. In Section 3, ASM is introduced and QCM framework based on ASM (QCM-ASM) is explained. In Section 4, QCM-ASM framework applying Adaptive QoS Class Selection (AQCS) mechanism is presented. Performance assessments by using ns-2 are given in Section 5. Finally Section 6 presents conclusions.

2   QCM Framework Description

To achieve efficient end-to-end delivery over heterogeneous networks we present a simple interoperable QoS Class Mapping (QCM) framework shown in Fig. 1. The QCM framework is included into the following important two building blocks (boxes in Fig. 1.): 1) parameter mapping accurately for determining the best suitable QoS class at the entrance; and 2) class mapping for interworking at the Ingress Edge Router (IER) of each network. We will describe the framework and function of these blocks in detail.

Each application has its own specific Service Level Agreement (SLA). To guarantee the SLA, packets of the application have the QoS requirement as network performance; The QoS requirement is represented by QoS parameters such as delay, loss and jitter. The packets determine their own QoS class through parameter mapping block. And whenever the packet incomes a different network, an Ingress Edge Router (IER) determines appropriate QoS class to the network through class mapping block at this time. All IERs is generating, managing its own class mapping table and updating it through network management information such as monitoring. The IER executes not only packet transformation but also QoS conversion. Here, we do not

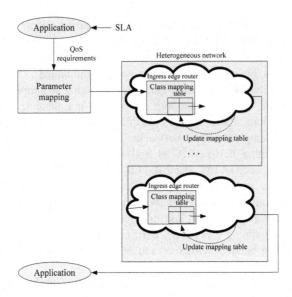

Fig. 1. A schematic of a QoS Class Mapping framework over heterogeneous networks

deal with packet transformation and only focus on QoS class conversion. And we assume PQ scheduler to all nodes for showing a differentiated service based on class.

2.1 Parameter Mapping

A network defines its own QoS class and stipulates upper bounded values of each QoS class by standard organization or network provider. Parameter mapping chooses the best suitable QoS class about QoS requirements of a packet. Here, we study how optimally to map a packet with QoS requirement to one of the priority class. Here, we assume that lower number QoS class has the higher priority.

Mapping parameters to QoS class can be formulated as follows where N is the QoS parameters number and M_α is the QoS class number of α network. Each packet having $\mathbf{X}=[x_0, x_1,\dots, x_{N-1}]^T$, QoS requirements, will be mapped to a y QoS class of a α network. $\mathbf{P}^\alpha =[\ \mathbf{P}^\alpha_0, \mathbf{P}^\alpha_1,\dots, \mathbf{P}^\alpha_{M-1}]$ is the bounded requirements of α network and i-class' bounded requirements values of α network has $\mathbf{P}^\alpha_i=[\ p^\alpha_{i0}, p^\alpha_{i1,\dots}, p^\alpha_{iN-1}]^T$. Here this formulations, $\mathbf{F}_\alpha$, how that a packet with $\mathbf{X}$ requirement be mapped to class y having least difference between $\mathbf{X}$ and the bounded requirements $\mathbf{P}^\alpha$ of α network.

$$y = F_\alpha(\mathbf{X}). \tag{1}$$

$$c_i = \mathbf{W} \cdot (\left\| \mathbf{X} - \mathbf{P}^\alpha_i \right\|) = \sum_{j=0}^{N-1} w_j \left| x_j - p^\alpha_{ij} \right| \quad \forall i = 0,1,\dots M_\alpha - 1 \cdot \tag{2}$$

$$y = \left\{ k \mid c_k = \min_i \{c_i\} \right\} \quad \forall i = 0,1,\dots M_\alpha - 1. \tag{3}$$

The $\mathbf{W} = [w_0, w_1, \ldots, w_{N-1}]^T$, as weight factor can reflect application characteristics, in case of strong delay sensitive; weight of delay parameter can increase relatively.

Here we can obtain the optimal parameter mapping by this formulation. This implies several meanings. First, it can map certain application requirement to more exact QoS class in any networks. Secondly we can control weight factor $\mathbf{W}$ according to application characteristic for example, error intolerant application, jitter-sensitive application etc., and in last this formulation is scalable so many parameters can be considered. And optimal parameter mapping prevents resource waste through effectively distributing resource and provide end-to-end QoS on the basis.

2.2 Class Mapping

Using on the concept of QoS classes for a network is recommended by Global Standardization Organizations i.e. ITU, IETF, 3GPP, IEEE which define a set of QoS classes of the most relevant networks. And they prefer the approach of QoS classes instead of individually specified performance parameters for avoiding complicated implementation for the network. Interworking by QoS class mapping can also provide simple implementation.

In this framework, class mapping module coverts QoS class based on class mapping table. The class mapping table is QoS class matching table between different networks. This table can be made based on the previous works [8][9][11][12] and is also managed and updated by a network provider according to the network state information. However, this method must consider the relation with all networks and the different granularity QoS class can bring out uncertain mapping.

3   QCM-ASM Framework

To solve these problems as stated above about class mapping we define Application Service Map (ASM) and QCM-ASM framework is proposed, a preliminary version of which was presented in [6].

3.1 ASM

Each application has its own suitable performance requirements which are not dependent on any specific network for its validity. [2] classifies performance requirements into QoS categories shown in Fig. 2 and has axes of important QoS parameters such as loss and delay. And the size and shape of the boxes show a general indication of the limit of delay and loss tolerable for each application. The Application Service Map (ASM) is an enlarged form of Fig. 2. The various applications are mapped on the ASM considering QoS requirements and each application has Location Information (LI) on the ASM. The LI represents 2-tuple (loss, delay) by marking its ToS (Type of Service) field with integer, respectively.

The Fig. 3 is ASM sample in which loss is divided into 4 and delay is divided into 16. Each area can express from (0, 0) to (3, 15) that starts from lower left. The specific applications are located in the LI. Other applications can be located in the

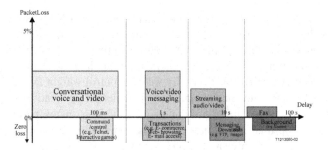

Fig. 2. Mapping of performance requirement [2]

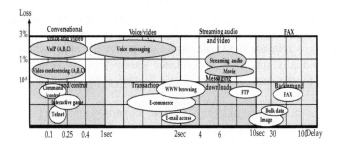

Fig. 3. ASM example having axes which are delay and loss

schema by their similarity to these exemplars. The new emerging applications can be mapped onto ASM according to QoS parameter requirements. The underlying requirements of loss and delay of ASM can be grouped appropriately. Namely, this can be used as the basis for deriving realistic and meaningful network QoS classes for differentiated service. It provides an indication of the upper and lower boundaries for applications as essentially acceptable. And because ASM has finer granular than any other networks, it can executes perfect QoS class mapping.

3.2   Overall QCM-ASM Framework

For the end-to-end delivery, packets of an application explicitly determine a LI of ASM according to its QoS requirement. The LI of ASM is stamped in packets in ToS field. Packet's QoS class is determined based on LI written in ToS field at the IER. IERs of a network generate and manages LI-to-class mapping table. A network provider has information on delay and loss on per-class basis through monitoring the network state and updates the mapping table between LI of ASM and QoS class through class mapping based on the monitoring information. The packets served as QoS class determined by LI in the network.

Class mapping. The mapping between LI and QoS class is executed by the parameter mapping formulation as early mentioned. However, LI of ASM has the upper and lower bounded values. So, the parameter mapping formulation needs to be partially modified to find most suitable mapped class in two requirement set. When each LI has X_{max} and X_{min} as the upper and lower QoS values, each LI chooses QoS class z in a α network as

$$z = \max(y_{min}, y_{max}) = \max(F_\alpha(\mathbf{X}_{min}), F_\alpha(\mathbf{X}_{max})). \tag{5}$$

LI-to-class mapping is generated by the above equation (5) and each LI can select optimal QoS class.

QCM-ASM framework can offer the differentiated service as much as the number of LI degree while QCM framework can provide the differentiated service.

4   AQCM-ASM Framework

In this section, we propose Adaptive QoS Class Selection (AQCM)-ASM framework based on QCM-ASM framework. It has all advantages of QCM-ASM and it additionally can guarantee delay requirement of more packets by adaptively changing the served QoS class.

4.1   AQCS

Providing end-to-end delay has been an important issue in QoS network and we focus on delay of QoS requirement parameters in this part. In QoS enabled network, higher priority packet should be guaranteed rather than lower priority. But, high priority packets do not always need to serve firstly if only their own requirement satisfy. To achieve it, we should know how much delay the packet experience until now. However, it is not easy that a packet knows its own state. Namely, if a packet knows the elapsing delay and loss at that time determining QoS class, end-to-end requirement of as many packets as possible can be guaranteed.

AQCS mechanism literally means that QoS class of a packet is adaptively selected according to the existing state. Here the existing state does not mean a network state but the state of a packet and the state can represent the elapsing time of a packet until now.

We should know an elapsing time; the elapsing time (t_e) is the experienced time until now. The elapsing time consists of propagation delay (t_p), transmission delay (t_t) and queuing delay (t_q) of a node and the elapsing time until the prior node like (6).

$$
\begin{aligned}
t_e &= t_p + t_t + t_q + t_e \\
&= t_p + \frac{P_{size}}{C} + (t_{out} - t_{in}) + t_e.
\end{aligned}
\tag{6}
$$

To get the queuing delay we check the arrival time of a packet (t_{in}) and the departure time of the packet (t_{out}). The queuing delay of a packet is driven as ($t_{out} - t_{in}$) in the node, transmission delay is driven by packet size (P_{size}) information in a header and link capacity (C). And the propagation delay of a link can be obtained from configuration information of a node.

The elapsing time is determined by packet by packet. All packets have this information and update from to time to time. So, the elapsing time should be encoded in option field of a packet header and it is updated at every node.

The AQCM mechanism, selecting the best suitable QoS class that not overestimated high class, based on estimation of the elapsing time, can utilize the network resource more efficiently.

4.2 Overall AQCM-ASM Framework

The AQCM-ASM framework is based on QCM-ASM framework so it has all characteristics in QCM-ASM. And it has the same operation with QCM-ASM except calculation of elapsing time and temporal LI. The LI in a packet represents the requirement of its own packet and the elapsing time represents the experienced delay of the packet. This information shows the remaining delay that subtracts the elapsing time from LI, the requirement. In this framework we select the QoS class base on the remaining delay. All nodes update the elapsing time of every packet and each IER selects QoS class based on temporal LI made of the remaining delay. Here, temporal LI is calculated on IEG at time decision of QoS class. QoS class of a packet is determined by temporal LI.

AQCM-ASM can support fine granular application like QCM-ASM. And it guarantees not only high priority class but also low priority even if it is relative QoS not absolute QoS. So, it can lead effective resource distribution.

5 Experimental Results

NS-2 is employed to compare the performance. We present a multi-node simulation to see if the frameworks can support the differentiated service. For this experiment we simulated a network with a topology as shown Fig. 4. The topology consists of 4 different domains (WLAN→DiffServ→IP→UMTS) and have 9 nodes connected by 1Mbps links having a propagation delay (10msec or 5msec). And there are 7 kinds of traffic summarized in Table 1. We are required to get a moderate end-to-end delay under the mentioned configuration. We run simulation of 0.95-loaded input traffic. To setup more realistic traffic environment, cross traffic exists; input traffic from source (1) to destination (9) is 0.56-loaded and cross traffic is 0.39-loaded, having respectively same specification(offered load is 0.39).

Fig. 5, Fig.6 and Fig. 7 show us the overall end-to-end delay and delay difference of the proposed frameworks. Here, a delay difference subtracts the experienced end-to-end delay from the delay requirement of a packet. Above zero value represents

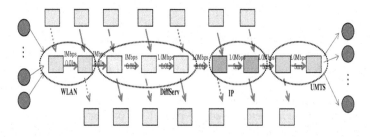

Fig. 4. Simulation Topology Model

Table 1. Specification of input traffic

Delay Requirement	traffic	Rate *offered load	number	size
80msec	exp	0.001Mbps*0.56	100	64
100msec	exp	0.001Mbps*0.56	100	128
120msec	exp	0.0012Mbps*0.56	100	128
140msec	exp	0.13Mbps*0.56	1	210
200msec	exp	0.15Mbps*0.56	1	210
250msec	exp	0.2Mbps*0.56	1	210
350msec	exp	0.2Mbps*0.56	1	210

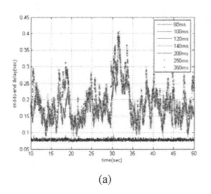
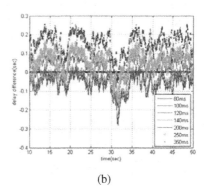

(a)   (b)

Fig. 5. QCM framework characteristics (a) end-to-end delay (b) delay difference

satisfying the requirement; below zero represents dissatisfying the requirement. Three frameworks are tested, i.e. Fig.5 are QCM framework, Fig.6 QCM-ASM framework and Fig. 7 AQCM-ASM framework.

As shown in Fig.5, QCM framework is based on IP QoS class. Packets of requirement below 100msec are high priority and packets of delay requirement from 100msec to 350msec represent same QoS class. Because of the 0.95-loaded the QoS violation is seen little in packets of delay requirement from 100msec to 350msec. The detail Fig.5 (b) show delay difference of all application from 100msec to 350msec have same pattern. It means the QCM can support differentiated service as much as very coarse degree but it is hard to support finer-granular application.

As shown in Fig.6, QCM-ASM framework is based on ASM. It is seen only packets having the lowest requirement have many violations because of PQ scheduler. The detail Fig. 6 (b) tell me all packets except the lowest requirement are satisfactory about each its requirement. However, because of the 0.95-loaded, the applications having of loose requirement have serious QoS violation. As a result, QCM–ASM framework can support a more differentiated service than QCM framework.

As shown in Fig.7, AQCM-ASM framework is applied with AQCS. It is seen only packets having the lowest requirement have little violation because of PQ scheduler. The detail Fig.7 (b) tell me the most packets are satisfactory about each its requirement.

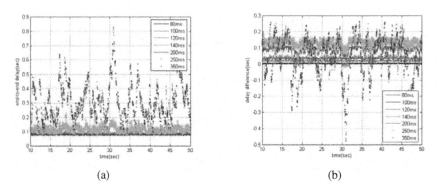

(a)   (b)

Fig. 6. QCM-ASM framework characteristics (a) end-to-end delay (b) delay difference

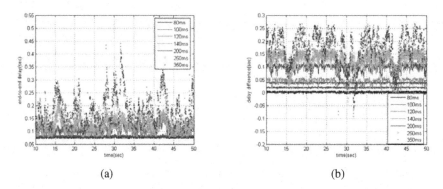

(a)   (b)

Fig. 7. AQCM-ASM framework characteristics (a) end-to-end delay (b) delay difference

Though 0.95-loaded, the most of applications meet their QoS requirement. As a result, we know that AQCM–ASM framework can support fine granular application and distribute the resource well.

6 Conclusions

This paper recognizes an interoperability of class over heterogeneous networks and the well-interoperability means the exact mapping between classes. So we propose QCM frameworks; The QCM-ASM is based on ASM and AQCM-ASM applies AQCS mechanism to QCM-ASM. The QCM can be easily implemented because of the established studies. However, it has a limitation to provide end-to-end QoS for each fine-granular application because class mapping has many flaws. So QCM-ASM based on LI of ASM is proposed. Each application requirement represents globally LI and the mapping table between LI and its own QoS class is made and managed in each network. It can support finer applications that emerge in the future and a specific network's QoS class change can have no effect on the framework. AQCS-ASM is proposed to guarantee of delay requirement in high priority traffic and low priority traffic simultaneously. Although this framework cannot absolutely guarantee the QoS requirement of all class packets, it can reduce end-to-end delay of the relative lower priority packets by handovering spare resource of the high class packet.

Acknowledgments. This research was supported by the MKE (Ministry of Knowledge Economy), Korea, under the ITRC(Information Technology Research Center) support program supervised by the IITA(Institute of Information Technology Advancement (IITA-2008-(C1090-0801-0036)).

References

1. ITU-T NGN-GSI Release I, http://www.itu.int/ITUT/ngn/release1.html
2. ITU-T Recommendation: G.1010
3. Skorin-Kapov, L., Huljenic, D., Mikic, D., Vilendecic, D.: Analysis of end-to-end QoS for networked virtual reality services in UMTS. Communications Magazine, IEEE 42(4), 49–55 (2004)
4. Znati, T.F., Melhem, R.: Node delay assignment strategies to support end-to-end delay requirements in heterogeneous networks. IEEE/ACM Transactions On Networking 12(5) (2004)
5. Fu, X., et al.: NSIS: a new extensible IP signaling protocol suite. Communications Magazine, IEEE 43(10), 133–141 (2005)
6. Ryu, M.S., et al.: QoS class mapping over heterogeneous networks using Application Service Map. ICNICONSMCL (2006)
7. DaSilva, L.A.: QoS mapping along the protocol stack: discussion and preliminary results. IEEE ICC 2, 713–717 (2000)
8. Tasaka, S., Ishibashi, Y.: Mutually compensatory property of multimedia QoS, Communications. IEEE International Conference, 1105–1111 (2002)
9. Garibbo, A., Marchese, M., Mongelli, M.: Mapping the quality of service over heterogeneous networks: a proposal about architectures and bandwidth allocation. IEEE International Conference on Communications 3, 1690–1694 (2003)
10. Marchese, M., Garibbo, A., Davoli, F., Mongelli, M.: Equivalent bandwidth control for the mapping of quality of service in heterogeneous networks. In: IEEE International Conference Communications, vol. 4, pp. 1948–1952 (2004)
11. Chahed, T., Hebuterne, G., Fayet, C.: Mapping of loss and delaybetween intserv and diffserv. In: ECUMN 1st European Conference, pp. 48–55 (2000)
12. Dixit, S.S., Kumar, S.: Traffic descriptor mapping and traffic control for frame relay over ATM network. IEEE/ACM Transactions on Networking 6(1), 56–70 (1998)
13. 3GPP: TS 29.207-640
14. ITU-T Recommendation: Y.1541
15. Park, S.-Y., et al.: Collaborative QoS Architecture between DiffServ and 802.11e Wireless LAN. IEEE VTC (2003)

A Fair Utility Function for
Incentive Mechanism against Free-Riding
in Peer-to-Peer Networks

Yuhua Liu[1], Chun Yang[1], Kaihua Xu[2], and Hongcai Chen[1]

[1] Department of Computer Science, Huazhong Normal University, Wuhan, 430079, China
[2] Research Center of the Digital Space Technology, Huazhong Normal University,
Wuhan, 430079, China
yhliu@mail.ccnu.edu.cn

Abstract. There are two problems, unfair and too strict, with the utility function of the proposed incentive mechanisms for restraining the free-riding in peer-to-peer (P2P) networks. This paper is devoted to establish a new fair utility function to solve those problems. We absorb both the absolute contribution value and the physical performance of peers in our utility function and the absolute contribution value is divided into supply value and profit value. Besides, we take the time the user is in the system as a factor of a peer's utility. The mathematical modeling, Analytic Hierarchy Process (AHP), is used in this paper to construct the function. The simulations compare the fair utility function with the proposed three functions and the results verify that this function is effective to solve those two problems.

Keywords: peer-to-peer, free-riding, utility function, Analytic Hierarchy Process.

1 Introduction

Peer-to-peer (P2P) systems are distributed systems consisting of interconnected nodes able to self-organize into network topologies with the purpose of sharing resources such as content, CPU cycles, storage and bandwidth [1] and it has gained great popularity for it provides us many fast and convenient services such as file searching, file downloading, etc. However, the free-riding [2] is growing crazily while P2P systems are developing rapidly.

Free-riding means exploiting P2P network resources (through searching, downloading objects or using services) without contributing to the P2P network at desirable levels [3]. In 2000, an extensive analysis of user traffic on Gnutella in [2] shows a significant amount of free-riding in the system that nearly 70% of Gnutella users share no files, and nearly 50% of all responses are returned by the top 1% of sharing hosts.

Free-riding could lead to degradation of the system performance and add vulnerability to the system [2]. As a result of it, the number of files in the system becomes limited and the number of popular files may become even smaller as the time goes by, which could affect user's interest in the system and they eventually pull

S. Balandin et al. (Eds.): NEW2AN 2008, LNCS 5174, pp. 222–233, 2008.

out of the system. When users who share popular files pull out of the system, the system becomes even poorer in terms of the amount of files shared. This is a vicious cycle and it may eventually lead to the collapse of the system [3].

Free-riding is a grave threat against the existence and efficient operation of P2P networks. To ensure P2P systems developing continually and healthily, some incentive schemes would be required [4]. The most important point for incentive schemes is to establish the utility function. Utility function estimates the utility of the user to the system. The difference between incentive mechanisms lies in the definitions of utility functions. Our work in this paper is to propose a utility function that can evaluate every peer objectively and fairly.

2   Related Work

There have been tremendous interests in free-riding since it was found out in [2] and the incentive mechanism is the earliest and most effective method to restrain free-riding [4].

Ramaswamy L. and Liu L. introduced the concept of utility function to measure the usefulness of peers to the system as a whole in 2003 for the first time, and described a scheme based on that concept to control free-riding [5].

$$U(P_i, T) = \alpha \times | ULimit(P_i, T) | . \tag{1}$$

The utility function (1) they designed was based on number of files shared by each user. This utility function could ensure that users sharing larger number of files would have a higher utility value than those users sharing smaller number of files.

$$U(P_i, T) = \beta \times \sum_{\forall F_j \in ULimit(P_i, T)} size(F_j) . \tag{2}$$

The utility function (2) was a variant of (1) that took into account the file size along with the number of files shared.

$$U(P_i, T) = \sum_{\forall F_j \in ULimit(P_i, T)} \gamma(k + DCount(F_j, T)) \times size(F_j) - \sum_{\forall F_l \in DList(P_i, T)} size(F_l) . \tag{3}$$

The utility function (3) accounted for the popularity of files as measured by the number of times it was downloaded.

In [6], the authors considered the problem of fair resource sharing to optimize the performance of resource sharing in peer-to-peer systems. Resource sharing systems currently face rational peers which may exhibit a variety of strategies including: free-riding and greedy behavior. To estimate the fairness, they defined the function as:

$$u(p,t) = \int_{t'=0}^{t} b(p, r(t'), t') al(p, t') - \sum_{r \in R} c(p, r) * part(p, r, t) . \tag{4}$$

Ahsan Habib and John Chuang in [7] established their utility function based on the streaming session quality Q and the contribution cost C,

$$U_i(x_i) = a_i Q(x_i) - b_i C(x_i) . \tag{5}$$

where a_i and b_i defined the values of streaming quality and contribution cost to user i.

The paper [8] gave out its utility function (6), where $\frac{x_i}{b_i}$ was the ratio of the allocated resource to the bidding resource.

$$U_i(x_i) = \log(\frac{x_i}{b_i} + 1) \quad (x_i \in [0, b_i]) \; . \tag{6}$$

Figueiredo D., Shapiro J. and Towsley D. in their work [9] used the utility function (7) to represent the total cost,

$$u_i(l_i, c_i, p_i) = (l_i p_i - \lambda_i c_i) + \alpha_i c_i - \beta_i l_i \; . \tag{7}$$

Analyzing all of the above utility functions, we can find out that they are not fair for all the users in P2P systems. Almost all of the utility functions calculate the user's absolute contribution value such as the number of sharing files, the total size of files downloaded by others, or other criterions, but they didn't consider the differences of users' hardware i.e. the heterogeneity of peer-to-peer network. Because not all the system users' computers are of the same well performance (including computational speed, storage, I/O speed, etc.), and their Internet connection types are also not the same. For example, if there is a user willing to share same popular files but he is using Intel 386 and connecting to internet by modem (33.6Kbps), apparently, he can't share very many files and the total size of files downloaded by others may be few because of his bad physical performance, so with the proposed utility functions, the system may take it as a free-rider. Therefore, it is not fair for these users.

Besides, the utility functions are too strict for the free-riders, and free-riders may drop the system since they can get few resources from it [4]. Consequently, the number of users would decrease rapidly. But the amount of users is the key point to P2P systems, and too few users would lead to the failure of the systems. Especially in [9], the authors found that moderate degree of free-riding would not do harm to the system, but could increase the throughput and improve the efficiency of the network. Therefore, the utility function should be moderate for the free-riders so that it can restrain free-riding but won't bring forth the negative effects.

This paper is devoted to establish a new utility function to solve the two problems, unfair and too strict. For fair, both the absolute contribution value and physical performance of users are adopted in our utility function and the absolute contribution value is divided into supply value and profit value; For moderate, we take the time the user is in the system into account and allow user to change for utility with its accumulated online time. Besides, we use the mathematical modeling, Analytic Hierarchy Process, to weigh each factor to construct the function.

3   Fair Utility Function

In our minds, the utility of a user is composed of its absolute contribution value and its physical performance. To establish the utility function, we should first estimate the absolute contribution value and the physical performance. Before doing our further analysis, we give out the necessary notations:

T: The period of updating;

Log_i: The times that the user i has logged on during T hours;

n_{i_share}: The total number of files that the user i has shared in T hours;

n_{i_down}: The total number of files that the user i has downloaded in T hours;

n_{i_up}: The total number of files that the user i has uploaded (downloaded by others) in T hours;

$t_i=(t_{i\_1}, t_{i\_2}, \cdots, t_{i\_f}, \cdots, t_{i\_ni\_share})$, where t_{i_f} is the time the file f has been shared by user i during T hours, so $|t_i|= \sum_f t_{i\_f}$ is the total time all the files have been shared by user i;

$S_{i\_share}=(s_{i\_1}, s_{i\_2}, \cdots, s_{i\_f}, \cdots, s_{i\_ni\_share})$, where s_{i_f} is the size of the file f shared by user i, and $|S_{i\_share}|=\sum_f s_{i\_f}$, which is the total size of the files shared by user i;

$S_{i\_down}=(\chi_{i\_1}, \chi_{i\_2}, \cdots, \chi_{i\_d}, \cdots, \chi_{i\_ni\_down})$, where χ_{i_f} is the size of the file d downloaded by user i, and $|S_{i\_down}|=\sum_d \chi_{i\_d}$, which is the total size of the files downloaded by user i;

$S_{i\_up}=(v_{i\_1}, v_{i\_2}, \cdots, v_{i\_u}, \cdots, v_{i\_ni\_up})$, where v_{i_u} is the size of the file u uploaded by user i, and $|S_{i\_up}|=\sum_u v_{i\_u}$, which is the total size of the files uploaded by user i;

T_clock_i: The clock rate of CPU for user i;

W_i: The word of CPU for user i;

V_RAM_i: The speed of RAM for user i;

S_RAM_i: The size of RAM for user i;

S_HD_i: The size of hard disks for user i;

B_up_i: The upload bandwidth of user i;

ξ_i: The absolute contribution value of user i;

Γ_i: The physical performance value of user i;

$U(i, h)$: The utility function which is the utility value of user i during h hours and h is the total time since user i first enters the system.

3.1 Absolute Contribution Estimation

The absolute contribution of a user during a period is relative to eight factors: the number of shared files, the number of downloaded files, the number of uploaded files, the size of shared files, the size of downloaded files, the size of uploaded files, the time each file has been shared, the logged times, and the online time. In notations, they are n_{i_share}, n_{i_down}, n_{i_up}, S_{i_share}, S_{i_down}, S_{i_up}, t_i, Log_i, T.

A user's contribution can be divided into four parts, sharing, downloading, uploading and online time, and they should be all estimated. It is unfair to just calculate downloading and uploading for users with bad hardware performance, because these peers may have few connections for their poor physical performance even though they share many popular files, and it is harmful to just calculate sharing and downloading for some users may share uninteresting files that no peers would visit, and also harmful to just calculate sharing and uploading because the greedy users can download files crazily without any restriction. In addition, it is useful to take the online time into account. If a free-rider shares no file, its sharing value and uploading value would be zero, so it can't download any file if the utility function doesn't absorb the factor online time and this free-rider may drop the system. Reversely, if the utility function takes the online time as its factor, this free-rider can earn utility by staying in the system. Thereby, the second problem, too strict, could be solved. Although the free-riders do no significant contribution to the system, the system can obtain profit all the same as long as the free-riders stay in the system, for example, using P2P system to do advertisement doesn't care the behavior of free-riding but just care the number of online users.

Both the number and size of files for sharing, downloading and uploading can impact on the user's contribution, and there would be bugs available for free-riders if we consider only one side, so they also should be adopted in estimation together. Since it is possible that some files may be stopped sharing during a period and the stability of a P2P system suffers a lot from peers' frequent login and logout, we take the file's sharing time and peer's logged times as two factors into our estimation.

The absolute contribution is the subtraction of supply value (φ_i) and profit value (ψ_i), i.e.,

$$\xi_i = \alpha\varphi_i - \psi_i . \tag{8}$$

where α is a normalizing constant which is variable. The supply value is the benefit that the system has gained from the user and the profit value is the benefit that the user has gained from the system.

Supply Value. The mathematical modeling, Analytic Hierarchy Process (AHP) [10], is used to resolve this problem.

Among the four parts, only sharing, uploading and online time are related to the supply value. The sharing can be divided into two sub-elements: the total number and the total size of all shared files. Because not all the files are shared for T hours during a period, we revise the two sub-elements into relative number and relative size, in which the number and the size are calculated by scalar product of n_{i_share} and t_i and scalar product of S_{i_share} and t_i. The uploading can also be divided into two sub-elements: the total number and the total size of all uploading files. As indicated before, the logged times is also an important factor, and the supply value should be inverse to it, so we let it divide the four sub-elements. Since the utility function is to calculate the utility in a period, the online time is just the period T and it doesn't need to be divided by logged times for the online time have nothing to do with it.

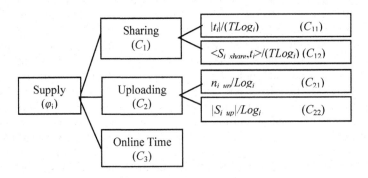

Fig. 1. Hierarchy structure of supply value

Table 1. Scale for pairwise comparison

Degree of importance	Definition
1	Equal (no preference)
2	Intermediate between 1 and 3
3	Moderately preferable
4	Intermediate between 3 and 5
5	Strongly preferable
6	Intermediate between 5 and 7
7	Very strongly preferable
8	Intermediate between 7 and 9
9	Extremely strongly preferable
1/2, 1/3, …, 1/9	Reciprocal of above numbers

According to the analysis above, we build the hierarchy structure model as shown in Fig.1.

The next step in the AHP is the establishment of priorities by using of pairwise comparison procedure (Table 1). We construct three pairwise comparison matrixes (A, B_1, B_2) based on the Table 1 and the reality of supply value, where A is the relative importance of C_1, C_2, C_3 to φ, B_1, B_2 are that of $C_{11}, C_{12}, C_{21}, C_{22}$ to C_1, C_2.

$$A = \begin{bmatrix} 1 & \frac{1}{3} & 3 \\ 3 & 1 & 9 \\ \frac{1}{3} & \frac{1}{9} & 1 \end{bmatrix},$$

$$B_1 = \begin{bmatrix} 1 & \frac{1}{2} & 0 & 0 \\ 2 & 1 & 0 & 0 \\ 0 & 0 & 0 & 0 \\ 0 & 0 & 0 & 0 \end{bmatrix}, \quad B_2 = \begin{bmatrix} 0 & 0 & 0 & 0 \\ 0 & 0 & 0 & 0 \\ 0 & 0 & 1 & \frac{1}{2} \\ 0 & 0 & 2 & 1 \end{bmatrix}$$

In A, we take the sharing (C_1) strongly prior to uploading (C_2) for supply (φ_i) in order to resist the behavior that just sharing uninteresting files to escape serving for others and take online time (C_3) the lowest importance to inspire sharing. In B_1 and B_2, the size of files is moderately prior to the number for sharing and uploading because the difference between their contributions is not strong, but C_{21}, C_{22} have no actions on C_1 and C_{11}, C_{12} have no actions on C_2, so their importance are 0.

Using formula $Aw=\lambda w$, we can get the biggest eigenvalues and their eigenvectors of A, B_1 and B_2.

$$\lambda^{(1)}=3, \ w^{(1)}=[0.23 \ 0.69 \ 0.08]^{T};$$

$$\lambda_1^{(2)}=2, \ w_1^{(2)}=[0.333 \ 0.667 \ 0 \ 0]^{T};$$

$$\lambda_2^{(2)}=2, \ w_2^{(2)}=[0 \ 0 \ 0.333 \ 0.667]^{T}.$$

So the combination weight is,

$$w=[\ w_1^{(2)}, \ w_2^{(2)}]w^{(1)}=[0.08 \ 0.15 \ 0.23 \ 0.46 \ 0.08]^{T} . \tag{9}$$

Table 2. Weights for the factors of supply value

Factor	C_{11}	C_{12}	C_{21}	C_{22}	C_3
Weight	0.08	0.15	0.23	0.46	0.08

Because A, B_1 and B_2 are consistent matrixes, the consistency checking and combination consistency checking need not to be done. Therefore, the combination weights w could be taken as the weights of four sub-elements (Table 2).

Now, we can establish the calculation formula of supply value as

$$\varphi_i = 0.08|t_i|/TLog_i + 0.15\langle S_{i\_share}, t_i\rangle/TLog_i$$
$$+ 0.23\, n_{i\_up}/Log_i + 0.46|S_{i\_up}|/Log_i + 0.08T \tag{10}$$

and

$$\langle S_{i\_share}, t_i\rangle = \sum_{f=1}^{n_{i\_share}} (s_{i\_f} \cdot t_{i\_f}) . \tag{11}$$

Profit Value. The profit value just has two factors, the number of downloaded files and the size of downloaded files. Since they are similar to C_{11}, C_{12} or C_{21}, C_{22}, we can give out their weights directly from $w_1^{(2)}$ or $w_2^{(2)}$ (Table 3).

Table 3. Weights for the factors of profit value

Factor	$n_{i\ down}$	$S_{i\ down}$
Weight	0.333	0.667

The calculation formula of profit value could be

$$\psi_i = 0.333 n_{i\_down} + 0.667|S_{i\_down}| . \tag{12}$$

Since both the supply value and the profit value are all given out, with equations (8), (10) and (12), we can get the calculation formula of absolute contribution value,

$$\xi_i = \alpha\varphi_i - \psi_i = \alpha(0.08|t_i|/TLog_i + 0.15\langle S_{i\_share}, t_i\rangle/TLog_i$$
$$+ 0.23 n_{i\_up}/Log_i + 0.46|S_{i\_up}|/Log_i + 0.08T) - (0.333 n_{i\_down} + 0.667|S_{i\_down}|) . \tag{13}$$

3.2 Physical Performance Estimation

The quantification of a user's physical performance is very complex for it could be influenced by lots of factors. In order to simplify the problem, we just choose the important factors that would influence peer's sharing behaviors seriously, the clock rate of CPU, word, the speed and size of RAM, the size of hard disk and upload bandwidth (we use it to evaluate the internet connection type because the bandwidth

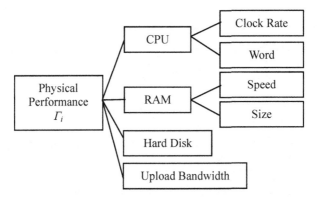

Fig. 2. Hierarchy structure of physical performance

is composed of download bandwidth and upload bandwidth and just the latter would impact on peer's contribution to the system).

Similar to the calculation of supply value, we also use the AHP mathematical modeling to establish the physical performance. The physical factors can be divided into four classes: CPU, RAM, hard disk's size and upload bandwidth. Thereby, we can build a three levels' hierarchy structure model (Fig. 2). The first level is the physical performance, and the second is the five classes, in which the CPU and RAM should be divided into two sub-elements respectively, so the clock rate of CPU, word, the speed and size of RAM make the third level.

The pairwise comparison matrix of the second level to the first level is,

$$A' = \begin{bmatrix} 1 & 2 & 7 & 1 \\ \frac{1}{2} & 1 & 5 & \frac{1}{2} \\ \frac{1}{7} & \frac{1}{5} & 1 & \frac{1}{7} \\ 1 & 2 & 7 & 1 \end{bmatrix}$$

The matrixes of the third to the second, clock rate and word to CPU and speed and size to RAM, could be,

$$B_1' = \begin{bmatrix} 1 & 3 \\ \frac{1}{3} & 1 \end{bmatrix}, \ B_2' = \begin{bmatrix} 1 & \frac{1}{3} \\ 3 & 1 \end{bmatrix}$$

From the introducing the calculate process of AHP in section 3.2, we can get the calculation result of the second and the third levels in Table 4 and Table 5, and the final result in Table 6.

Table 4. Calculation result of the second level

Factor	CPU	RAM	S_HD_i	B_up_i
Weight	0.373	0.205	0.049	0.373
Checking	λ_A=4, CI=0.05, CR=0.06			

Table 5. Calculation result of the third level

Factor	T_clock_i	W_i	V_RAM_i	S_RAM_i
Weight	0.75	0.25	0.25	0.75
Checking	$\lambda_{B1} = \lambda_{B2} = 2$			

Table 6. Final result of six factors for physical performance

Factor	T_clock_i	W_i	V_RAM_i
Weight	0.280	0.093	0.051
Factor	S_RAM_i	S_HD_i	B_up_i
Weight	0.154	0.049	0.373
Checking	$CR'=0.06$		

In this part, the consistency checking and the combination consistency checking are done because the matrix A' is not a consistent matrix. In Table 4, CI is the consistency index and CR is the consistency ratio. The CR' in Table 6 is the combination consistency ratio. Both CR and CR' are all less than 0.1, so the weights in Table 6 are rational.

Therefore, we establish the estimation formula for physical performance as,

$$\Gamma_i = 0.28T\_clock_i + 0.093W_i + 0.051V\_RAM_i \\ + 0.154S\_RAM_i + 0.049S\_HD_i + 0.373B\_up_i \qquad (14)$$

3.3 Utility Function

The utility function is used to measure the usefulness of every peer in the system to the community as a whole. We define it as

$$U(i, h) = U(i, h\text{-}T) + U(i, T) . \qquad (15)$$

where $U(i, h\text{-}T)$ is the accumulated utilities of user i since its first entry until the last period, and $U(i, T)$ is the earned utility of user i in current period which should be the ratio of absolute contribution value to physical performance.

$$U(i, T) = \frac{\xi_i}{\Gamma_i} . \qquad (16)$$

With (15),(16), utility function could be established as

$$U(i,h) = U(i,h-T) + \frac{\xi_i}{\Gamma_i} . \qquad (17)$$

and ξ_i, Γ_i are established in (13) and (14).

4 Simulation Results and Discussions

Our simulation is based on Gnutella protocol and Flooding searching algorithm. We use the BA model [11] to construct the topology of 5000 nodes and simulate the

Table 7. Measurement units of each factor

Factor	T	t_i	$S_{i\ share}$	$S_{i\ down}$
Unit	hour	hour	M	M
Factor	$S_{i\ up}$	T_clock_i	W_i	V_RAM_i
Unit	M	GHz	bits	MHz
Factor	S_RAM_i	S_HD_i	B_up_i	
Unit	M	G	Kb/s	

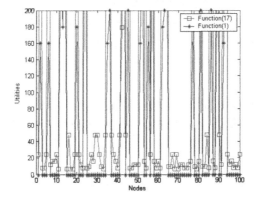

Fig. 3. Comparison between function (17) and function (1) on 100 nodes

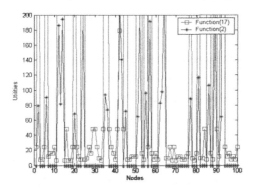

Fig. 4. Comparison between function (17) and function (2) on 100 nodes

operation of P2P system on machine. Each node in the simulation system is assigned with virtual hardware, uploading bandwidth and sharing files randomly. The measurement units of each factor in the calculation of utility are given out in Table 7. In addition, we suppose $\alpha=3$ and the period $T=2$ hours. In our simulation, we calculate the value of utility function (17) for every node during each period and we also calculate the values of utility functions (1), (2) and (3) to get the comparison data.

Fig. 3 to Fig. 5 are the comparisons of utility values between utility function (17) and functions (1), (2) and (3) respectively on the first 100 nodes of 5000 nodes. From

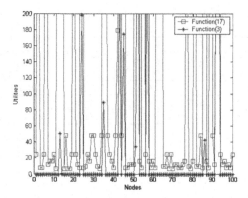

Fig. 5. Comparison between function (17) and function (3) on 100 nodes

them, we can see that the utilities of free-riders (such as the nodes 1, 3, 4, 5 and so on.) with functions (1), (2) and (3) are zero, but with our utility function (17), their utilities are more than zero but not too high, most of which are less than 50 while other nodes' utilities are much bigger than theirs. Therefore, it verifies that our utility function is not too strict for the free-riders but it also can distinguish them from the users willing to share.

Additionally, in Fig. 5, the nodes assigned with poor physical performance in the simulation (such as nodes 14, 35, 93 etc.) among the nodes willing to share have much bigger utilities with our function than function (3) and the opposite nodes assigned with better physical performance (such as nodes 13, 24, 43 etc.) have much lower utilities. Thereby, the simulation verifies that our function is fair.

In sum, the above analysis demonstrates that the two problems presented in section 2, unfair and too strict, could be resolved by our utility function (17).

5   Conclusions

Free-riding is growing crazily in P2P systems and it could lead to degradation of the system performance and add vulnerability to the system. Many incentive mechanisms were proposed for restraining the free-riding. However, there are two common problems with them, unfair and too strict. In order to solve them, we propose a new fair utility function to calculate each peer's utility value. The utility function is established by Analytic Hierarchy Process. For fairness, the absolute contribution value and physical performance of peers are adopted in the function and the absolute contribution value is divided into supply value and profit value. To solve the problem of too strict, we take the time the user is in the system as a factor of the supply value. The simulations verify that our utility function can resolve the two problems and it also can distinguish the free-riders from the users willing to share. This fair utility function can be accompanied with resource allocation or authority management or other incentive measures to restrain the free-riding in P2P systems.

Acknowledgments. This work is supported by the Natural Science Foundation of China (Grant No. 60673163).

References

1. Stephanos, A., Diomidis, S.: A Survey of Peer-to-Peer Content Distribution Technologies. ACM Computing Surveys 36(4), 335–371 (2004)
2. Adar, E., Huberman, B.: Free Riding on Gnutella. First Monday 5(10), 134–139 (2000)
3. Karakaya, M., Korpeouglu, I., Ulusoy, O.: A distributed and measurement based framework against free riding in Peer-to-Peer networks. In: 4th International Conference on Peer-to-Peer Computing Zurich, pp. 276–277 (2004)
4. Yijiao, Y., Hai, J.: A Survey on Overcoming Free Riding in Peer-to-Peer Networks. Chinese Journal of Computers 31(1), 1–15 (2008)
5. Ramaswamy, L., Ling, L.: Free Riding: A New Challenge to Peer-to-Peer File Sharing Systems. In: 36th Annual Hawaii International Conference on System Sciences (HICSS-36 2003), pp. 220–229. IEEE Computer Society, Big Island (2003)
6. Anceaume, E., Gradinariu, M., Ravoajia, A.: Incentive for P2P fair resource sharing. In: 5th IEEE International Conference on Peer-to-Peer Computing, Konstanz, pp. 253–260 (2005)
7. Ahsan, H., John, C.: An Incentive Mechanism for Peer-to-Peer Media Streaming. In: International Workshop on Quality of Service (IWQoS), Montreal, pp. 171–180 (2004)
8. Ma, R.T.B., Lee, S.C.M., Lui, J.C.S.: A Game Theoretic Approach to Provide Incentive and Service Differentiation in P2P Networks. In: ACM SIGMETRICS/Performance 2004, New York, pp. 189–198 (2004)
9. Figueiredo, D., Shapiro, J., Towsley, D.: Incentives to Promote Availability in Peer-to-Peer Anonymity Systems. In: 13th IEEE International Conference on Network Protocols (ICNP 2005), pp. 110–121. IEEE Computer Society, Boston (2005)
10. Saaty, T.L.: The Analytic Hierarchy Process. McGraw Hill Company, New York (1980)
11. Yuhua, L., Chun, Y., Liansheng, T., Shaohua, T.: A Generalized B-A Model for the Recognition of Intermediate Vertex Effect. Dynamics of Continuous, Discrete and Impulsive Systems (DCDIS), Series B. 14(S7), 23–27 (2007)

Algorithm for Selecting either an Overlay or Flat Route Based on the Amount of the Delay Measurement Load on the Home Agent in a Hierarchical Mobile IPv6 Network

Hidetoshi Kobayashi and Kazumasa Takami

Graduate school of Engineering, Soka University
1-236 Tangi-cho, Hachiouji-shi, Tokyo, 192-8577 Japan
k_takami@t.soka.ac.jp

Abstract. In a Hierarchical Mobile IPv6 (HMIPv6), two alternative routes are available: a Home Agent (HA) route (overlay route), and a direct communication route (flat route), which does not pass through the HA and is determined using the Return Routability procedure. The flat route is not always optimal when handover frequencies are taken into consideration. Therefore, it is necessary to select either the overlay or flat rate based on the particular communication situation. This paper proposes a communication route selection algorithm that is suitable of an HMIPv6 configuration, in which the route to be used is selected based on the packet delay measurement of each route. The algorithm's effectiveness has been verified through a simulation of its operation, and the evaluation of the reduction in the number of delay measurement packets arriving at the HA.

Keywords: Hierarchical Mobile IPv6, Packet delay, Route Selection.

1 Introduction

Recent remarkable advances in networking technology have given rise to new network environments, as exemplified by the Next Generation Network (NGN) [1]. In such a network, mobile communication will gain in importance. The high data communication speed becoming available to mobile phones and the all-IP implementation of mobile networks are expected to increase the variety of multimedia services. For example, continuous communication, such as videoconferencing and other real-time communication and streaming delivery, is expected to increase. An increase in this mode of communication will heighten the importance of communication quality. In particular, packet delay will be critical in real-time communication or streaming delivery. To reduce the degradation in quality caused by intermittent disruptions of a video or a delay in speech signals, it is necessary to select a route with a small packet delay.

In the Internet, Mobile IPv6 [2] has been proposed to handle terminal mobility. Furthermore, Hierarchical Mobile IPv6 [3] has been proposed to achieve efficient location management and fast handover in a wide-area mobile communication network.

S. Balandin et al. (Eds.): NEW2AN 2008, LNCS 5174, pp. 234–245, 2008.

This paper proposes a packet delay measuring method suitable for hierarchical terminal mobility control in a Hierarchical Mobile IPv6 network environment, which includes a mix of flat and overlay routes. It also proposes a routing algorithm based on the proposed packet delay measuring method. This paper verifies the operation of the algorithm, and measures the effect of the algorithm on reducing the load to the Home Agent (HA), and the routing error rate. Section 2 describes Hierarchical Mobile IPv6, the two types of routes, and associated routing issues. Section 3 discusses the proposed routing algorithm. Section 4 presents the evaluation model used and the evaluation results. Section 5 gives the conclusions and future issues.

2   Types of Routes in Hierarchical Mobile IPv6, and Issues for the Selection of Routes

Hierarchical Mobile IPv6 (HMIPv6) aims to distribute the location management function over the network that uses Mobile IPv6 (MIPv6), which is a protocol applicable to mobile communication environments, as shown in Fig. 1. HMIPv6 makes it possible to reduce location registration (or binding update) time, by introducing MAPs (Mobility Anchor Points) and separating the management of local movements of a Mobile Node (MN) (movements of the MN within the same MAP) from the management of global movements (movements of the MN across different MAPs).

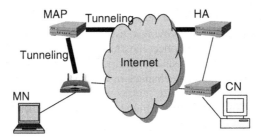

Fig. 1. Hierarchical Mobile IPv6

In HMIPv6, two types of communication routes exist between an MN and the Corresponding Node (CN):

- Overlay route
- Flat route.

The overlay route is a communication route via the HA. The flat route is an optimal route selected by a procedure called Return Routability, which informs the corresponding node (CN) of the location of the MN.

As far as the geographical distance between nodes is concerned, packets should experience a smaller delay on the flat route than on the overlay route. However, since delays are also affected by the frequency of MN handovers and the geological layout of network facilities, there may be cases where the delay on the geographically longer route is smaller than that on the geographically shorter route. Also, there have been

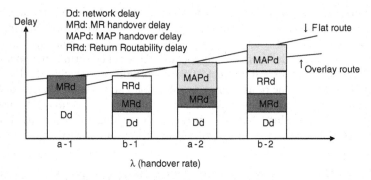

Fig. 2. Handover rate and delays of each route

reports that packets on an overlay route experience a smaller packet delay and a smaller packet discard rate than packets on a route selected using IP routing [4]. Cases have also been found in a theoretical analysis of an HMIPv6 environment that the delay on the overlay route was smaller than that on the flat route.

In our study, we assume delays in different HMIPv6 routes as shown in Fig. 2. We consider not only a network delay but also MR handover delay (MRd), which is the time needed to complete a handover when an MN moves to a different Mobile Router (MR) under the same MAP, MAP handover delay (MAPd), which is the time needed to complete a handover when an MN moves to a different MAP, and Return Routability delay (RRd), which is the time needed to complete Return Routability, a procedure for routing optimization. We also assume that, at a high handover rate (λ), the delay on the overlay route can become smaller than that on the flat rate. Since MNs are mobile, it is better to measure delays periodically than measuring them only once in order to be able to adapt to changes in network bandwidths. However, if the delay measurement interval is short, delay measurement packets can impose a significant load not only on the network but also on the HA. Therefore, it is necessary to consider how to reduce this load.

3   Routing Algorithm

Network bandwidths change continuously. To keep up with such changes, it is necessary to measure delays in the network periodically and determine the optimal routes, without generating excessive loads on the HA and the network. Because of the nature of Mobile IP, when algorithms so far proposed are applied to measuring the delay on the overlay route, delay measurement packets must pass through the HA, and thus become a significant load not only on the network but also on the HA. To solve this problem, we propose an algorithm shown in Fig. 3. Here, T1 is a delay measurement interval, and T2 is the threshold of the time during which an MN stays within the MAP link (MAP link staying time threshold). The proposed algorithm is described in detail below.

Step 0: When a movement of the MN is detected, one of the following is selected. If the movement is to a different MR within the same MAP area, Step 6 is executed. If the movement is to a different MAP area, Steps 1 to 6 are executed.

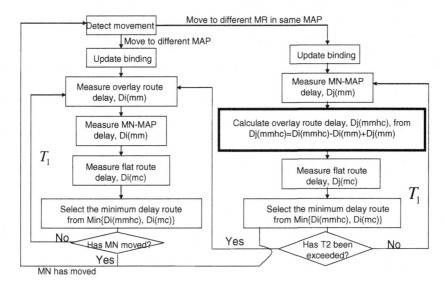

Fig. 3. Flowchart of the algorithm

Step 1: The MN detects its movement to a different MAP area (outside the home link), generates LCoA (on-Link Care-of Address) and RCoA (Regional CoA), and registers its new location.

Step 2: The delay of the overlay route (MN-MAP-HA-CN), Di(mmhc), is measured using reverse tunneling, and stored.

Step 3: The delay between the MN and the MAP (MN-MAP), Di(mm), is measured and stored.

Step 4: The delay of the flat route (MN-CN), Di(mc), is measured and stored.

Step 5: The route with the minimum delay, Min{Di(mmhc), Di(mc)}, is selected from the measurements. Steps 2 to 5 are repeated at an interval of T_1.

Step 6 (This is the case of the movement to a different MR within the same MAP area.)

Step 6-1: MN generates new LCoA and registers the LCoA as the new location.

Step 6-2: The delay of MN-MAP, Dj(mm), is measured using reverse tunneling based on LCoA, and is stored.

Step 6-3: Dj (mmhc) is calculated from the new measurements.

$$Dj(mmhc)=Di(mmhc)-Di(mm)+Dj(mm)$$

Step 6-4: The delay of the flat route (MN-CN), Dj(mc), is measured and stored.

Step 6-5: The route with the minimum delay, Min{Dj(mmhc), Dj(mc)}, is selected. Steps 6-2 to 6-5 are repeated at a fixed interval of T1. If the time that the MN stays within the same MAP area (the "staying time") exceeds a certain threshold, T_2, Step 2 is executed.

The proposed algorithm is expected to reduce the number of measurement packets arriving at the HA for cases where an MN moves to a different MR area within the same MAP area. This reduction is achieved by measuring the delay only up to the

MAP, and by calculating the delay on the overlay route from past overlay route delays and the delay between MAPs.

Pings are used to measure delays in two types of routes existing in HMIPv6. Pings are easy to use for measuring the round trip time (RTT) [5], and the ping capability is implemented in almost all PCs, routers and other devices. The packet size of a ping is not much larger than a packet in streaming data. Even with headers and tunneling and other overheads added, the total packet size of a ping is not so large. Therefore, the load on HAs is considered to be small.

4 Evaluation

4.1 Routing Algorithm

We have built a simple environment using a network simulator, OPNET, and evaluated the measurement and calculation parts on the right-hand side of Fig. 3, as well as the entire algorithm. The evaluation environment is shown in Fig. 4, and the evaluation conditions used are summarized in Table 1.

The simulation starts immediately after the MN moves from the home link to an external link. The same is true for the simulation that focuses on the handover rate. The size of the packet sent for delay measurement is that used in IPv6. This size is that of a packet with an IPv6 header and a routing control header added in actual Mobile IPv6. Reference [6] reports that the minimum and average handover delays are 100 msec and several hundred milliseconds, respectively. Since the handover delay may vary depending on the particular situation of the particular network environment, we have varied the handover delay from 100, 200 to 300 msec in the simulation.

Table 1. Evaluation conditions in the simulation

Item		Condition, value
Numbers of MNs, HAs and CNs		1
Number of MNs registered in the HA		1
Communication between MN and CN		1-to-1 bidirectional communication
Delay measurement interval (T1)		60 sec
MAP link staying time threshold (T2)		100, 200, 300, 500, 1000 sec
MN's movement speed (V)		3.11, 2, 1, 0.5, 0 km/h
Size of delay measurement packet	Overlay route	256 Bytes
	Flat route	136 Bytes
	MN-MAP	152 Bytes
MAPd (MAP handover delay)		100, 200, 300 msec
MRd (MR handover delay)		100, 200, 300 msec

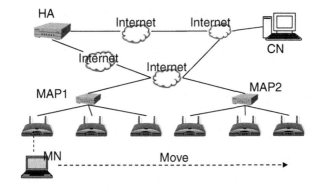

Fig. 4. Network environment used for verification

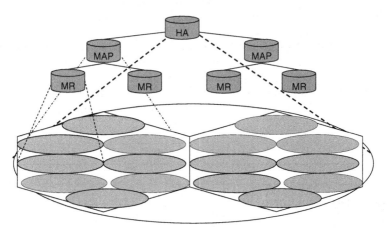

It is assumed that MAP is a regular hexagon, and MR is circular.
V: MN's movement speed (3.11km/h), A: Total area (100 km^2)
A_{MAP}: MAP area (2.83 k m^2), A_{MR}: MR Area (0.28 km^2)
D_{MR}: Diameter of MR (0.6 km), M_{HA}: Number of MNs registered in the HA

Fig. 5. Assumed urban environment model

The MAP handover rate (λ_{MAP}) and the MR handover rate (λ_{MR}) are calculated using Eqs (1) and (2) based on the assumed urban environment model shown as Fig.5, respectively, which were given in Reference [7][8].

$$\lambda_{MAP} = M_{HA} \times V \big/ (2\sqrt{3}/3)^{0.5} \times \sqrt{A_{MAP}} / A \qquad (1)$$

$$\lambda_{MR} = M_{HA} \times V \times A_{MR} \times \frac{1}{A_{MAP} \times D_{MR}} \qquad (2)$$

The RRd is defined as the Return Routability completion time in our study. RRd is calculated using Eq. (3), which considers the number of messages sent between the MN and CN in accordance with Return Routability.

$$RR_d = k(Dd_{OR} + Dd_{FR}) + BU_{CN} \qquad (3)$$

Dd_{OR}: Measured delay on the overlay route (msec)
Dd_{FR}: Measured delay on the flat route (msec)
BU_{CN}: Time for updating the binding to CN after RR
k: Number of RR messages sent (k=1)

The delay on the overlay route (ORd) and that on the flat route (FRd) are calculated using Eqs (4) and (5), respectively.

$$ORd = Dd_{OR} + \lambda_{MR} \cdot MRd + \lambda_{MAP} \cdot MAPd \qquad (4)$$

$$FRd = Dd_{FR} + \lambda_{MR}(RRd + MRd) + \lambda_{MAP} \cdot MAPd \qquad (5)$$

The variation in the bandwidths of the network environment and the processing delay in tunneling are taken account of by randomly generating a packet delay of 0.01 to 0.1 sec at the Internet cloud node (cloud in Fig. 4). The movement of the MN is simulated by equipping the MN with a timer, which turns on and off. It is assumed that the MN moves across 3 MR areas under the same MAP before it goes to a different MAP, and that delay measurement packets are never discarded.

Figure 6 shows a simulation result for the case where T2=200 sec, handover rate, λ_{MR}=0.385, and handover rate, λ_{MAP}=0.0285. The lower chart in Fig. 6 is an enlarged

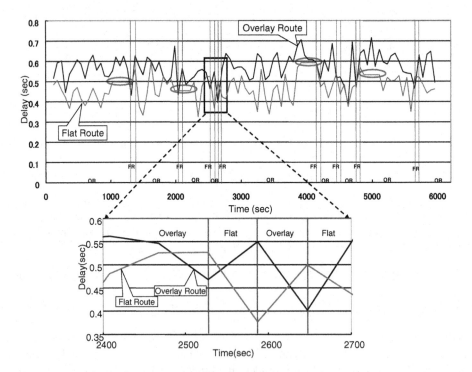

Fig. 6. Simulation results

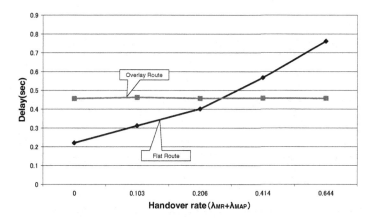

Fig. 7. Relationship between the handover rate and the delay in each route

chart for the rectangle in the upper chart. This simulation has allowed us to confirm that delays can be measured correctly by using delay measurement packets, and that routing is also carried out correctly. We have also confirmed that, for the MN's movement with the same MAP, the algorithm was started, the delay was measured, and routing was performed, all within the MAP link staying time threshold (T2) (the straight lines within ellipses in Fig. 6.).

We have varied the handover rate, and examined the relationship between the handover rate and the delay in each route. The results are shown in Fig. 7. The figure shows that, when the MN does not move, the flat route is selected, but that, as the handover rate increases, the delay on the flat rate exceeds that on the overlay route.

4.2 Reduction in the Load on the HA

The extent to which the proposed algorithm reduces the load on the HA depends on the measurement interval, T1, and the MAP link staying time threshold, T2. We have compared, by simulation, the number of delay measurement packets arriving at the HA in the proposed method, which takes account of T1 and T2, with that in a method of always making measurements via the HA. The simulation was performed in the same network environment as that in Section 4.1. Detailed simulation conditions were as follows:

- The MN moves at a speed of 2 km/h on a straight line towards the MR 0.5 km away.
- The simulation time was 11, 400 sec (approximately 3 hours).
- The measurement interval in the via-HA measurement method was 60 sec.
- The measurement interval, T1, in the proposed method was 60 sec.
- The MAP link staying time threshold, T2, of the proposed method was varied from 200, 300, 500 to 1000 sec.
- No measurement packets were discarded.

Figure 8 shows the probability (arrival rate) of measurement packets arriving at the HA before the expiry of the MAP link staying time interval threshold, T2. The

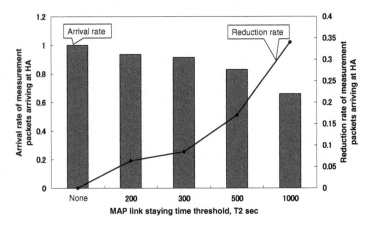

Fig. 8. Arrival rate of measurement packets arriving at HA

measurement packet arrival rate is a ratio of the number of measurement packets arriving at the HA to the number of measurement packets transmitted during the simulation time. The total number of measurement packets arriving at the HA was 188 in the via-HA measurement method. We used this 188 as a reference, i.e., 188 packets were regarded as "1", when we examined the number of measurement packets arriving at the HA in the proposed method for a variety of T2. The result is shown in Fig. 8. When T2=200 sec, the number of measurement packets arriving at the HA was 176; i.e., the arrival rate was about 0.94. Similarly, when T2=300 sec, the number was 172 and the arrival rate was about 0.91. When T2=500sec, the number was 156, and the arrival rate was about 0.82. When T2=1000sec, the number was 124, and the arrival rate was about 0.65. So, as the value of T2 increases, the arrival rate decreases. This reduction in the arrival rate verifies the effectiveness of the proposed method.

The line chart in Fig. 8 shows the relationship between T2 and the reduction rate of measurement packets arriving at the HA. This chart indicates that the reduction rate is proportional to T2.

4.3   Routing Error Rate within the MAP Link Staying Time Threshold, T2

We have examined the routing error rate within the MAP link staying time threshold, T2. For the MN's movement within the MAP, the proposed algorithm measures only the inter-MAP delay for the overlay route, and calculates the delay on the overlay route from past delays on the route using an equation. It does so until T2 is exceeded, and after T2 is exceeded, the algorithm measures the delay on the overlay route in the normal way. We have chosen to do so in order to reduce the number of delay measurement packets arriving at the HA.

However, since network bandwidths vary continuously, the MN's movement within the MAP changes the bandwidths available on the route between the MAP and the HA and the route between the HA and the CN. This may cause the calculated value to differ from the value measured within T2. Therefore, using a simulation, we have examined the routing error rate in the calculated value for the case where the MN moves within the MAP. The conditions for this simulation were as follows:

- The simulation environment is the same as was used for the verification of the operation in Section 4.1.
- The simulation time was 11,400 sec (approximately 3 hours).
- The time during which an MN moves in each MR was 1860 sec (approximately 30 minutes).
- MN's handover rate, λ_{MR}, was 0.194, and λ_{MAP} was 0.014.
- MAPd and MRd were varied from 100, 200 to 300 msec.
- The measurement interval, T1, was 60 sec.
- The MAP link staying time threshold, T2, was varied from 200, 300, 500 to 1000 sec.
- No measurement packets were discarded.

There were two reasons why we chose the handover rates, λ_{MR}=0.194 and λ_{MAP}=0.014. The first reason was that routing errors are likely to occur frequently because the difference between the delay on the flat route and that on the overlay route is small. The second reason was that, when the handover rate is either extremely high or extremely low, the difference between the delay on the flat route and that on the overlay route is large, and one type of route would always be selected, which excludes the possibility of any routing error occurring. We varied T2 greatly to examine how a change in T2 affects the routing error rate.

The routing error rate was derived from the number of times that the route selected in the simulation was different from the route determined from measurements within a period of T2.

Figure 9 shows the result. It shows the routing error rate when MAPd and MRd were varied within T2. At T2=200 and 300 sec, the routing error rate varied with changes in MAPd and MRd. It ranged from the minimum of about 0.15 (15%) to the maximum of about 0.5 (50%). At T2=500 and 1000 sec, the routing error rate did not

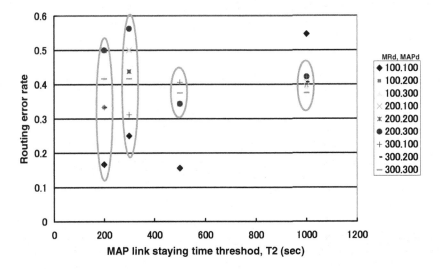

Fig. 9. Relationship between the MAP link staying time threshold, T2, and the routing error rate (λ_{MR}=0.194, λ_{MAP}=0.014)

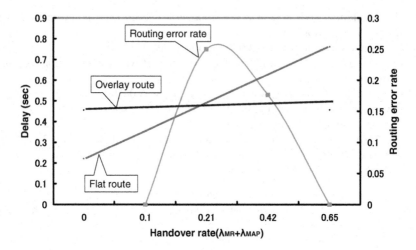

Fig. 10. Routing error rate

vary so much with changes in MAPd and MRd. The average routing error rate was about 0.35 (35%) at T2=500 sec, and as high as about 0.4 (40%) at T2=1000 sec.

These results seem to indicate that, if the network delay does not vary greatly, the routing error rate is small for a short T2. That is, the greater T2 is, the higher the routing error rate. In this simulation, a delay ranging from 0.01 to 0.1 sec was added randomly to represent the delay in the Internet-cloud. This made measured delays unstable. This was the reason why the routing error rate was varied greatly at T2=200 and 300 sec. This implies that, in a real network environment in which the network delay varies greatly, if T2 is small, the number of delay measurements within T2 will be small, which will tend to make the routing error rate unstable.

If T2 is too large, the routing error rate goes up. Therefore, particularly when an MN stays within the MR link for a long time, it is necessary to determine T2 carefully.

Figure 10 shows the routing error rate, and the delays on the overlay route and the flat rate for different handover rates. The figure indicates that, as the difference between the delay on the overlay route and that on the flat route increases, the routing error rate decreases.

From the overall point of view taking all of the delay on the overlay route, the delay on the flat route, MAP link staying time threshold, T2, the handover rate and the routing error rate into consideration, we have confirmed that, as the difference between the delay on the overlay route and that on the flat route decreases, the routing error rate tends to increase irrespective the value of T2, and that, as this difference increases, the routing error rate becomes high if the value T2 is large.

The routing error rate seems to be inversely proportional to the absolute difference between the delay on the overlay route (ORd) and the delay on the flat route (FRd). Consequently, as the difference between the delay on the overlay route and that on the flat route decreases, the routing error rate within T2 becomes high. Therefore, it is necessary to study how to determine the MAP link staying time threshold, T2.

5 Conclusions and Future Issues

This paper has proposed the method of measuring packet delays and the algorithm for selecting either an overlay or flat route based on that method in a Hierarchical Mobile IPv6 network, verified the operation of the algorithm, and confirmed its effectiveness by evaluating the extent to which the load on the HA is reduced.

Future issues include how to optimize the delay measurement interval, T1, and MAP link staying time threshold, T2. We believe that these issues can be resolved by considering an indicator, such as the moving speed of an MN, that leads to the optimization of T1 and T2. Therefore, it is necessary to find out what can be the appropriate indicator. It is also necessary to study what header should be added to a communication packet when its route has been determined, and what negotiation should be made with the corresponding node. We believe that possible solutions to the above are the use of messages similar to Binding Update and Return Routability used in Mobile IPv6, and the addition, to each communication packet, of a header that can identify the communication packet. If such messages are to be used, it is necessary to consider the processing time required when switching routes.

References

1. ITU-T Recommendation (2004): Y. 2001, General overview of NGN
2. Johnson, D., Perkins, C., Arkko, J.: Mobility Support in IPv6, IETF Internet Draft, draft-ietf-mobileip-ipv6-21.txt (February 2003)
3. Soliman, H., Castelluccia, C., El-Malki, K., Bellier, L.: Hierarchical MIPv6 Mobility Management (HMIPv6), IETF Internet Draft, draft-ietf-mipshop-hmipv6-00.txt (June 2003)
4. Murata, M.: Masauki Murata: Highly Reliable Services by Overlay Networks. IEICE Journal 89(9), 792–795 (2006)
5. Lee, S., Zhang, Z.-L., Nelakuditi, S.: Exploiting as hierarchy for scalable route selection in multi-homed stub networks. In: Proceedings of the 4th ACM SIGCOMM conference on Internet measurement, pp. 294–299 (2004)
6. Ogawa, T., Itoh, T.: A Study of DHCP-Based Seamless Handover Method. IEICE Trans. Commun J88-B(11), 2228–2238 (2005)
7. Takagi, Y., Ohnishi, H., Sakitani, K., Baba, K.-i., Shimojo, S.: Route Optimization Methods for Network Mobility with Mobile IPv6. IEICE Trans. Commun. 87(3), 480–489 (2004)
8. Takagi, Y., Ihara, T., Ohnishi, H.: Mobile IP Route Optimization Method for Next Generation Mobile Networks. IEICE Trans. Commun. 84(3), 344–353 (2001)

Application of Wavelet Packet Transform to Network Anomaly Detection

Christian Callegari, Stefano Giordano, and Michele Pagano

Università di Pisa, Dipartimento di Ingegneria dell'Informazione,
Via Caruso 16, I-56126 Pisa, Italy
{christian.callegari,s.giordano,m.pagano}@iet.unipi.it

Abstract. In the last few years, the number and impact of security attacks over the Internet have been continuously increasing. Since it seems impossible to guarantee complete protection to a system by means of the "classical" prevention mechanisms, the use of Intrusion Detection Systems (IDSs) has emerged as a key element in network security. In this paper we address the problem considering different methods, based on the Wavelet Packet Transform, for detecting anomalies in the network traffic, taking into account both the best basis and the value of transformed coefficients.

The performance comparison among the different solutions shows that very little information about network anomalies is carried by the best basis selection, while the "distance" between the transformed coefficients leads to very interesting results, highlighting the effectiveness of the proposed approaches.

Keywords: IDS, Anomaly Detection, Wavelet Packet, Best Basis.

1 Introduction

In the last few years Internet has experienced an explosive growth. Along with the wide proliferation of new services, the quantity and impact of attacks have been continuously increasing. The number of computer systems and their vulnerabilities have been rising, while the level of sophistication and knowledge required to carry out an attack have been decreasing, as much technical attack know-how is readily available on Web sites all over the world.

Recent advances in encryption, public key exchange, digital signature, and the development of related standards have set a foundation for network security. However, security on a network goes beyond these issues. Indeed it must include security of computer systems and networks, at all levels, top to bottom.

Since it seems impossible to guarantee complete protection to a system by means of prevention mechanisms (e.g., authentication techniques and data encryption), the use of an Intrusion Detection System (IDS) is of primary importance to reveal intrusions in a network or in a system.

State of the art in the field of intrusion detection is mostly represented by misuse based IDSs. Considering that most attacks are realized with known tools,

S. Balandin et al. (Eds.): NEW2AN 2008, LNCS 5174, pp. 246–257, 2008.

available on the Internet, a signature based IDS could seem a good solution. Nevertheless hackers continuously come up with new ideas for the attacks, that a misuse based IDS is not able to block.

This is the main reason why our work focuses on the development of an anomaly based IDS, based on the Wavelet Packet Transform (WPT). In more detail we have developed several algorithms which aim at revealing the anomalies in the network traffic, taking into account both the transformed coefficients and the information about the best basis.

The remainder of this paper is organized as follows: Sect. 2 presents the related work and Sect. 3 recalls the theoretical backgrounds (WPT), while Sect. 4 describes the implemented system, detailing the detection algorithms. Then in Sect. 5 we provide the experimental results and finally Sect. 6 concludes the paper with some final remarks.

2   Related Works

Due to its properties, the Wavelet transform is quite a "classical" approach to detect irregular traffic patterns in traffic traces. As an example, in [1] and [2], the authors respectively present the application of the Discrete Wavelet transform (DWT) to the detection of changes caused by flashcrowds, outages, and attacks, and the detection of network problems affecting dominant Round Trip Times. Moreover, in [3] the authors explore the use of the Continuous Time Wavelet Transform (CWT), for detecting DoS attacks in network traffic.

Our idea is to generalize the existing approaches, by taking into account the WPT, which provides, with respect to the Wavelet approaches, more degrees of freedom in the choice of thebasis for representing the signal.

After an extensive survey, to the best of our knowledge, there is no work directly related to the application of the WPT to the detection of network anomalies.

3   The Wavelet Packet Transform

The WPT can be seen as a generalization of the more traditional Wavelet decomposition, since it permits to define a covering of the time–frequency plane, tuned to the features of the analyzed signal. In the following subsections we will briefly introduce the dyadic wavelet decomposition and then its generalization through wavelet packets, mentioning the algorithms for the best basis selection.

3.1   Wavelet Decomposition and Multiresolution Analysis

The Wavelet decomposition [4] is based on the representation of any finite–energy signal $x(t) \in L^2(\mathbb{R})$ by means of its inner products $\{x_{m,n}\}_{m,n\in\mathbb{Z}}$ with a set of functions, $\{\psi_{m,n}(t)\}_{m,n\in\mathbb{Z}}$, which are scaled and translated versions of an adequately chosen *mother wavelet* $\psi(t)$:

$$\psi_{m,n}(t) \;=\; a_0^{-m/2}\psi\left(a_0^{-m}t - nb_0\right)$$

Under quite stringent constraints on the choice of the mother wavelet, the functions $\{\psi_{m,n}(t)\}_{m,n\in\mathbb{Z}}$ may define an orthonormal dyadic wavelet basis (corresponding to $a_0 = 2$ and $b_0 = 1$). In this case, $\psi(t)$ can be represented in terms of the so–called *scaling function* $\phi(t)$

$$\psi(t) = \sqrt{2}\sum_n g_n \phi(2t - n)$$

where $\phi(t)$ itself is defined by a two–scale difference equation

$$\phi(t) = \sqrt{2}\sum_n h_n \phi(2t - n)$$

with the additional constraint that $g_n = (-1)^{n-1} h_{-n-1}$. In this work, we will consider the well–known Daubechies bases family of compactly–supported mother wavelets, introduced by the Belgian mathematician Ingrid Daubechies in 1988. The number of non null coefficients h_n determines the regularity of the mother wavelet and the number of its vanishing moments: for the Daubechies basis $_N\psi$ of order N (with N vanishing moments) only $2N$ coefficients are non zero, so that both $_N\phi$ and $_N\psi$ have compact support of width $2N - 1$.

The Multiresolution Analysis represents the theoretical framework for the efficient calculation of the wavelet decomposition [5]. Let $\boldsymbol{x} = (x_1, x_2, \ldots)$ denote the approximation of a finite–energy signal $x(t)$ at a given resolution; the wavelet coefficients $\{x_{m,n}\}$ at lower resolutions can be obtained considering the filter bank shown in Fig. 1, where the coefficients h_n and g_n depend on the chosen mother wavelet. In particular, the outputs of the high-pass filter h_n give the detail coefficients (at the given resolution), while the outputs of the low-pass filter give an approximation at a lower resolution, which is further decomposed in a similar way.

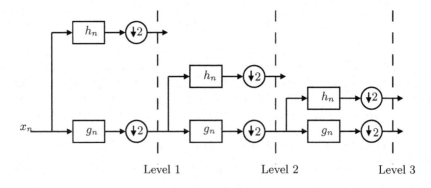

Fig. 1. Wavelet Decomposition

3.2   Wavelet Packets

In the Wavelet decomposition, the filtering procedure is applied, at each resolution level, only to the low-pass component of the signal; a more general scheme (although computationally more expensive) is depicted in Fig. 2, where the decomposition is carried out recursively for both components of the signal. For instance, if we consider the outputs of the filters at a specific level, we obtain the traditional subband decomposition, widely used in signal processing. The idea behind Wavelet Packets decomposition is that a *more compact* representation of the signal (for instance, with a lower number of significative coefficients) might be achieved by picking up the transformed coefficients at different resolution levels. In other words, since the Wavelet Packet decomposition provides an over-complete expansion of the original signal (i.e., it gives more information than is strictly necessary for its reconstruction), it is possible to choose a more adequate basis (according to some information cost criterium) in order to represent the signal. For instance, in [6], a general algorithm for the best basis selection has been described and a few different cost criteria (number above a threshold, concentration in l^p norm (with $p < 2$), entropy and logarithm of energy) have been presented. Roughly speaking, the idea of the algorithm is to visit the tree, corresponding to the complete Wavelet Packet decomposition, starting from its leaves, which are assumed to be part of the best basis. Then the following procedure is applied recursively: if the cost of the parent node is

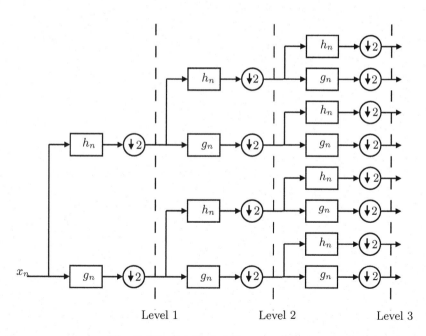

Level 1   Level 2   Level 3

Fig. 2. Wavelet Packet Decomposition

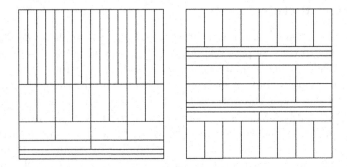

Fig. 3. Wavelet and Wavelet Packet bases tile the time-frequency plane

not higher then the sum of the costs of its children, the latters are replaced by their parent in the basis.

The same concept can be better emphasized considering the representation of the transformed coefficients on the time–frequency plane [7]. It is worth mentioning that this representation permits to highlight the significant advantage of the Wavelet Transform over the Windowed Fourier Transform: the shape of the resolution cell is not fixed (although its area is lower bounded by the Heisenberg product) and permits a better time resolution at higher frequencies (i.e., where it really matters). In particular, the wavelet basis corresponds to an octave-band decomposition of the time–frequency plane (see the left part of Fig. 3), while the wavelet packets permit a more general covering: for instance, the one on the right in Fig. 3 is suitable for a signal containing two almost pure tones near 1/3 and 3/4 of the Nyquist frequency.

4   System Design

In this section we provide a description of the algorithms used by the implemented system to reveal anomalies in the network traffic. In more detail, we describe four different algorithms, which aim at detecting anomalies, exploiting both information regarding the best basis and the transformed coefficients.

The system analyzes the given traffic traces, by means of a sliding window mechanism. At each step it computes the WPT of the samples currently in the window (and the corresponding best basis) and compares the obtained coefficients (and basis) with the previous ones. Since we are interested in a precise localization of the anomalies and the width of the mother wavelet grows with N, among the Daubechies wavelets, we have considered in our texts only $_1\psi$ (i.e., the Haar wavelet) and $_2\psi$. However, the experimental results were very similar and so, for sake of brevity, in the following we will present only those obtained with $_1\psi$. As far as the best basis selection is concerned, among the different cost functions described in [6], we have considered the number of transformed coefficients above a threshold, the simplest from a computational point of view.

Thus, to reveal an anomaly the system computes a "distance" (its definition depends on the used algorithm as discussed in the following subsections) and compares it with a threshold, set by means of Monte Carlo simulation.

Algorithm 1. The first method is based on the idea that the best basis should significantly change during an anomaly. Thus, the system reveals an anomaly if the "distance" between the bases corresponding to two subsequent temporal windows overcomes a threshold, set by means of Monte-Carlo simulation. It is worth noticing that such comparison is performed between the current samples and the previous ones, with the exception of the case when an anomaly is detected. Indeed, when the system reveals an anomaly, the comparison for the new data is made with the samples related to the last "anomaly-free" window. In more detail such distance is computed, assigning a distinct number to each node of the coefficients tree; Figs. 4 and 5 show three distinct node numbering, which will be referred to as numbering 0, 1, and 2 respectively. At this point, the best basis is written as an array of fixed length, equal to the size of the analysis window. For example, the base shown in Fig. 4 (thick circles) corresponds to $[3, 3, 8, 7, 13, 13, 10, 10, 24, 24, 24, 24, 21, 21, 19]$. Thus, the distance is computed as the Euclidean distance between two arrays.

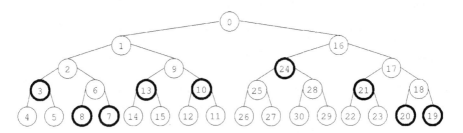

Fig. 4. Example of a Best Basis (Numbering 0)

Algorithm 2. The second method is a variation of the first one. In this case the distance is simply given by the difference between the number of nodes composing the best basis.

Algorithm 3. This algorithm, differently from the previous ones, does not take into account any information regarding the best basis, but it just considers the values of the transformed coefficients belonging to a given level of the tree (the choice of the level is a parameter of the algorithm). In this case the system computes the Euclidean distance between the actual coefficients and the previous ones, generating an alarm if such distance overcomes a threshold (set by means of Monte-Carlo simulation).

Algorithm 4. Finally this algorithm tries to exploit both the information regarding the base and the transformed coefficients, computing the Euclidean distance between the current coefficients in the best basis, written as a vector, and the corresponding coefficients for the previous attack-free window.

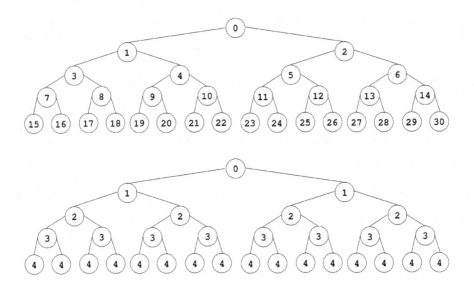

Fig. 5. Coefficient Tree Numbering (Numbering 1 and 2)

5   Experimental Results

In this section we describe the simulations carried out to check the effectiveness of the heuristics proposed in the previous section. In more detail we first present an overview of the dataset and the anomalies used in the simulation; then we focus on the comparison between the different methods and the discussion of the achieved results.

5.1   Simulation Data

The simulation dataset is composed of three distinct datasets [3], so as to have both simulated and real traffic data. The first dataset is extracted from the dataset provided by the 1999 DARPA evaluation project [8,9], which is a partly emulated and partly simulated dataset representing the *standard de facto* for IDS evaluation, and consists of fifteen distinct traces. These traces represent the number of IP packet received by the internal and the external gateways, measured every two seconds (for the first five traces, corresponding to week 1) and every five seconds (traces 6 to 15, corresponding to week 2). The second dataset is composed of three traces collected at the University of Naples "Federico II", representing the number of packet (with destination port 80) received by the gateway over intervals of five seconds. Finally the third dataset consists of five traces collected at UCLA [10], corresponding to the number of received packets, measured every five seconds. To be noted that the traces are composed of 3600 samples, thus they correspond to five-hour periods, with the exception of traces related to week 1 of the DARPA dataset, which correspond to a two-hour period.

To perform the simulations, these traces have been processed, so as to add some anomalies over them. The anomalies have been partly realized by means of publicly available software tools for the generation of DoS and DDoS attacks, as like as TFN2K [11] and Stacheldraht [12], and partly synthetically generated, according to known profiles that have been considered in [13] (we have taken into account "Constant Rate", "Increasing Rate", and "Decreasing Rate" anomalies). To be noted that such anomalies have been added to the traffic traces at different "time" position and with different amplitude, obtaining a total of 10752 different test traces.

5.2   Results

The simulations, carried out to validate the proposed methods, have shown that the results strongly depend on the information we want to exploit. The figures in this section present the ROC curves corresponding to the implemented algorithms.

In more detail Fig. 6 shows the results obtained by applying the first algorithm. The different curves correspond to distinct parameters settings, namely:

- Curve 1: Window size = 16 pkts, Overlap = 8 pkts, Numbering 0
- Curve 2: Window size = 16 pkts, Overlap = 8 pkts, Numbering 1
- Curve 3: Window size = 16 pkts, Overlap = 8 pkts, Numbering 2
- Curve 4: Window size = 32 pkts, Overlap = 8 pkts, Numbering 1
- Curve 5: Window size = 64 pkts, Overlap = 60 pkts, Numbering 1
- Curve 6: Window size = 64 pkts, Overlap = 8 pkts, Numbering 1

As we can notice from the graph, the method does not provide any acceptable results, presenting an hit rate almost equal to the false alarm rate, which means that it is equivalent to randomly decide if there is an anomaly or not. Moreover, the algorithm performance are almost independent of the choice of the simulation parameters; for this reason in the following graphs we only show the results corresponding to a window size of 16 packets and an overlap between consecutive windows of 8 packets.

As shown by Fig. 7, the performance of algorithm 2 are only slightly better. Thus, it is easy to conclude that the first two algorithms are ineffective for the anomaly detection.

Figure 8 presents the results for the third algorithm, considering its application to the different levels of the WPT tree (level 0 corresponds to the original trace samples). As expected, the direct application of the algorithm to the original samples slightly improves the previous methods, but still it is not acceptable. Instead, the results obtained applying the algorithm to the transformed coefficients are quite good, and get better the deeper we move in the wavelet packet tree. It is worth noticing that we have only explored level from one to four, because considering deeper levels would take to a bigger window size and to a computational effort, unacceptable for on-line anomaly detection systems.

The subsequent figure, Fig. 9, shows the results obtained with algorithm four. From an accurate comparison with Fig. 8 it turns out that the fourth algorithm performs slightly worse then the third one. Thus we can conclude that the information carried out by the best basis is misleading for the detection of network anomalies. On the other hand the experimental results show the effectiveness of

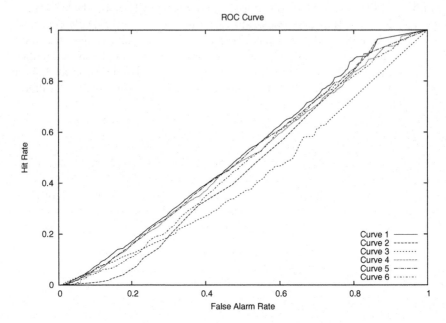

Fig. 6. Performance of algorithm 1

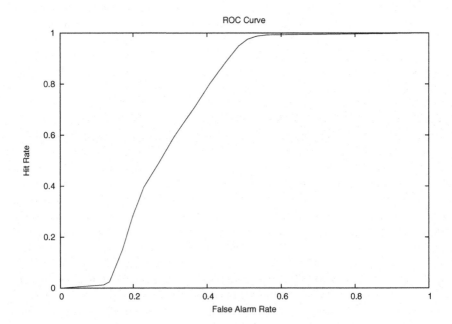

Fig. 7. Performance of algorithm 2

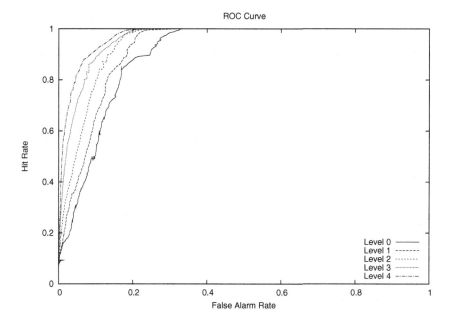

Fig. 8. Performance of algorithm 3

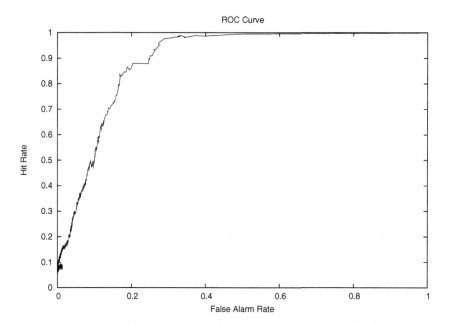

Fig. 9. Performance of algorithm 4

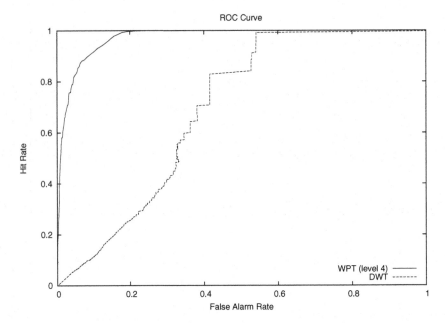

Fig. 10. Performance comparison: WPT and DWT

the method based on the analysis of the WPT coefficients, as highlighted by the performance of algorithm 3.

Finally Fig. 10 compares the ROC curve obtained by the WPT (algorithm 3), with the one obtained by the DWT, when an analogous heuristic is used. The figure shows that the use of WPT significantly improves the results, justifying the approach considered in this paper.

6   Conclusions

In this paper we have investigated the suitability of the WP decomposition of traffic traces for anomaly detection purposes. In more detail, we have presented four distinct algorithms, which exploit both the information carried by the transformed coefficient and by the best basis. To compare the proposed solutions, we have analysed their performance over more than 10000 traffic traces, taken from three distinct reference datasets.

The performance analysis has highlighted that some of the implemented systems obtain, for a proper choice of their parameters, very good results. In more detail we have shown that the best basis does not carry any relevant information for the anomaly detection, while the methods based on the analysis of the transformed coefficients take to very good results, already for relatively small sizes of the observation windows, showing the suitability of such methods for on-line detection. Moreover we have shown that the use of WPT significantly improves the results achievable when the same heuristic is applied by means of wavelet transform.

Acknowledgments

The authors would like to thank Prof. A. Pescapè and A. Dainotti of the University of Naples "Federico II" for having provided some of the traffic traces used in this paper and A. Ciambelli for his contribution to this work.

This work was partially supported by the RECIPE project funded by MIUR.

References

1. Barford, P., Kline, J., Plonka, D., Ron, A.: A signal analysis of network traffic anomalies. In: IMW 2002: Proceedings of the 2nd ACM SIGCOMM Workshop on Internet measurment, pp. 71–82 (2002)
2. Huang, P., Feldmann, A., Willinger, W.: A non-instrusive, wavelet-based approach to detecting network performance problems. In: IMW 2001: Proceedings of the 1st ACM SIGCOMM Workshop on Internet Measurement, pp. 213–227 (2001)
3. Dainotti, A., Pescapè, A., Ventre, G.: Wavelet-based detection of DoS attacks. In: Proceedings of GLOBECOM 2006, pp. 1–6 (2006)
4. Daubechies, I.: Ten lectures on Wavelets. CBMS-NSF Series in Applied Mathematics, vol. 61. SIAM, Philadelphia (1992)
5. Mallat, S.: Multifrequency channel decompositions of images and wavelet models. IEEE Transactions on Acoustics, Speech and Signal Processing 37, 2091–2110 (1989)
6. Wickerhauser, M.: Lectures on wavelet packet algorithms, November 18 (1991)
7. Hess-Nielsen, N., Wickerhauser, M.: Wavelets and time-frequency analysis. In: Proceedings of the IEEE, vol. 84, pp. 523–540 (April 1996)
8. MIT, Lincoln laboratory, DARPA evaluation intrusion detection(accessed on December 12, 2008) (2008), http://www.ll.mit.edu/IST/ideval/
9. Lippmann, R., Haines, J., Fried, D., Korba, J., Das, K.: The 1999 DARPA off-line intrusion detection evaluation. Computer Networks 34(4), 579–595 (2000)
10. UCLA Data Traces, http://lever.cs.ucla.edu/ddos/traces
11. CERT Coordination Center. Denial-of-service tools - advisory, 1999-17, http://www.cert.org/advisories/CA-1999-17.html
12. CERT Coordination Center. DoS Developments - advisory ca-2000-01, http://www.cert.org/advisories/CA-2000-01.html
13. Yuan, J., Mills, K.: Monitoring the macroscopic effect of DDoS flooding attacks. IEEE Trans. Dependable Secur. Comput. 2(4), 324–335 (2005)

Towards the Dynamic Semantic Web

Ian Oliver

Nokia Research Center
Itämerenkatu 11-13
Helsinki
Finland

The Semantic Web is emerging as the future platform for computation. The ability to represent information, its structure and some of its semantics in a canonical form that is readable and understandable by machine forms the key infrastructure for truly ubiquitous computing. On top of this structure we are seeing the development of more sophisticated services, reasoning and applications and a re-emergence of machine intelligence in the mainstream environment.

The Semantic Web in its current form is expanding from its current static focus: that of a method of representing all information world- (universe-) wide. Coupled with its ability to grow and link any piece of information with any other piece of information, this forms a true *web* of information that can be queried, modified, added-to, searched and reasoned about by any number of 'agents'.

This then brings us to the point to where we must start considering how this information is conceptually perceived in terms of its dynamicity, locality, persistence, access and interpretation. It is true to say that the current information sets presented to the users are relatively static in form and this is typically true for 'web-wide' structures. As smaller, more local and personal structures such as personal information evolve the dynamicity increases with the non-monotonicity and volatility of information and meaning.

One problem with the Semantic Web is ironically the notion of semantics, which in the static case is 'easily' defined - everyone can agree on the semantics and thus interpretation of a given structure that conforms to a given ontology. As more users or agents have access to the web-wide information both from static and dynamic sources, interpretation will vary and deeper representations of the semantics and the grounding of the semantics of the information will be required.

The semantics of a given ontology or structure for the more static web-wide structures will be (and is already) generally accepted either by standardisation or by de-facto standard. As ontologies are used by a greater number of agents or users the ontologies' semantics becomes more fixed and less open to interpretation. When applied to the personal and localised information structures adherence to any given ontology and its semantics becomes less and ad hoc in nature.

Logics for processing and reasoning about this information will need to develop from the strict, monotonic, ontology-conformant processing to those which can deal with variable interpretation, non-monotonicity, incompleteness, incorrectness, inconsistency etc. Sophisticated conflict resolution, truth and belief revision and search strategies will be required in order to successfully reason, interpret, gather and search this information.

S. Balandin et al. (Eds.): NEW2AN 2008, LNCS 5174, pp. 258–259, 2008.

The almost ubiquitous nature of mobile computing provided through highcomputing power mobile devices (mobile phones, mobile tablets, sensors etc) and high-speed data connections (UMTS/3G, WLAN etc) means that it is possible to perform complex computations locally and distribute the results rather than centralised via some service through some light-weight interface. This leads to the situation where the multiplicity of devices can interact with each other to form very dynamic, ad hoc and distributed computation platforms - a semantic web 'social' network.

The nature of applications in this dynamic, web-wide environment will change dramatically from the current monolithic-style applications to more highly distributed, mobile and agent-like entities. This leads us to how information should be made available: we believe that tuple-space structures, distributed computation and ubiquitous communication facilities provide a solution or platform to this. The nature of the application however still requires development and investigation.

The Semantic Web future is arriving, however, it will be more dynamic and more ubiquitous encompassing everything from the World- and Universe- wide structures to local and personal structures more than anyone has envisaged.

From Smart Homes to Smart Cities: Opportunities and Challenges from an Industrial Perspective

Cornel Klein and Gerald Kaefer

Siemens AG
Corporate Research and Technologies
Software & Engineering – Architecture
Otto-Hahn-Ring 6, 81730 Munich
Cornel.Klein@siemens.com, Gerald.Kaefer@siemens.com

Abstract. Driven by the advances in hardware technologies, smart environments (or "pervasive computing") already penetrate many spaces of our daily live. Smart Homes, Smart Buildings and larger ensembles like airports, hospitals or university campuses are already equipped with a multitude of mobile terminals, embedded devices as well as connected sensors and actuators. Some activities already envision the "Smart City" which uses the opportunities provided by pervasive computing technologies to the benefits of their inhabitants. In such a setting, smart environments are expected to play a crucial role for coping with the challenges of urbanization and demographic chance e.g. regarding sustainability, energy distribution, mobility, health or public safety/security.

While "smartness" is often centered on a user perspective, we give the business perspective of a large industrial supplier of infrastructures and solutions. This includes application scenarios for smart cities and an outline of the involved business- and research challenges. A particular focus will be the setup and energy efficient operation of smart infrastructures and data centers. While "autonomic computing principles" like self-configuration, self-healing, self-protection, and self-optimization are well understood for enterprise IT infrastructures, their application to highly-distributed, heterogeneous pervasive computing systems is less straightforward. A particular concern is the incorporation of energy efficiency in such settings. This involves a reduction of the rapidly raising IT energy costs and concepts for making software applications and services more aware of their energy consumption. It is therefore the basis for the identification of IT energy hotspots in software and IT system architecture and therefore for a sustainable development and operation of smart environments.

Keywords: Smart Cities, Pervasive Computing, Sustainability, System Management, Green IT, Cloud Computing, Platforms, Infrastructure.

S. Balandin et al. (Eds.): NEW2AN 2008, LNCS 5174, p. 260, 2008.

Smartness of Pervasive Computing Systems through Context-Awareness

Arkady Zaslavsky

Luleå University of Technology, Sweden
arkady.zaslavsky@ltu.se

Emerging pervasive computing and communications technologies evolved into ample pioneering initiatives, leading towards a world in which computing systems are distributed, mobile, intelligent, supportive, unobtrusive, invisible and cooperative. Central to the notion of a pervasive systems is context-awareness. Context-aware computing endeavours to make systems aware of specific, relevant circumstances in the computing environment, and enable them to adapt their behaviour accordingly. Context-aware systems can process (intelligently) the context information acquired by any type of a sensor (either physical, computational or virtual). This, in turn, enhances services provided to users (including service personalization), makes pervasive systems smart by reacting (and possibly pro-acting) to changing circumstances, and enables adaptability and autonomy of systems, freeing users from avoidable sometimes routine interactions.

From stand-alone applications, which have demonstrated the benefits of using context, research community is now looking at modelling and reasoning about context in uncertain, distributed, open and heterogeneous computing environments. The inherent pervasiveness, heterogeneity and information uncertainty in such systems challenge computational methods and flexible implementable architectures that would support diverse clients, services, systems and applications. Realizing this vision requires suitable approaches in modelling, reasoning and architecting effective context-aware systems that can handle reasoning in uncertain and rapidly changing environments. In recent years, research efforts have focused on various aspects of context, including context middleware and toolkits for context discovery and acquisition, ontologies that provide vocabularies to describe, interpret and share context information, different approaches to reason about context and a variety of context models. Of particular interest in dealing with complex and open pervasive systems are: (1) context models that represent context in a general way, including reasoning algorithms that can handle varying degrees of uncertainty, and (2) architectures that promote agility and autonomy of individual computing entities.

We have developed the Context Spaces (CS) model to describe context and apply reasoning over modelled information under uncertainty. The Context Spaces model overcomes some of the shortcomings of logic-based modelling and lack of unifying properties of sensor data fusion approaches for context-awareness. Context Spaces aims towards a general context model to aid thinking and describing context, and to design operations for manipulating and utilizing context. The concepts use insights from geometrical spaces and the state-space model, hypothesizing that geometrical metaphors such as states within spaces are useful to support reasoning about context.

S. Balandin et al. (Eds.): NEW2AN 2008, LNCS 5174, pp. 261–262, 2008.
© Springer-Verlag Berlin Heidelberg 2008

The model provides a unifying way to represent context and enables effective reasoning to be applied over the modelled information. Context Spaces approach distinguishes between the concepts of context and situations, as illustrated in Fig 1. Context is the information used in a model for representing real world situations. Thus, situations are per-

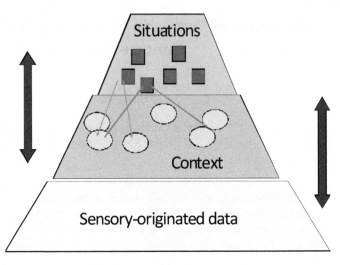

Fig. 1. Context-situation pyramid

ceived as a meta-level concept over context and algorithms are designed to assess the association and mapping between context and situations (i.e., determining occurrences of situations based on information in a model). This relationship between context and situations is represented in a general way by the concepts of state and space. An *application space*, i.e., the universe of discourse in terms of available contextual information for an application, is determined by the types of information, deemed relevant and discoverable/inferrable by the system designers. It is a multi-dimensional space made up of domains of values for each relevant information type, in which context can be sensed. Within it we model subspaces (possibly defined in fewer dimensions), which reflect real-life situations. We call these subspaces *situation spaces*. Situation spaces are defined over regions of values in selected dimensions and represent collections of values that reflect real-life situations. For example, "fire in the building" situation can be modelled with fire alarm sensors, smoke detectors, human presence sensors, etc. The actual values of sensory originated information are defined by the context state, e.g., the collection of current sensor readings representing a specific context.

The CS modelling approach is further extended with Context Spaces algebra, comprising operators that enable distributed context reasoning, including compositions of situations and reasoning about logically conditioned situations, in cooperative context-aware systems. The algebra facilitates merging between the perspectives of different entities in the pervasive computing environment, enhancing the reasoning outcome under uncertain conditions, enabling adaptive behaviour, and computing a numerical measure of confidence in the validity of situation expressions under uncertainty. A range of applications, including healthcare, intelligent transportation systems, etc demonstrates the usefulness, validity and practicality of the CS approach.

Home Automation with ZigBee

Maxim Osipov

OOO Siemens Corporate Technology, St.Petersburg
maxim.osipov@siemens.com

Abstract. This article discusses a topic of home automation, also called domotics, and provides an overview of state-of-the-art in communication protocols for this application area. Comparitive analysis for different networking standards and communication media is provided. Special focus is given to wireless technologies and advantages provided by ZigBee wireless networking standard for home automation solutions. ZigBee protocol implementation details are covered together with examples of possible applications in conjunction with Linux Open Source software platform.

Keywords: Home automation, domotics, smart home, connected home, home networking, ZigBee, KNX, BACnet, X10, UPnP, HomePlug, Z-Wave, LonWorks, Linux, Open Source.

1 Introduction

1.1 Traditional Home Automation Overview

Traditionally, home automation solutions were developing mostly in the following three directions:

1. HVAC (heating, ventilating and air conditioning)
2. Security, including fire alarm and detection systems, access control, intrusion detection, as well as intercoms
3. Lightning control, automation for electricity and water metering

There are several application fields for these solutions, starting with traditional area - building automation, followed by automation for cottage's and finally high-end systems for flats. Only a subset of solutions is applicable for each area, mapping between system functionality and application is presented in Table 1.

These types of solutions were often implemented as separate products, using different technologies, having little overlap between each other and with very limited influence from traditional information and communication areas, such as personal computers and IP networks. Specialized protocols and standards were created for almost every field of home automation and often equipment from different manufacturers was totally incompatible.

Home electronic appliances were not considered as a part of home automation solutions and were developing separately, including audio / video systems integration, universal remote controls and kitchen appliances.

S. Balandin et al. (Eds.): NEW2AN 2008, LNCS 5174, pp. 263–270, 2008.

Table 1. Application areas of home automation

Functionality	Buildings	Cottages	Flats
HVAC	Yes	Yes	No
Security	Yes	Yes	Yes/No
Intercom	No	Yes	Yes/No
Lightning	Yes	Yes	Yes
Garage doors	No/Yes	Yes	No
Metering	Yes/No	Yes/No	Yes/No
Boiler & Pool	No	Yes	No

1.2   Technological Basis

From a functional point of view, traditional home automation solution was a simple control system, including set of sensors, switches and actuators, connected together using direct wiring or specialized bus. Programming of automation tasks for a system was done using PLC (Programmable Logic Controllers) and specialized programming languages, such as:

- Ladder diagram (LD)
- Sequential Function Charts (SFC)
- Function Block Diagram (FBD)
- Structured Text (ST)
- Instruction List (IL)

Similar, specialized communication protocols were developed in order to establish interaction between simple electronic devices, constituting home automation system. Development of home automation technologies was accompanied by introduction of more and more advanced communication protocols, described in Table 2.

1.3   Next Steps

As we can see from Table 2, in year 1999 and foreword home automation experienced very significant interest from different companies, working on standardization of inter-device communication.

This boom of home automation was mostly caused by the progress of information and communication technologies (ICT). It became clear, that integration of home appliances into one network together with traditional home automation will create more possibilities for external control and development of smart management strategies. This would bring significant benefits, including:

- More featureful and convenient lifestyle
- Advanced functions for home electronics
- Environmental care by optimized energy consumption

This is how "smart home" concept was born, which we will discuss in the next section.

Table 2. Home Automation Protocols

Name	Year	Media	Description
X10 [6]	1975	Powerline	Noise, attenuation, 256 devices, speed 20 bit/s
BACnet [3]	1995	ARCNET, Ethernet, BACnet/IP, RS-232, RS-485, LonWorks	Higher level protocol, provides abstractions for control functionality
LonWorks [7]	1999	Twisted pair, Powerline	Widely accepted, 78/5.6/3.4 kbit/s
KNX [4]	1999	Twisted pair, Powerline, RF, Infrared, Ethernet	Widely accepted, speed depends on media
UPnP [5]	1999	IP networks	Higher level protocol abstracting control functionality
HomePlug [8]	2001	Powerline	Standard for PC networking over powerline
ZigBee [10]	2004	Wireless	Open standard for wireless mesh communication
Z-Wave [9]	2005	Wireless	Proprietary alternative to ZigBee

2   Smart and Connected Home Concept

2.1   Importance of Home Networking

The key concept of "smart home" technologies is integration of home control, entertainment and computer networks into one environment. In order to do so, it is necessary to solve several problems:

1. Establish communication links between all devices
2. Make sure, that protocols are understood by all members of the network

This is rather difficult, taking into account amount of different tasks which are to be solved by elements of "smart" living environment and contradictory requirements imposed. For example, for multimedia it is necessary to have a good communication channel, and power requirements are not an issue. But for a wireless smoke sensor, installed on a ceiling, the most important feature is power efficiency, as it should live for years on one battery, and communication spead is not very important.

2.2   Types of Networks

Question of communication links deserves more attention. There are several possibilities to establish such links [13]:

Traditional direct wired connections. Such connections have long history and are widely used in home automation - for example by intercoms. More complex links, providing possibility for digital information exchange are also available. However this approach requires a lot of wiring and thus extremely inefficient.

Dedicated wired networks. Dedicated networks for home automation already exist and are rather wide spread. Good examples are BACnet and KNX standards, which are supported by major automation product suppliers. However, with massive introduction of small sensors and actuators in houses, usage of wired networks becomes economically infeasible, and wires are not possible to install in some interiors.

Usage of powerline for networking. Usage of powerline for networking of home automation devices is a very natural approach, since most of devices are connected to power network. Development in this area has begun with X10 protocol. X10 was rather slow, unstable and tricky to use, and this direction was not developing until recently, when HomePlug and IEEE P1901 standards came into the game. These standards provide possibility for high speed broadband networking over regular powerlines. However, powerline modems are still rather expensive.

Wireless networks. Wireless networks are very attractive for home automation tasks for the reason of installation simplicity and receive major attention from industry now. Many existing standards evolve towards usage of wireless media. But in order to be economically beneficial, wireless modules shall be cheap and highly energy efficient.

2.3 Benefits of Wireless Technologies

While more and more devices become connected to a network, problem of network complexity, installation and management arises. Most of traditional approaches, like direct wiring, are already economically infeasible and gave up positions to dedicated communication buses. But when smart home technologies come to every flat, necessity for wiring at all becomes a huge overhead.

Currently developed home networking technologies are concentrating on two types of media - powerline and wireless. Both do not require additional wiring (not always the case for powerline) and in its modern state can provide enough bandwidth. Among these options, wireless technology has significant benefits, including simplicity of device installation (no wiring required), possibility to install modules in remote locations, flexibility and reduced costs. Another important advantage, comparing to powerline, is higher communication reliability (many electric devices in houses are a source of significant noise).

Second problem of home automation networking is necessity for network self-organization, including identification of devices and capabilities, creation of communication paths between modules. One cannot expect technical skills from regular user, so devices for smart home shall be intelligent enough to allow zero-configuration. Such functionality is much more simple to implement with wireless networks - in this case problem of correct physical connections, which could affect network behavior, can be avoided completely.

2.4 Types of Wireless Networks

Different wireless protocols are targeted to different application areas. This is mostly due to contradictory requirements to speed, power consumption, communication model etc. Following types of networks are usually identified and described, based on territory coverage [14]:

- Wireless Personal Area Network (WPAN)
- Wireless Local Area Network (WLAN)
- Wireless Metropolitan Area Network (WMAN)
- Wireless Wide Area Network (WWAN)

Home automation uses WLAN and WPAN families. These types of network connect devices around some person or in small area. Normal reach of WPAN is around few meters. Special requirements, as we already discussed above, include little configuration and maintenance, dynamic network formation and reconfiguration.

The most typical representative of WPAN's, conforming to all above requirements to home automation networking protocols, is ZigBee standard, which is developed by ZigBee Alliance.

2.5   ZigBee Technology Overview

ZigBee was designed to be low cost, extremely power efficient and provide effective communication with low data rate, needed for most home automation scenarios. Key technical data is provided in Table 3.

ZigBee is based on IEEE 802.15.4-2003 WPAN (Wireless Personal Area Network) specification. It defines physical (PHY) and medium access control (MAC) layers of protocol. Higher layers, including network (NWK), application support sub-layer (APS) and ZigBee Device Objects (ZDO) are described by ZigBee standard (see Figure 1).

One of the most interesting features of ZigBee is possibility of mesh networking, i.e. dynamic network formation from existing nodes. This extends network range and provides higher network reliability by creation of new paths in case of network configuration changes (see Figure 2).

ZigBee specification defines so called Profiles of devices, which allow equipment from different manufacturers to interoperate. Usage of ZigBee for standardized products makes it possible to provide a unified integration solution, which will create a link between different types of home automation and serve as a basis for "smart home".

Table 3. ZigBee technical data [2]

Frequencies	868MHz, 915MHz, 2.4GHz
Data rates	20kb/s, 40kb/s, 250kb/s
Typical distance	10-75 meters
Network	Mesh
Battery life	Several years

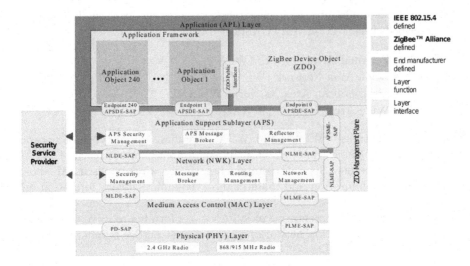

Fig. 1. ZigBee stack [1]

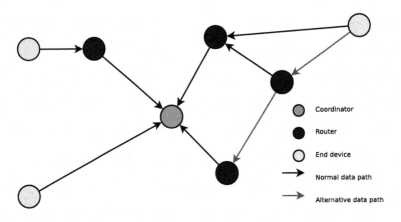

Fig. 2. Simple ZigBee network structure

3   Linux Platform and ZigBee Technology

3.1   Platforms for Home Automation

If we look at control systems hardware, during last years it could not support modern operating systems - 8/16 bit autonomous controllers did not have enough performance and memory for such task. Also, required functionality did not prove usage of advanced processors.

However, currently more and more complicated solutions are being implemented, with high connectivity, configurability and certain intelligence. For such devices more powerful hardware is necessary, which also requires to use more advanced system software in order to support networking and application functionality.

3.2 Linux Platform Position

From a wide range of possible solutions, most of manufacturers today choose Linux. It has reliable support for most of existing processor architectures, including:

- ARM - handhelds and other power-efficient applications
- MIPS - platforms for network equipment
- PowerPC - used in high-performance applications, such as game consoles
- SH - widely used for set-top-boxes

For the home entertainment solutions, Linux already has taken a leading role as a platform for set-top-boxes - multimedia devices, providing pay-per-view television, digital video recording and other services. Most famous examples include products like TiVo [11] and others. Linux serves as a software platform for some modern TV's and is ported to most of game consoles. So we can see, that from multimedia systems perspective, home automation is pretty well covered by Linux devices.

From traditional IP networks perspective, it is not unusual to find Linux inside network infrastructure and projects like OpenWrt [12] provide alternative firmware with extended functionality to substitute original platform on such devices. Other role which Linux plays, often invisibly, is an embedded platform for various control panels, automation controllers, etc.

3.3 Linux and ZigBee

With support for most of existing networking protocols, Linux becomes a natural choice for implementation of almost any kind of modern home automation system. However, there are some gaps in technological position of this operating system, connected mostly with specific technology areas. For example, home networking protocol stacks implementation is currently not complete.

In order to fill the gap and put this platform into other, more traditional home automation areas, OOO Siemens CT ELIN has implemented ZigBee protocol stack for Linux. This protocol stack provides a basis for integration of traditional networking technologies with home automation network and opens possibility for easy creation of smart home applications.

References

1. ZigBee Specification. ZigBee Alliance (2008)
2. EEE Std. 802.15.4TM-2003. IEEE Computer Society (2003)

3. American Society of Heating, Refrigerating and Air-Conditioning Engineers, Inc.,
   http://www.ashrae.org
4. KNX Association, http://www.knx.org
5. UPnP Forum, http://www.upnp.org
6. X10. Wikipedia,
   http://en.wikipedia.org/wiki/X10_industry_standard
7. LonWorks, http://www.echelon.com
8. HomePlug Powerline Alliance, http://www.homeplug.org
9. Z-Wave, http://www.zen-sys.com
10. ZigBee Alliance, http://www.zigbee.org
11. TiVo, http://www.tivo.com
12. OpenWrt, http://www.openwrt.org
13. Home Automation. Wikipedia,
    http://www.wikipedia.org/wiki/Home_Automation
14. Wireless Network. Wikipedia,
    http://en.wikipedia.org/wiki/Wireless_computer_network

The Smartest Space of All: A Global Space of (Machine-Understandable) Knowledge

Reto Krummenacher

Semantic Technology Institute, University of Innsbruck, Austria
reto.krummenacher@sti-innsbruck.at

Abstract. Internet computing is entering a new era, and the traditional human targeted World Wide Web slowly but surely moves towards a Web of billions of services, where the Web resources become entities in a global service-oriented architecture. In such a world of services, it is the service that counts for a customer and not the software or hardware components that implement the application. Current service-oriented architectures are still very much restricted in their application context to corporate solutions, while a Web with billions of heterogeneous and distributed service and content providers and consumers has requirements that call for specialized solutions.

With our work on Triple Space Computing we aim at solutions to the Service Web challenges by integrating technology from tuplespace computing (blackboards), the Semantic Web and Web services. This unique synergy of technology is researched in the EC STREP TripCom and is seen as the fundamental layer for the coordination and communication of services in future Internet systems. Triple Space facilitates the reusability and integration of heterogenous knowledge, the lightweight orchestration of services, and thus enables large scale collaborative solutions in a scalable, robust and easily maintainable way.

The World Wide Web is one of the ground-breaking achievements of modern computer science. The core technology, so simple to use, allows billions of users to publish, share and consume information from all over the world. The addition of so-called Web services that enable the remote execution of applications delivered a new dimension of communication, in particular in B2B and EAI systems. The same counts for the recent achievements around the Semantic Web. In this talk we present our work towards a scalable semantic middleware that further enables the integration, communication and coordination of billions of autonomous, distributed and heterogeneous service providers and consumers. Such middleware is the fundamental building block for the realization of Web services and the Semantic Web in the true Web sense: scalable, open and very simple to use.

The main objective of the Service Web is to automate as much of the service computing life-cycle management as possible, as maintaining millions or even billions of services that are implemented and invoked in changing and diverse environments and contexts is not feasible with the current human-centered XML-based Web service technologies [1]. Tasks such as service discovery, composition or process mediation can be automated to a certain extent by use of semantic technologies. For example semantics enables the authoring of meaningful service descriptions, as opposed to the XML-based end point descriptions that deliver a pure syntactical view point of the remote application. The main element that such Semantic Web services however lack, is

S. Balandin et al. (Eds.): NEW2AN 2008, LNCS 5174, pp. 271–272, 2008.

a semantics-aware service bus that supports the communication and coordination needs of the services. The middleware bus is thus the main accessor to large scale integrated and interoperable service computing systems, and must provide reusability, decoupling in reference and time, abstraction, composability, statelessness, and pay tribute to the inherit Web principles of openness, decentralization, dynamicity, n:m interactions, and last but not least scalability.

In consequence, with the project TripCom we provide a semantic middleware called Triple Space that abstracts from the characteristics of data and service providers through a transparent integration mechanism for various sources of information and services - no matter if human or machines. As mentioned previously, the concept of Triple Space is a convergence of space-based computing [5] and semantic technology with a clear focus on service computing; the original intention was to bring Web service communication to rely on persistent publication of data, as it is the paradigm of the World Wide Web, instead of being mainly dependent on transient message exchanges [2,6,7]. Still, in a more generalized view the Triple Space could conceptually be seen as a database that delivers communication services. In fact, the need to abstract relational database management systems into data spaces to guarantee uniform access to heterogeneous, highly distributed and interrelated data sources has been acknowledged recently by the database community [4], while on the other hand spaces were also compared to integration platforms or enterprize service buses, where a space would be modeled by a n:m message queue [3].

In that sense Triple Space provides a large scale integration framework for semantic data and services and hence enables a global knowledge infrastructure that brings advantages both to providers and consumers of the Service Web, and that goes far beyond corporate solutions or simple applications.

In this talk we will present Triple Space Computing as novel communication and coordination paradigm that brings Internet computing to the world of services. In detail, we will consider Triple Space in the context of large scale, distributed scenarios and show how space principles together with semantics are the true enabler of smart spaces, i.e. how we see Triple Space being the 'Smartest Space of All'.

References

1. Domingue, J., Fensel, D.: Toward a Service Web: Integrating the Semantic Web and Service Orientation. IEEE Intelligent Systems 23(1), 86–88 (2008)
2. Fensel, D.: Triple-Space Computing: Semantic Web Services Based on Persistent Publication of Information. In: IFIP Int'l Conf. on Intelligence in Communication Systems, pp. 43–53 (2004)
3. Fensel, D., Kühn, E., Leymann, F., Tolksdorf, R.: Queues Are Spaces - Yet Still Both Are Not The Same? Technical report, TripCom Project (2007)
4. Franklin, M., Halevy, A., Maier, D.: From Databases to Dataspaces: a New Abstraction for Information Management. ACM SIGMOD Record 34(4), 27–33 (2005)
5. Gelernter, D.: Generative communication in Linda. ACM Transactions on Programming Languages and Systems 7(1), 80–112 (1985)
6. Krummenacher, R., Hepp, M., Polleres, A., Bussler, C., Fensel, D.: WWW or What is Wrong with Web services. In: 3rd European Conf. on Web Services, pp. 235–243 (2005)
7. Simperl, E., Krummenacher, R., Nixon, L.: A Coordination Model for Triplespace Computing. In: 9th Int'l Conference on Coordination Models and Languages (2007)

A Definition Approach to Smart Logistics

Dieter Uckelmann

Abstract. Having the right product at the right time at the right place and in the right condition – these are the well-known requirements for logistics and transportation in general. But fulfilling these requirements is getting more and more complex in a dynamically changing logistic environment. There is a shift from traditional supply chains to open supply networks. Long-lasting business relationships are overrun by short-term business connections. The highly dynamic logistic markets and the advancing complexity of logistic networks require new methods, products and services. Aspects such as flexibility, adaptability and proactivity gain importance and can only be achieved by integration of new technologies. While problem initiated approaches usually only lead to minor improvements, technology driven approaches can evoke more radical changes. The technology driven approach that is used to define Smart Products and Smart Services is utilized and extended to define "Smart Logistics". Within the paper a definition of Smart Logistics is given. Additionally current technical components of Smart Logistics are specified.

1 Introduction and Motivation

In 1996 Mark Weiser, generally seen as the father or ubiquitous computing, published a paper in which he described a dispute with Nicholas Negroponte about what the third wave in computing would look like. Mark Weiser forecasts that *the defining words will not be "intelligent" or "agent", but rather "invisible" and "calm" and "connection"* [1]. When thinking about Smart Logistics this dispute is still open. While Weiser until today still proves to be right, there is a huge amount of ongoing research about intelligent and agent based logistics [2], [3], [4]. As the term "Smart" in combination with products and services can be deducted from technology driven innovations, the definition of Smart Logistics will be subject to change dependant on the respective current technology developments. Weiser even thinks the term "Smart" is misleading and criticizes its time dependency: *The "Smart House" of 1935 had an electric light in every room. The "Smart House" of 1955 dared to put a TV and a telephone in every room. And the "Smart House" of 2005 will have computers in every room* [1]. These examples show that "Smart" is defined by deployment of innovative and commercially available state-of-the-art technology. A definition of Smart Logistics is reasonable to group the corresponding research and to accomplish a common understanding. The definition has to be flexible and adaptive though, to incorporate future technology developments.

In literature a definition for Smart Products and Services can be found. This definition is described in the next chapter. A good definition of Smart Logistics instead is still missing. The definition is based on an analogy to the terms Smart

S. Balandin et al. (Eds.): NEW2AN 2008, LNCS 5174, pp. 273–284, 2008.

Products and Smart Services. First the paper defines characteristics of Smart Logistics. In the next step the technical components supporting the growth of Smart Logistics are listed. Among these components are radio-frequency and other means of automatic identification, locating and sensing technologies, networking and data-processing as well as billing capabilities.

2 Characteristics of Smart Logistics

As described within the introduction the definition of Smart Logistics is based in the same conditions as Smart Products and Services. A good definition of Smart Products and Smart Services is given by Fleisch et al. [5].

Smart Products and Services according to Fleisch et al. [5] describes that humans dispense some of their control activities to products and services. This specification resembles Weiser's explanation of ubiquitous computing. *Ubiquitous computing just might help to free our minds from unnecessary work...*[1]. Unnecessary work in this context could be defined as everything that can be delegated to Smart Products and Smart Services. It enables people to focus on subjects that can not be delegated, thus requiring more "smartness" than Smart Products and Services can provide.

Smart Products and Services usually are evolving from new technologies in a technology driven approach. Smart products in this respect offer additional functionality based on increased visibility which has been achieved through ubiquitous computing technologies. Their function is based on proximate environments and relations including assets, spare parts, tools related to Smart Products as well as product lifecycle history and neighbourhood to other products. Fleisch et al. [5] lists several practical examples for these kinds of Smart Products. These examples can be grouped as follows:

- Identify item and perform simple if-then-else algorithm
  - Handgun, which only works if a chip at the wrist of the owner is in close proximity
  - Consumer goods which only work if original spare parts are used (e.g. printer and toner cartridge)
  - Error message is generated, if spare parts are misplaced during assembly
  - Hazardous goods generate an alarm, if other goods are in close proximity that could lead to catastrophes
  - Toolbox checks its own completeness
- Identify item, sense environment
  - Packaging of cold chain goods indicates, if cold chain has been disrupted
- Identify item, sense environment, perform simple if-then-else algorithm, communicate
- Machine (tool, car, airplane) only works if original parts are used; if usage cycle ends, new spare parts are ordered automatically
- Lot senses its location, communicates to production machines, machine generates warning, if production lot does not fit
- Vending machine senses its inventory as well as its service needs and generates a message, if replenishment or service is required

Fleisch et al. [5] does not consider more sophisticated "Smart" functionality such as software agents. Even the "Internet-of-Things" is not required in order to offer Smart Products or Smart Services in this context. The above mentioned examples are mostly based on one-to-one relationships instead.

Fleisch et al. [5] relates the need for communication of products to the grade of functionality. Today, a hammer does not need light or sound notification nor a screen, whereas machines, electronic equipment or vehicles offer more functionality, thus requiring more means of communication. Reversing this theory, integration of ubiquitous computing raises the communication ability as well as the noticeable functionality. Fleisch et al. [5] deducts two questions from this thesis:

1. Which additional functionality can be offered to the customer?
2. Which additional functionality offers the production company benefits?

Thus a win-win relationship between customer and production company is needed in order to develop a successful Smart Product.

According to Fleisch et al. [5] Smart Services offer the ability to measure what could not be measured before. They enable operating, pricing and trading of previous untradeable services. Usage based billing is only possible, if usage can be quantified. Jonkers et al. [6] sees two elements within usage data acquisition – metering and collecting. While metering registers the usage of resources in real-time, collectors aggregate data from one or more meters and forward these data for accounting.

Benefits that can not be quantified will remain to be general expenses. They can not be delegated to external service providers. Examples are the introduction of GPS which lead to electronic navigation services. Ubiquitous computing leads to new Smart Services for tracking and tracing of goods and authenticity checking of brand products and pharmaceuticals [5].

Even though Fleisch et al. [5] has recognized the chances of new billable Smart Services, they have not influenced the development of the "Internet-of-Things" to integrate billing into the proposed EPC network. The billing capabilities are a prerequisite though, to enable Smart Services. Surely there are ways of integrating existing billing solutions into the EPC network, but a comprehensive approach is missing.

Fleisch et al. [5] sees five important types of Smart Services:

- Control services – delegating of control tasks such as tracking and tracing, theft protection, counterfeiting, reordering
- Leasing services – high visibility enables conversion of the basis of calculation from owner based information to usage based information
- Risk services – insurances changes their pricing model from estimates to usage based fees
- Information services – instant online access to ubiquitous computing related information
- Complex services – they offer a combination of the above mentioned services

Smart Products and Services are technology driven. Problem oriented approaches that lead to relatively small process optimizations, whereas technology driven Smart Services and Products may lead to larger, more radical changes. Smart Logistics incorporates both – Smart Services and Smart Products.

Smart Logistics shall include and is defined by the following characteristics:

- Smart Logistics embraces Smart Service as well as Smart Products within Logistics
- Smart Logistics is derived from a technology driven approach, and thereby subject to change
- Smart Logistics frees humans from (control) activities that can be delegated to Smart Products and Services
- Smart Logistics are invisible and calm and can therefore be described as transparent
- Smart Logistics are connected, thus communicating and possibly interacting with their environment
- Smart Logistics facilitate state-of-the-art data processing (which may include, but do not require software agents)
- Smart Logistics integrates existing logistic technologies, such as material handling systems, and enable these to react and act in a correspondingly smart manner
- Smart Logistics include state-of-the-art billing, payment or licensing as integral component

The above given criteria shall be used to define Smart Logistics. As this definition is based on state-of-the-art technology and does not require autonomy or decentralization, it is easy to distinguish from autonomous cooperation and control within logistics [7].

3 Technical Components of Smart Logistics

Even though technical components on Smart Logistics may change over time, as new technologies involve, there are still some basic components that foster Smart Logistics. Among these are technologies to identify, locate, sense, process and act.

3.1 Identification

Provokingly physicist and Professor Neil Gershenfeld, author of *When Things Start to Think*, claimed in his Bill of Things' Rights, things have a right to have an identity [8]. As real and virtual world are moving closer together, logistic items that can not be identified are not existent. No controlled action is possible until identification is enabled again.

RFID offers a convenient way of identifying logistic objects. The "right product" logistic requirement can be easily checked by using RFID. Usually "right time" and "right place" are confirmed together with the identification number. Therefore three of the basic logistic requirements can be covered by automatic identification. There are other means of automatic identification besides RFID, but radio-frequency identification offers the following advantages:

- bulk-reading
- no line of sight required
- robust compared to barcode concerning dirt etc.
- large memory

- re-writable / changeable data storage
- ease of use
- speed of data entry
- cheaper than other solutions

Bulk reading and no line of sight for reading are two of the basic advantages when it comes to RFID. Robustness might be an issue in certain projects where readability of barcodes is disturbed, i.e., through dirt. Even though the current RFID discussion is mainly focussed on a relatively small 96bit electronic product code (EPC), the chip can hold much more data which may be added or changed if needed. The ease of use which may be described as "transparent identification" – once the RFID-system is installed - and the speed of data entry adds to the convenience and may influence a pro-RFID decision. Finally, with prices for tags and readers going down, we may see RFID being the cheaper solution compared to other systems in certain cases.

3.2 Locating

Event though identification is often associated with recording the place of identification, real-time locating systems (RTLS) have to be distinguished from identification. RTLS offers locating close to real-time with sub-minute updates [9]. While RFID is on its way to being ubiquitous, real-time locating systems (RTLS) are still seldom to find. One reason for this may be the cost for the corresponding infrastructure. Implementing a real-time locating system has the capability to lead to considerably better supply chain visibility. RTLS combine the functions to identify objects at a certain time and to locate them. While radio frequency based identification has gained wide acceptance in the market, using RF-technologies for real-time locating is still limited to branch-specific approaches, such as healthcare management, car yards or container terminals.

Traditionally active transponders or ultra-sound systems are used for real-time locating, requiring expensive asset tags and an extensive network of base-stations. Numerous real-time locating systems are available in the market. There are several criteria to distinguish between different RTL systems:

- Physical / symbolic location
- Absolute / relative location
- Transmission media
- Locating method
- Location computed by object / infrastructure
- Accuracy
- Scalability
- Communication intervals
- Costs for infrastructure and beacon (e.g. tags)

Some of the real-time locating systems provide physical position (e.g. GPS) while others refer to symbolic locations such as parking lot or goods issue [10]. GPS is a good example for absolute location. A GPS position is unique worldwide. Relative positions may be of importance, if distance and direction relative to an object are relevant. For example the distance of a workpiece to a workstation may provide valuable information for managing a production line. Especially for indoor locating systems relative location systems are used. Transmission media may include radio frequency, infrared, ultrasound, magnetic or optical technologies. The following different locating methods and combinations for RTLS are known:

- Cell-of-origin (transponder-of-origin)
- Amplitude (RSSI, received signal strength indicator) triangulation
- Time of flight ranging systems
- Time difference of arrival (TDOA)
- Angle of arrival (AoA)

The cell-of-origin approach only indicates that the transponder is within read range. It is easy to set up and to maintain. Basically every RFID-reader or tag attached to a fixed locating represents a cell-of-origin locating system. As an example, low frequency readers hooked to forklifts have been used for pallet tracking. Transponders in the floor are indicating a position, while the forklift passes them.

Signal strength is another easy method for real-time locating. Any hardware offering the possibility to measure signal strength may be used. Some newer RFID-UHF-readers offer to measure the received signal strength indicator (RSSI). If a tag is within reading range of multiple antennas, amplitude trilateration may be used to locate the tag. The strength of an electromagnetic signal decreases with the square of the distance. However reflection and absorption may lead to false results. Consequently existing systems based on the analysis of signal strength use a map of received signal strength as a reference for locating. Another disadvantage of this system is, that it is vulnerable to changes within the environment.

Time of flight ranging and time difference of arrival (TDOA) systems are less influenced by environmental changes. Using triangulation algorithms an estimation of the location may be given. These are more accurate than the above mentioned systems, while a higher degree of synchronization between the readers is needed to measure exact timings.

Angle of arrival systems for locating purposes are well known for example at airports where rotating radar antennas are using this method to locate planes. Antennas are rotated to find the direction of the highest signal strength. Antenna arrays are also used to measure at different angles.

Hybrid approaches combine any of the above mentioned technologies. One possibility to provide a cost effective solution of location tracking is to combine GPS and passive RFID in a hybrid personal data terminal used for time and location recording [11].

Location may be computed by the object (e.g. beacon, transponder) as well as by infrastructure. Angle of arrival is typically calculated through infrastructure.

Accuracy is one of the most important issues for RTL systems, whereas it is not always necessary to adopt the most accurate solution. Instead sufficient accuracy for the individual process should be used. With some systems, accuracy may be improved by setting up a tighter infrastructure. Other systems may be more accurate in limited spaces but lack the possibility to scale to larger areas. Generally cost and scalability are trade-offs with accuracy.

Communication intervals should be optimized if active transponders are used in order to maximize battery life while providing enough data to achieve the required accuracy. Besides static definition of communication intervals (e.g. once every 15 minutes) also dynamic intervals may be set for certain RTLS. For example motion detectors or proximity sensors may be used to activate communication.

As mentioned above, relatively high prices hinder the diffusion of this technology within supply chains. There would be an increased request for RTLS, if lower prices could be achieved.

But lower prices for RTLS may not be the only way to capture a broader market. If potentials of RTLS can be achieved for multiple stakeholders within the supply chain, and billing across the supply chain for increased visibility would be possible, then service providers could enter the market. RTLS service providers for "Smart Logistics" would offer a usage based fee across the supply chain to charge for usage information and not for owning a technology. As described within the introduction, corresponding services for GPS are already available, whereas other RTL solutions are still based on selling infrastructure.

3.3 Sensing

The remaining basic requirement of logistics is to provide logistic goods in the "right condition". Gershenfeld includes the right *to detect the nature of their environment* in his Bill of Things' Rights [8]. The condition of goods, for example, is of uttermost importance with cold chain and fresh food logistics. Temperature and humidity sensors are well known examples of sensors that are used in these scenarios. For fruits and vegetables ethylene sensors could be interesting, because the ethylene concentration within a closed room such as a container provides a good indication about the ripening process [12]. Sensors may be based on active RFID transponders, thus facilitating the existing battery to power the sensor.

The three technologies RFID, RTLS and sensors in combination fulfil a control function for the "4R" within logistics. The "right product" has been identified, „right time" and „right location" have been recorded and the "right condition" has been checked through sensors. The three described technologies communicate with a state-of-the-art IT infrastructure.

3.4 State-of-the-Art Networking and Data-Processing

Smart logistic objects are able to communicate with each other and with their environment such as machines and transportation. Data-processing can be done centralized or de-centralized. This leads to a coalescence of material flow and information flow and enables every item to manage and control its logistic process. De-centralized data-processing can be supported by software agents. Today, software agents have still failed to gain a relevant market share within logistics. Therefore, in a lot of cases, software agents fail to meet the defined criteria of being a commercially available state-of-the-art technology. But software agents may be a common technology within the future.

The "Internet-of-Things" is another example of a technology that has not yet reached the status of being commercially available. Certain parts are available, but still there are many open questions left.

3.5 Billing

The general problem with Smart Products and Services is that they need to pay off for the customer as well as for the manufacturer. Much too often a return-on-investment for the product and service provider can not be achieved from one single stakeholder as there is a move from closed technology environments to open loop installations. Thus new concepts to share benefits and costs need to be implemented.

The concept of the "Internet-of-Things" provides most of the necessary infrastructure to overcome this problem. Products are uniquely identified through automated identification such as RFID. It is a question of time, when the "Internet-of-Things" will be able to handle locating and sensor data as well. Through an internet like infrastructure, a link points to product related information. The presented paper focuses on integration of a billing system to the "Internet-of-Things", thus enabling individual and independent payments for products and information. As the value of individual information such as best-before date or storage conditions during shipment is very small, existing micro-payment solutions are suggested in order to accumulate tiny payments to billable amounts. The billed amount has to exceed the cost of billing.

Existing solutions usually are proprietary, fail to scale and are not reasonable for micro payments. Yet, as ubiquitous computing is growing the transferred amount of information per transaction decreases thus requiring minimal payments on usage basis. Today, telecom providers and some online billing services are closest to fulfil these requirements. But telecom providers have to adapt to the changing requirements within ubiquitous computing. *The upcoming paradigm shift towards ubiquitous computing is highly likely to change the IT and Telecommunications business considerably, probably more than what the Internet did to telecommunications a few years ago. Telcos today are unfortunately not well prepared for this, in particular due to lack of technological involvement* [13]. Online-payment services are more familiar with ubiquitous computing. Still, these service providers are not positioned to handle micropayments in business-to-business (B2B) relations. B2B requirements include service level agreements, different charging schemes, and clearance of mutual bills before payment [6].

Different pricing schemes can be applied. Examples are flat-fees, usage-based, time-based, transaction-based, and volume based pricing schemes. In order to achieve a high visibility within billing, it is desirable to use a transaction-based pricing model. It should be possible to charge even the smallest valuable information generated by "Smart Devices" such as the best-before-date. Unfortunately the costs of billing are most often too high with a rate of up to nearly 50 percent (Table 1) of the individual transaction.

There are three typical strategies to support facilitation of new technologies. Firstly, mandating may be seen as the main force to spread new technologies such as RFID in markets dominated by large players such as retail, aviation or automotive. Examples are mandates from the Department of Defense, Metro and Walmart. But there are limits to mandates in non-dominated, distributed markets.

Secondly, cost and technology sharing is another method of enforcing new technologies between business partners. Cost and technology sharing models are

Table 1. Comparison of different payment systems (based on [14])

System type	Pay before			Pay later				
System	Micro Money	Yahoo Pay-Direct	PayPal	First-Gate Click & Buy	Infin Micro-payment	Net 900	Pay-safe-key	T-Pay
Set-up fee	500 €	-	-	25 – 5,000 €	-	150 €	1.000 €	116 €
Monthly cost	0–50 €			5 €	50 – 250 €	Small	500 €	>11,60 €
Trans-action fee	20–30%	2,2–2,5% + 0,3 US-$	0,7–2,9 % + 0,3 US-$	< 30-40 %	20–40%	20–47 %	Nego-tiable	>0,12 €
Regis-tration required	Yes	Yes	Yes	Yes	No	Yes	Yes	Yes
Pervasive-ness at providers	Medium	High	High	High	Low	Low	Medium	High
Pervasive-ness at customers	Low	n/a	High	High	n/a	n/a	n/a	n/a

limited to scenarios with few players involved. The model fails to scale in large environments.

The third option is an open and market driven and usage based model across the supply chain to invoice Smart Products and Smart Services. Asymmetries concerning the distribution of costs and benefits between Smart Service and Products providers and users may only be overcome, if standard market rules can be applied.

The "Internet-of-Things" can be used as the fundament for an open and scalable infrastructure, if billing and invoicing capabilities are added. The necessary additional components to the "Billing Integrated Internet-of-Things" requires a subscription to a billing service and the billing awareness of the components of the "Internet-of-Things" (Fig. 1.)

The billing service shall be between the Object Name Service and the EPC Information Service and could be integrated with the EPCIS Discovery Services. It provides information about the availability and the price of Smart Products and Services based on the automatic identification of logistic objects. It ensures that billing level events (BLE) are generated, if these service and products are used. Billing level events could be standardized in order to forward billing events to proprietary billing systems or to the financial systems of companies.

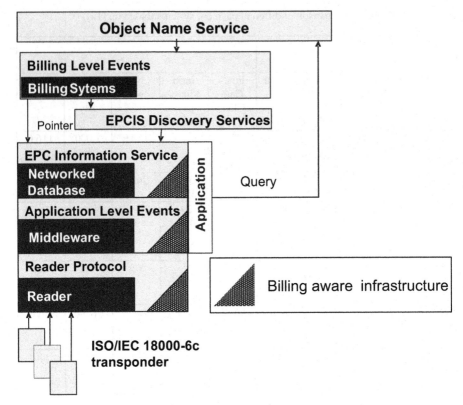

Fig. 1. The Billing Integrated Internet-of-Things (based on: [15], [16])

The concept of the Billing Integrated Internet-of-Things allows multi-directional cost-sharing and profit generation across supply chains. It allows the accelerated distribution of Smart Logistics, as the invoicing would be an integral part of an existing infrastructure. Providers of Smart Services and Products are released of setting up their own billing infrastructure. Therefore a Billing Integrated Internet-of-Things not only fosters the development of Smart Logistics – it is a Smart Service in itself.

4 Conclusion and Future Work

Within this paper a definition of Smart Logistics has been given. Based on Smart Products and Smart Services, characteristics of "Smart" have been elaborated. These characteristics have been enhanced by state-of-the-art data processing and billing. While software agents may be used to enable autonomous and decentralized decision making, they are not a prerequisite to Smart Logistics. Centralized data-processing may be used as well.

The billing integration plays an important role within the given definition of Smart Logistics. Billing is always a part of Smart Logistics. But up to now, there has not been a comprehensive approach for an open and integrated billing solution. The integration of billing capabilities into the concept of the "Internet-of-Things" allows a multi-directional open billing structure. Smart Logistics in a lot of instances only pays off, if multiple members of the supply chain pay for the individual benefits they receive through Smart Logistics. The bi-directional approach allows every member of the supply chain to pay for received services and products as well as to charge for own offerings. The integrated billing capability would lower the barrier to enter the market with new offerings thus accelerating the introduction and dissemination of new Smart Logistic offerings.

Future work includes building a test scenario within a lab environment followed by feasibility studies within real environments. Currently the EPC network is set up and integration of billing solutions are evaluated. A "Smart Truck" including RFID, real-time locating, sensors and sophisticated data-processing is under development at the LogDynamics Lab to be used for a technical feasibility study.

Beyond technical feasibility, the acceptance of bi-directional payments has to be verified within logistic networks.

References

1. Weiser, M.: Open House. In: ITP Review, the web magazine of the Interactive Telecommunications Program of New York University, New York (1996)
2. Moore, M.L., Reyns, R.B., Kumara, S.R.T., Hummel, J.R.: Distributed intelligent agents for logistics (DIAL). In: Systems, Man, and Cybernetics, 1997. Proceedings of the IEEE International Conference on Computational Cybernetics and Simulation, vol.3, pp. 2782–2787 (1997)
3. Davidsson, P., Henesey, L., Ramstedt, L., Törnquist, J., Wernstedt, F.: An Analysis of Agent-Based Approaches to Transport Logistics. Transportation Research Part C: Emerging Technologies 13(4), 255–271 (2005)
4. Graudina, V., Grundspenkis, J.: Technologies and Multi-Agent System Architectures for Transportation and Logistics Support: An Overview. In: Rachev, B., Smrikarov, A. (eds.), Proceedings of the International Conference on Computer Systems and Technologies and Workshop for PhD Students in Computing, pp. IIIA.6–1 – IIIA.6–6 (2005)
5. Fleisch, E., Christ, O., Dierkes, M.: Die betriebswirtschaftliche Vision des Internets der Dinge. In: Fleisch, E., Mattern, F. (eds.) Das Internet der Dinge, pp. 3–37. Springer, Berlin (2005)
6. Jonkers, H., Hille, S.C., Tokmakoff, A.: M. Wibbels: A functional architecture to support commercial exploitation of Internet-based services. In: Winiwarter, W., Bressan, S., Ibrahim, I.K. (eds.), Third International Conference on Information Integration and Web-based Applications and Services (iiWAS), pp. 277–288, Linz, Austria (2001)
7. Windt, K., Hülsmann, M.: Changing Paradigms in Logistics – Understanding the Shift from Conventional Control to Autonomous Cooperation and Control. In: Hülsmann, M., Windt, K. (eds.) Understanding Autonomous Cooperation & Control – The Impact of Autonomy on Management, Information, Communication, and Material Flow, pp. 4–16. Springer, Berlin (2007)
8. Gershenfield, N.: When Things Start to Think. Henry Holt and Company, New York (1999)
9. ISO/IEC 24730-1: Real-time locating systems (RTLS) - Part 1: Application program interface (API) (2006)

10. Hightower, J., Borriello, G.: Location systems for ubiquitous computing. Computer 34(8) (August 2001); ISSN: 0018-9162
11. Scholz-Reiter, B., Uckelmann, D.: Tracking and Tracing of Returnable Items and Pre-Finished Goods in the Automotive Supply Chain. In: Miles, S., Hardgrave, B., Williams, J. (eds.) Proceedings of the 1st RFID Academic Convocation, Boston (2006)
12. Jedermann, R., Schouten, R., Sklorz, A., Lang, W., van Kooten, O.: Linking keeping quality models and sensor systems to an autonomous transport supervision system. In: Kreyenschmidt, J., Petersen, B. (eds.) Cold Chain-Management, Proceedings of the 2nd international Workshop Cold Chain Management, pp. 3–18 (2006)
13. Eurescom: Smart Devices When Things Start to Think. Project P946-GI, Deliverable 1, Strategic Study Eurescom, the European Institute for Research and Strategic Studies in Telecommunications GmbH, http://www.eurescom.de/~pub-deliverables/p900-series/ P946/ D1/p946d1.pdf
14. Dannenberg, M., Ulrich, A.: E-Payment und E-Billing: elektronische Bezahlsysteme für Mobilfunk und Internet. Gabler, Wiesbaden (2004)
15. Harrison, M.: EPC Information Service, http://www.m-lab.ch/auto-id/SwissReWorkshop/ papers/EPCinformationService.pdf
16. Uckelmann, D.: The Value of RF-based Information. In: Haasis, H.D., Kreowski, H.J., Scholz-Reider, B. (eds.) Dynamics in Logistics, LDIC 2007, pp. 183–197. Springer, Berlin (2008)

A Method of Constructing Personal Network for Ubiquitous Personal Services

Kazumasa Takami[1,2], Hajime Kusu[1], and Akira Ikeda[2]

[1] Graduate school of Engineering
[2] Department of Engineering, Soka University
1-236 Tangi-cho, Hachiouji-shi, Tokyo, 192-8577 Japan
k_takami@t.soka.ac.jp

Abstract. In a ubiquitous environment, the connections of a variety of devices, including electronic tags such as RFID (Radio Frequency Identification) tags to the network, together with an advance in Web services, are expected to give rise to a wide range of services, stimulating research into smart space services, which incorporate a sensing environment. With a view to building a convenient personal environment in such an environment, in this paper we propose a method of building a personal network (PN), which is a wide-area personal space and consists of a visited personal area network (PAN), in which the user can use devices available in a temporarily visited network for information input/output, and a home network built at home or in the office. In particular, we study a Touch&Select method, which allows the user to select an appropriate device in a visited PAN intuitively, and a monitoring protocol, which is designed to ensure connectivity of the PN. Evaluations through prototyping and simulation have verified the effectiveness of these proposals.

Keywords: Ubiquitous Network, Personal Agent, PAN, PN, RFID.

1 Introduction

When a ubiquitous environment is well in place, a variety of devices both at home and in the office will be connected to the network. In the future, people will be able to operate devices, whether owned by a public or private sector, as if they were their own. Such devices include large-screen monitors on the street normally used to display advertisements, Web cameras normally used for monitoring traffic or preventing crimes, and traffic signals capable of visible wavelength communication. In addition, when information of objects equipped with RFID (Radio Frequency Identification) tags can be read and collected easily and securely by wireless means, a true ubiquitous environment will become a reality, making ubiquitous services more and more convenient.

In such a well-developed ubiquitous environment, a personal area network (PAN) confined to a small area with 0 to 10 meters across (a pico space) is not sufficient to provide a ubiquitous service well tuned to individuals. If, instead, a personal network (PN) can be flexibly expanded via wide-area networks, it would be possible to provide personal services more easily and securely.

S. Balandin et al. (Eds.): NEW2AN 2008, LNCS 5174, pp. 285–296, 2008.
© Springer-Verlag Berlin Heidelberg 2008

There have been many studies on technologies for building a PN at a visited place using a mobile terminal (mobile node) and locally available devices temporarily [1]–[4]. The dramatic advance in the performance of terminals is allowing such technologies to be applied to a wide range of purposes. However, while such convenient and highly flexible technologies and services are being deployed, the systems involved are becoming so complex as to demand high skill of users. In fact, even engineers are sometimes unable to make full use of such services.

To solve this problem, in this paper we propose a method of building an Mobile IPv6-based PN consisting of a personal agent (PA), which provides a service to the user's mobile node in a way most useful for the user, and an agent that resides in the home PAN and another in the visited PAN. We also propose a Touch&Select method, which uses RFID to enable the user to intuitively select a device for input/out from among those connected to the LAN at the visited place, for integration into the Visited PAN. Section 2 outlines the configuration of the proposed personal network, and gives some service examples. Section 3 describes the method of building a Mobile IPv6-based PN, which is of the P2P-VPN type [5], and the monitoring protocol used. Section 4 describes the Touch&Select method, which uses RFID for intuitive selection of an appropriate device in a PAN. It uses a Touch&Share method and a Touch&Print method as specific examples. Section 5 evaluates the monitoring protocol by simulation and the Touch&Select method using prototyping.

2   Configuration of a Personal Network

In the ubiquitous communication environment it is expected that users should be able to readily access a network specifically configured for them, anytime and anywhere, regardless of the particular surroundings in which they find themselves at the time of communication. A personal area network (PAN) is build for an individual on the spot using devices available there. The PAN mainly used by the user is called a Home PAN (HPAN) while a PAN built at a temporarily visited place is called a Visited PAN (VPAN). A logical network consisting of an HPAN and multiple VPANs, overlaid on and connected to a wide-area network is defined as a personal network (PN). A PN consists of a personal agent (PA) implemented on a mobile node carried by an individual, and an agent in each location concerned. By standardizing the API of the

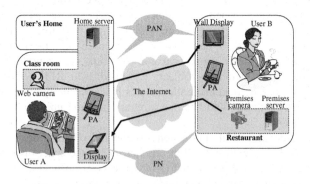

Fig. 1. Example of a ubiquitous personal service

PN, the user can personalize the currently available communication environment, enabling him or her to connect devices available there to his or her PN, and thus to receive services flexibly and seamlessly.

An example of ubiquitous personal service is shown in Fig. 1. The user integrates the Web camera and the large display in the classroom into his PAN, and establishes a visual communication with his friend in a café. This friend has also integrated the Web camera and display in the café into her PAN. The user and his friend set up connections to their HPANs, mutually authenticate themselves and establish a secure communication environment.

3   Method of Building a PN and Monitoring Protocol

Requirements for a method of building a secure personal network are to ensure: (a) the mobility of the personal agent (PA), which plays the central part in the Visited PAN, (b) dynamic establishment and release of the PN, (c) simultaneous establishment of multiple PNs in different locations, and (d) high security in PNs.

To satisfy these requirements, the method will use a Mobile IPv6-based network shown in Fig. 2. The home agent (HA) of the PA resides in the home network, and keeps the up-to-date IP address of the PA. Tunnel T1 is formed between the HA and the PA using IPsec. When the Correspondent Agent (CA) in an office LAN wants to access the PA, it does so via the HA. When the mutual authentication between the PA and CA is successful, a path for direction communication between the PA and the CA is established according to a routing optimization procedure. In parallel, P2P-VPN tunnels T2 and T3 are formed between the HA and the CA and between the PA and the CA using IPsec. The tunnels, T1-T2-T3, are managed as a temporary ring network, or a PN. The T1-T2-T3 configuration is released when the use of the PN is terminated, when the location of the PA is unknown, or when a fault is detected in the path.

To achieve this PN configuration management, it is necessary to be able to monitor when the PN is established or released, and to manage the PN configuration. We assume that a PN is established when a routing optimization procedure is executed between the PA and CA, and that the PN is released when the path between the PA and CA is no longer needed or when a fault is detected in the PN management

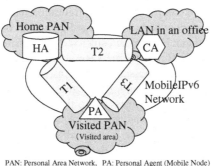

PAN: Personal Area Network, PA: Personal Agent (Mobile Node)
CA: Correspondent Agent ,   HA: Home Agent

Fig. 2. Establishment of a PN through dynamic interconnections of tunnels in different sections

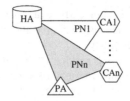

Fig. 3. Example of establishing multiple, independent PNs

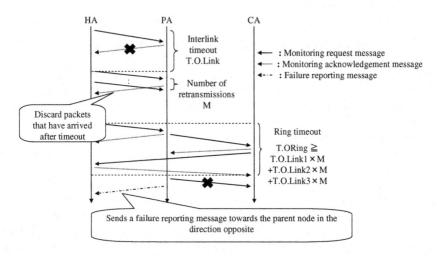

Fig. 4. Sequence of the PN monitoring protocol

mechanism. The PN management mechanism keeps track of the status of the tunnel in each section and the PN in the following manner. It circulates monitoring packets in the PN, and when a node (PA, HA or CA) receives one of these packets, it sends an acknowledgement packet to the upper node.

Multiple PNs in different locations can be established simultaneously as shown in Fig. 3. This allows for independence and security of each communication location.

We specifically propose a monitoring protocol, as shown in Fig. 4, particularly used in a single PN under the control of a PN configuration mechanism.

It is desirable that a PN is operated and managed by an HA that plays a central part in the PN configuration and can be assigned sufficient processing capacity. Therefore, the HA first sends a monitoring request message, and functions as a PN monitoring node. When a monitoring request message sent by a (parent) node reaches the next (child) node, the child node sends a monitoring request message to the next node, and sends a monitoring acknowledgement message to the parent node. If the parent node does not receive a monitoring acknowledgement message for a certain length of time, it re-sends a monitoring request message. Let us call the length of this waiting time a interlink timeout or T.O.Link, and let M be the maximum number of retransmissions. If the parent node receives no monitoring acknowledgement message even after M retransmissions, it assumes that the link concerned has been disconnected. This is how whether a link is alive or not is checked. This operation is repeated. The reception by the HA of a monitoring request message indicates that the message has been fully circulated. This is how whether the ring concerned is still alive or not is checked. Since the maximum delay on each link is T.O.Link×(1+M), the theoretical maximum delay of the ring can be determined as the total delay of three links. The above arrangement can detect any disconnection in a link or the ring, but cannot locate the link where the disconnection has occurred except for the link managed by the HA. To solve this problem and allow the HA to locate where a disconnection has occurred, the node that has detected a disconnection sends a failure reporting message towards the parent node in the direction opposite to the direction in which a monitoring request message has been circulated.

4 Selection of a Device by Touch&Select Method

As shown in Fig. 5, in the Touch&Select method, each device has an RFID tag, which keeps data used for access to that device. When the tag is read by a tag reader, the user can intuitively identify and select the device. Since the information about a device is attached to the device itself, the user can easily obtain information about the communication environment, such as network addresses, in an office he or she has temporarily visited. The user can easily use a device new to him or her.

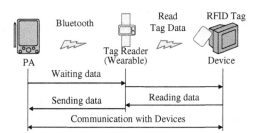

Fig. 5. Device selection using the Touch&Select method

The Touch&Select method can be applied, for example, to a Touch&Share method, which supports file transfer, a Touch&Print method, which supports printing, a Touch&Pay method, which supports shopping, and a Touch&Switch method, which switches input/output of devices. In this paper, we focus on the Touch&Share method and the Touch&Print method. We have built prototypes of these methods.

4.1 Touch&Share Method

File transfer between computers is an essential operation when using computer systems. The Touch&Share method allows the user to specify the device to which a file is to be transferred by simply touching it. A conceptual diagram of the system is shown in Fig. 6, in which the PA of the node that holds the file to be transferred is called PA1, and the PA of the node that is to receive the file is called PA2.

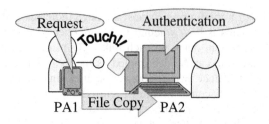

Fig. 6. Touch&Share method

The processing sequence of the Touch&Share method is as follows:

Step1: PA1 selects the file to be sent. PA2 selects the folder in which the file is to be saved, and begins to wait.

Step2: PA1 starts the wearable tag reader and receives data read from the tag attached to the node of PA2.

Step3: PA1 sends a file summary data. PA2 receives the file summary data and decides whether to receive the file.

Step4: After receiving PA2's acceptance to receive the file, PA1 sends the file.

We have built a prototype of the Touch&Share method with two PCs. It is a Windows application. PA1, which sends a file, is designated as the client while PA2, which receives the file, is designated as the server.

4.2 Touch&Print Method

When a user temporarily visits an office new to him or her, he or she cannot immediately understand the layout and connections of printers and PCs. If the user wants to connect his or her node to the network in the office, and if his or her node does not have appropriate network settings or drivers, he or she needs to make appropriate settings or install appropriate drivers. This is time consuming. With a Touch&Print method in place, the user can establish a PAN. By touching the tag attached to the printer, the user can have his or her wearable RFID tag reader read data, and use the printer to output his or her file. A conceptual diagram of the Touch&Print method is shown in Fig. 7. The sequence of using the system is shown in Fig. 8. The sequence, except for the functions of acquiring and installing the printer driver, has been implemented in the prototype system.

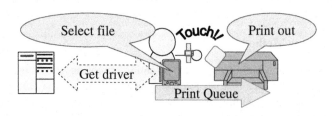

Fig. 7. Touch&Print method

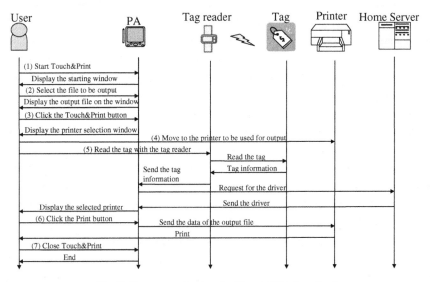

Fig. 8. Sequence of using the Touch&Print method

5 Evaluation

5.1 PN Monitoring Protocol

We implemented the proposed protocol on a network simulator, OPNET, to verify the operation of the protocol, and examined how the ring timeout detection time is affected when the interlink timeout T.O.Link, the maximum number of retransmissions, and the packet discard ratio were changed. The simulation model used is shown in Fig. 9.

The simulation conditions are as follows. We set the speed of the HA-PA link, including wireless communication, to 1.5 Mbps. Considering the handover processing in Mobile IPv6, we generated a load which was about 20% of the link speed. As a

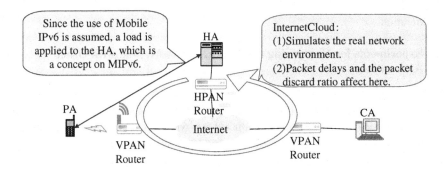

Fig. 9. Simulation model

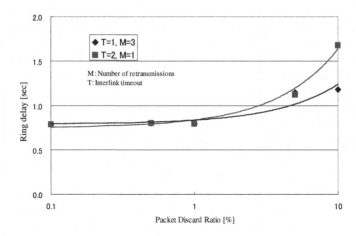

Fig. 10. Ring delay of a monitoring packet

load on the entire network, a packet delay of 0.01 to 0.5 sec in a uniform distribution was generated. We set the ring timeout to 12 sec. The reason for choosing 12 sec was that, if 3 retransmissions were made at an interval of 1 second, it took 4 seconds per link, and since there were 3 links, the total timeout length should be 12 sec. A monitoring packet was circulated from HA to PA and to CA for 10 minutes, and the ring delay was measured.

Figure 10 shows the relationship between the ring delay and the packet discard ratio for different values of T.O.Link and M. The figure shows that, as the packet discard ratio increases as a result of an increase in the load, the link delay increases, and consequently, it is necessary to extend the ring timeout value for the monitoring of the PN. Figure 10 shows that the value of ring timeout can be reduced if the ring delay is about 1 sec when the packet discard ratio is up to 1% and is about 2 sec when the packet discard ratio is higher than that. By selecting appropriate T.O.Link and M in the proposed protocol, it is possible to achieve a wide range of PN fault detection accuracy. However, since it is desirable to prevent instantaneous interruptions from causing erroneous detections of a fault too often, the ring timeout should be determined by taking the application and the network environment used into consideration.

5.2 Experiments for Comparing the Touch&Share Method with Other Alternatives

We experimented with copying a file from one computer to another computer nearby using three different methods: the proposed Touch&Share method, network sharing and the use of a USB memory stick. We conducted two types of experiments. In the first type of experiments, a file was copied using any appropriate procedure, and in the second type of experiments, specified procedures were used. The first type of experiments was intended to examine the extent of popularity and difficulty of each method while the second type of experiments was intended to purely measure the file copying time taken

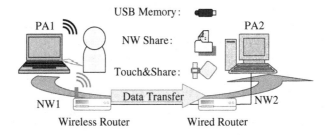

Fig. 11. Experimental environment for comparing the Touch&Share method with other alternatives

Table 1. Specified procedures for different alternatives

USB memory	Network sharing	Touch&Share
1. Insert a USB memory stick to PC1. 2. Open the source folder. 3. Copy the file to the USB memory stick. 4. Withdraw the USB memory stick. 5. Insert the USB memory stick to PC2. 6. Open the destination folder. 7. Complete copying.	1. Open the source folder. 2. Open the destination parent folder. 3. Set file sharing. 4. Open the destination folder. 5. Complete copying.	1. Start the server. 2. Open the source folder. 3. Start the application. 4. Complete copying.

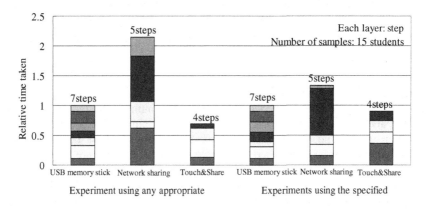

Fig. 12. Comparison of copying times taken by different alternatives

by each method. For the experimental environment, we used two PCs, which act as PAs, routers, RFID tags and a wearable tag reader, as shown in Fig. 11.

The two pairs of a PC and a router were considered two different networks. The Touch&Share client was implemented in PA1, while the Touch&Share server was implemented in PA2. An RFID tag containing the device data of PA2 was attached to the server PC. The detail of the procedure applied to each method is shown in Table 1.

We considered one step to be as follows. The operation of opening a folder was considered one step. The operation on an application to achieve its objective was also considered one step. When it comes to physical operations, an operation that results in some operation on the PC was considered one step. Figure 12 shows the average copying time taken by each method, experimented by 15 students.

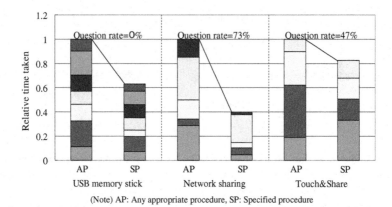

Fig. 13. Comparison of the copying time when no procedure was specified and that when a procedure was specified

The left-hand side of Fig. 12 shows the experiments using any appropriate procedures while the right-hand side shows experiments using the specified procedures. The relative copying time is compared using the time taken by the USB memory stick as one. The figure shows that the copying time taken by the proposed method is 70 to 90% of that taken by the USB memory stick method, and 30 to 70% of that taken by the network sharing method. In other words, the copying time was reduced by 10 to 30% compared to the USB memory stick method, and by 30 to 70% compared to the network sharing method. Figure 13 compares the copying time between the case where the procedure was not specified and the case where it was specified.

Figure 13 compares the results of the two types of experiments. The question rate shown at the top of each method is the ratio of the number of students who made questions about an appropriate procedure to the total number of students. The figure shows that there is no significant difference in the copying time between the two procedures. This can be considered to indicate that the students were able to perform the copying operation intuitively without any need to ask questions. Although nearly half of the students asked questions, it did not take as much time as the other method. This is because the questions posed were so simple that it did not take much time to explain the procedure.

The USB memory stick method excels in that it requires fewer steps and that it can handle large data. The network sharing method requires more knowledge and imposes more restrictions. Not many students succeeded in making perfect settings in the experiments. In contrast, the operation of the Touch&Share method is simple and involves fewer steps, and there are no restrictions similar to those of the network sharing method. As such, the Touch&Share method excels in being easy to use.

5.3 Experiments for Comparing the Touch&Print Method with other Alternatives

We experimented with actual printing of a file using two methods: the Touch&Print method and the conventional method. We compared the time required, the number of

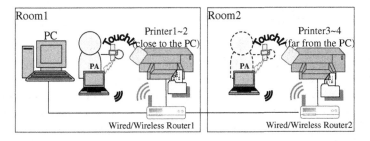

Fig. 14. Experimental environment for examining the advantage of the Touch&Print method

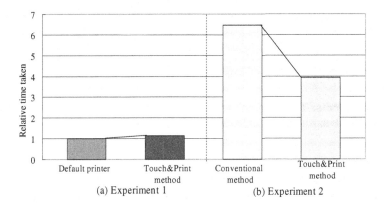

Fig. 15. Comparison of time spent

steps and features to investigate the advantages of the Touch&Print method. As shown in Fig. 14, we put two printers in one room, and two in another room, and ran a program we had developed. In order to determine which system was easy to use, we measured the length of time each of the 19 students spent. We conducted two experiments:

Experiment 1: Print a file on a default printer and a printer touched by the student. These printers are close to the PC.

Experiment 2: Print a file on an arbitrary printer selected by using a conventional method and the Touch&Print method. These printers are far from the PC.

Figure 15 shows the results of the above experiments. In Experiment 1, students spent slightly shorter time on average when using a default printer because the printer was explicitly specified and thus was easy to determine. However, in Experiment 2, in which students had to print a file on an arbitrary printer, they had to move to a printer, check its name, go to the PC and specify that printer on a printer selection window in the conventional method. In contrast, when using the Touch&Print method, students specified a printer by directly touching it, and no change in settings was required. As a result, it took a shorter time to print the file.

6 Conclusions and Future Issues

We have proposed a method of constructing a personal network (PN) to provide a personal space over a wide area in a ubiquitous environment, and a Touch&Select method, which allows an intuitive selection of an appropriate device in a temporarily visited environment. We evaluated a PN monitoring protocol specifically proposed for the construction of a PN, and confirmed its effectiveness. As specific examples of the Touch&Select method, we developed prototypes of a Touch&Share method and a Touch&Print method, and compared them with conventional methods. Experiments using the prototypes verified that, even in a newly visited place, these methods allow the user to select a device and incorporate it into the user's personal area network easily and quickly.

The issues that require further study include how to monitoring PNs when the user chooses to build multiple PNs, and how to determine whether to permit interconnections at visited places. It is also necessary to explore a further range of applications of the Touch&Select method and to evaluate how much convenience it provides.

Acknowledgments. This research was partially supported by TAF (The Telecommunications Advancement Foundation), 2007 fiscal year.

References

1. Baken, N., van Boven, E., den Hartog, F., Hekmat, R.: A Four-Tiered Hierarchy in a Converged Fixed-Mobile Architecture, Enabling Personal Networks. The Journal of the Communications Network 3, 98–104 (2004)
2. Louati, W., Zeghlache, D.: Network-Based Virtual Personal Overlay Networks Using Programmable Virtual Routers. IEEE Communications Magazine, 86–94 (August 2005)
3. Iida, I., Nishigaya, T., Murakami, K.: DUET: An Agent-based Personal Communications Network. IEEE Communications Magazine, 44–49 (November 1995)
4. Hirai, H., Mimura, N., Morikawa, H., Aoyama, T.: Realization of a Service-oriented Network Using a Host Grouping Framework. In: IEICE Gen. Conf. (2006) (in Japanese)
5. Yagi, K., Honda, O., Oosaki, H., Matsuda, K., Imase, M.: A Ring-based P2P-VPN Realizing the Cyber-Society, Technical report of IEICE (IN2005-12), pp. 61-66 (May 2005)

Conductive Inkjet-Printed Wireless Sensor Nodes on Flexible Low-Cost Paper-Based Substrates

Manos M. Tentzeris, L. Yang, A. Rida, R.Vyas, A. Traille, and C. Kruessi

GEDC/ECE, Georgia Institute of Technology, Atlanta, GA 30332, USA
etentze@ece.gatech.edu

Abstract. In this paper, inkjet-printed flexible antennas fabricated on paper substrates are introduced as a system-level solution for ultra-low-cost mass production of UHF Radio Frequency Identification (RFID) Tags and Wireless Sensor Nodes (WSN) in an approach that could be easily extended to other microwave and wireless applications. A compact inkjet-printed UHF "passive-RFID" antenna using the classic T-match approach and designed to match IC's complex impedance, is presented as a demonstrating prototype for this technology. In addition, the authors briefly touch up the state-of-the-art area of fully-integrated wireless sensor modules on paper and show the first ever 2D sensor integration with an RFID tag module on paper, as well as the possibility of a 3D multilayer paper-based RF/microwave structures.

Keywords: Inkjet-printed electronics, wireless sensor nodes, paper, UHF, RFID.

1 Introduction

RFID is an emerging compact wireless technology for the identification of objects, and is considered as an eminent candidate for the realization of a completely ubiquitous "ad-hoc" wireless networks. RFID utilizes electromagnetic waves for transmitting and receiving information stored in a tag or transponder to/from a reader. This technology has several benefits over the conventional ways of identification, such as higher read range, faster data transfer, the ability of RFID tags to be embedded within objects, no requirement of line of sight, and the ability to read a massive amount of tags simultaneously [1]. A listing of applications that currently use RFID are: retail supply chain, military supply chain, pharmaceutical tracking and management, access control, sensing and metering application, parcel and document tracking, automatic payment solutions, asset tracking, real time location systems (RTLS), automatic vehicle identification, and livestock or pet tracking.

The demand for flexible RFID tags has recently increased tremendously due to the requirements of automatic identification/tracking/monitoring in the various areas listed above. Compared with the lower frequency tags (LF and HF bands) already suffering from limited read range (1-2 feet), RFID tags in UHF band see the widest use due to their higher read range (over 10 feet) and higher data transfer rate [2]. The major challenges that could potentially hinder RFID practical implementation are: 1) Cost; in order for RFID technology to realize a completely ubiquitous network, the

S. Balandin et al. (Eds.): NEW2AN 2008, LNCS 5174, pp. 297–305, 2008.
© Springer-Verlag Berlin Heidelberg 2008

cost of the RFID tags have to be extremely inexpensive in order to be realized in mass production amounts 2) Reliability; and that extends to primarily the efficiency of the RFID tag antennas, readers, and the middleware deployed, 3) Regulatory Situation; meaning tags have to abide to a certain global regulatory set of requirements, such as the bandwidth allocations of the Gen2 Protocols defined by the EPC Global regulatory unit [3] and [4]) Environmentally-friendly materials, in order to allow for the easy disposal of a massive number (in the billions) of RFID's.

This article demonstrates for the first time how inkjet-printing of antennas/ matching networks on low-cost paper-based materials can tackle all four challenges enabling the easy implementation of ubiquitous RFID and WSN networks. It starts by discussing why paper should be used as a substrate for UHF/wireless inlays, how we can use conductive inkjet-printing technology for the fast fabrication of RF/wireless circuits, provides a design guideline for an inkjet-printed broadband antenna for UHF RFID tags which can be used globally, and eventually shows the capability of integrating sensors with RFID tags and discusses how added this functionality could revolutionize data fusion and real-time environmental cognition.

2 Paper and Inkjet-Printed Electronics

There are many aspects of paper that make it an excellent candidate for an extremely low-cost substrate for RFID and other RF applications. Paper; an organic-based substrate, is widely available;, the high demand and the mass production of paper make it the cheapest material ever made. From a manufacturing point of view, paper is well suited for reel-to-reel processing, as shown in Figure 1, thus mass fabricating RFID inlays on paper becomes more feasible. Paper also has low surface profile and, with appropriate coating, it is suitable for fast printing processes such as direct write methodologies instead of the traditional metal etching techniques. A fast process, like inkjet printing, can be used efficiently to print electronics on/in paper substrates. This also enables components such as: antennas, IC, memory, batteries and/or sensors to be easily embedded in/on paper modules. In addition, paper can be made hydrophobic as shown in Figure 2, and/or fire-retardant by adding certain textiles to it, which easily resolve any moisture absorbing issues that fiber-based materials such as paper suffer from [4]. Last, but not least, paper is one of the most environmentally-friendly

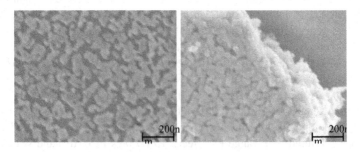

Fig. 1. SEM images of a layer of printed silver nano-particle ink, after a 15 minutes curing at 100°C (left) and 150°C (right)

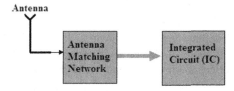

Fig. 2. Block diagram of a passive RFID tag

materials and the proposed approach could potentially set the foundation for the first generation of truly "green" RF electronics and modules.

However; due to the wide availability of different types of paper varying in density, coating, thickness, texture, and implicitly, dielectric properties: dielectric constant and dielectric loss tangent, dielectric RF characterization of paper substrates becomes an essential step before any RF "on-paper" designs. The electrical characterization of paper has already been performed [5] by the authors and results have shown the feasibility of the use of paper in the UHF and RF frequencies. Another note to mention here is that the low cost fabrication and even the assembly with PCB compatible processes can realize paper boards similar to printed wiring boards, that can support passives, wirings, RFID, sensors, and other components in a 3D multi-layer platform [5, 6].

Modern inkjet printers operate by propelling tiny droplets of liquid down to several pL [7]. This new technology of inkjet printing utilizing conductive paste may rapidly fabricate prototype circuits without iterations in photolithographic mask design or traditional etching techniques, that have been widely used in industry. Printing is completely controlled from the designer's computer and does not require a clean room environment. A droplet's volume determines the resolution of the printer, for e.g. a droplet of 10 pL gives ~ 25μm minimum thickness or gap size of printed traces/lines. The cartridge consists of a Piezo-driven jetting device with integrated reservoir and heater [7].

Inkjet Printing; unlike etching which is a subtractive method by removing unwanted metal from the substrate surface, jets the single ink droplet from the nozzle to the desired position, therefore, no waste is created, resulting in an economical fabrication solution. Silver nano-particle inks are usually selected in the inkjet-printing process to ensure a good metal conductivity. After the silver nano-particle droplet is driven through the nozzle, sintering process is found to be necessary to remove excess solvent and to remove material impurities from the depositions. Sintering process also provides the secondary benefit of increasing the bond of the deposition with the paper substrate [8]. The conductivity of the conductive ink varies from $0.4{\sim}2.5{\times}10^7$ Siemens/m depending on the curing temperature and duration time. Figure 5 shows the difference between heating temperature 100°C and 150°C after a 15 minutes curing. At lower temperature, larger gaps exist between the particles, resulting in a poor connection. When the temperature is increased, the particles begin to expand and gaps start to diminish. That guarantees a virtually continuous metal conductor, providing a good percolation channel for the conduction electrons to flow. To ensure the conductivity performance of microwave circuits, such as RFID modules, curing temperatures around 120oC and duration time of two hours were chosen in the following fabrication to sufficiently cure the nano-particle ink. Alternatively, much shorter UV heating approaches can achieve similar results.

The savings in fabrication/prototyping time that inkjet printing brings to RF/wireless circuits is very critical to the ever changing electronics market of today's verifying its feasibility as an excellent prototyping and mass-production technology for next generation electronics especially in RFID, wireless sensors, handheld wireless devices (e.g.4G/4.5G cell phones), flex circuits, and even in thin-film batteries.

The printing area is 122 mm × 193 mm. The text should be justified to occupy the full line width, so that the right margin is not ragged, with words hyphenated as appropriate. Please fill pages so that the length of the text is no less than 180 mm, if possible.

3 Benchmarking Prototypes

A major challenge in RFID antenna designs is the impedance matching of the antenna (Z_{ANT}) to that of the IC (Z_{IC}). For years, antennas have been designed primarily to match either 50 Ω or 75 Ω loads. However, RFID chips primarily exhibit complex input impedance, making matching extremely challenging.

It is to be noted that besides impedance matching, low cost, omnidirectional radiation pattern, long read range, wide bandwidth, flexibility, and miniaturized size are all important features that an RFID tag must acquire. Most available commercial RFID tags are passive due to cost and fabrication requirements. A purely passive RFID system utilizes the EM power transmitted by the reader antenna in order to power up the IC of the RFID tag and transmit back its information to the reader using the backscatter phenomena. A block diagram of a passive RFID tag is shown in the Figure 2 below. The antenna matching network must provide the maximum power delivered to the IC which is used to store the data that is transmitted to/received from the reader.

For a truly global operation of passive UHF RFID's, Gen2 protocols define different sets of frequency, power levels, numbers of channel and sideband spurious limits of the RFID readers signal, for different regions of operation (North America 902-928MHz, Europe 866-868MHz, Japan 950-956MHz, and China 840.25MHz-844.75 and 920.85MHz-924.75MHz). This places a demand for the design of RFID tags that operate at all those frequencies, thus requiring a miniaturized broadband UHF antenna. For instance, in a scenario where cargo/containers get imported/exported from different regions of the world in a secured RFID system implementation, an RFID tag is required to have a bandwidth wide enough to operate globally. This imposes very stringent design challenges on the antenna designers.

To achieve these design goals, while demonstrating the exceptional capabilities of the hereby reported paper-based inkjet-printed antennas technology, a T-match folded bow-tie half-wavelength dipole antenna was designed and fabricated on a commercial photo paper by the inkjet-printer mentioned above. The antenna used for this design was designed using Ansoft's HFSS 3-D EM solver. This design was used for the matching of the passive antenna terminals to the TI RI-UHF-Strap-08 IC with resistance R_{IC} = 380 Ohms and reactance modeled by a capacitor with value C_{IC}=2.8 pF. The RFID prototype structure is shown in Figure 3 along with dimensions, with the IC placed in the center of the T-match arms. The T-match arms are also responsible for the matching

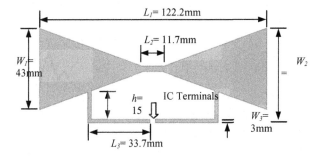

Fig. 3. T-match folded bow-tie RFID tag module configuration

of the impedance of the antenna terminals to that of the IC through the fine tuning of the length L3, height h, and width W3. A GS 1000µm pitch probe connected to a UHF balun to ensure the balanced signal between the arms of the T-match folded dipole antenna was used for impedance measurements. In order to minimize backside reflections of this type of antenna, the fabricated or inkjet printed antennas were placed on a custom-made probe station using high density polystyrene foam with low dielectric constant of value 1.06 resembling that of the free space. The calibration method used was short-open-load-thru (SOLT). Figure 4 shows the impedance plots. As shown in Figure 4(a), the simulated resistance for the antenna in the UHF RFID frequency range maintains a value close to 380 Ohms between the two successive peaks. The reactance part of the impedance, as shown in figure 4(b), features a positive value with a linear variation with frequency, pertaining to an inductance that conjugately matches or equivalently cancels the effect of the 2.8pF capacitance of the IC. Fairly good agreement was found between the simulation and measurement results. The distortion is possibly due to the effect of the metal probe fixture.

The return loss of this antenna was calculated based on the power reflection coefficient which takes into account the reactance part of the IC's impedance:

$$\left| s^2 \right| = \left| \frac{Z_{IC} - Z_{ANT}{}^*}{Z_{IC} - Z_{ANT}} \right|^2 \tag{1}$$

where Z_{IC} represents the impedance of the IC and Z_{ANT} represents the impedance of the antenna terminals with $Z_{ANT}{}^*$ being its conjugate. The Return Loss plot is shown in Figure 5 demonstrating a good agreement for both paper metallization approaches. The nature of the bow-tie shape of the half-wavelength dipole antenna body allows for a broadband operation, with a designed bandwidth of 190MHz corresponding to 22% around the center frequency 854 MHz which covers the universal UHF RFID bands. It has to be noted that the impedance value of the IC stated above was provided only for the UHF RFID frequency which extends from 850MHz to 960MHz; thus, the return loss outside this frequency region, shown in Figure 5, may vary significantly due to potential IC impedance variations with frequency.

In order to verify the performance of the ink-jet printed RFID antenna, measurements were performed on a copper-metalized antenna prototype with the same dimensions fabricated on the same paper substrate using the slow etching technique mentioned

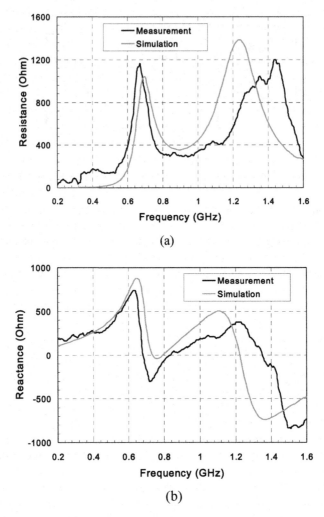

Fig. 4. Measured and simulated input resistance and reactance of the inkjet-printed RFID tag, (a) Resistance, (b) Reactance

before. The return loss results are included in Figure 5 and they show that the return loss of the inkjet-printed antenna is very slightly larger than the copper one. Overall a good agreement between the copper etched and the inkjet-printed antennas was observed despite the higher metal loss of the silver-based conductive ink.

The radiation pattern was measured by using the NIST Calibrated SH8000 Horn Antenna as a calibration kit for the measured radiation pattern at 915 MHz. The radiation pattern as shown in Figure 6 is almost uniform (omnidirectional) at 915 MHz with directivity around 2.1dBi. The IC strap was attached to the IC terminal with H2OE Epo-Tek silver conductive epoxy cured at 80°C. An UHF RFID reader was used to detect the reading distance at different directions to the tag. These measured distances are theoretically proportional to the actual radiation pattern. The

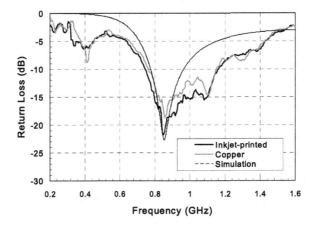

Fig. 5. Measured and simulated Return Loss of the inkjet-printed RFID tag

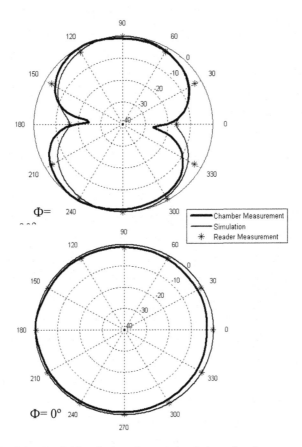

Fig. 6. Normalized 2D far-field radiation plots of simulation, chamber measurement and tag reading distance measurement

normalized radiation patterns of simulation, microwave chamber measurement and reader measurement are plotted in Figure 6, showing a very good agreement between simulations and measurements, which can be also verified for other frequencies within the antenna bandwidth.

4   RFID-Sensor Integration on Paper

In addition to the basic RFID automatic Identification capabilities along with the technologies and designs discussed above, the authors have demonstrated the capabilities of inkjet-printing technology in integrated wireless sensors on paper bridging RFID and sensing technology. The aim is to create a system that is capable of not only tracking, but also monitoring. With this real-time cognition of the status of a certain object will be made possible by a simple function of a sensor integrated in the RFID tag. The ultimate goal is to create a secured "intelligent network of RFID-enabled sensors." For this effort, the authors have developed the FIRST Sensor-enabled RFID on-paper ,that uses Gen2 protocols as means of communication on paper substrate, as shown in Figure 7. This module is easy to be extended to a 3D multilayer paper-on-paper RFID/Sensor module by laminating a number of photo-paper sheets (150um thickness/sheet). This is expected to drop the cost of the sensor nodes significantly and eventually made the "ubiquitous computing network" a possible reality with a convergent ability to communicate, sense, and even process information.

There are many aspects of paper that make it an excellent candidate for an extremely low-cost substrate for RFID and other RF applications. Paper; an organic-based substrate, is widely available; the high demand and the mass production of paper.

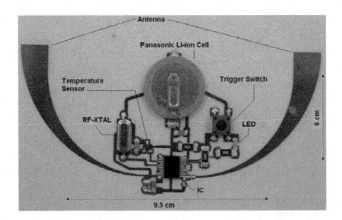

Fig. 7. Wireless Sensor transmitter prototype on paper substrate using silver inkjet printing technology

5   Conclusions

Paper which holds one of the biggest market share in the world can potentially revolutionize the electronics market and eventually take the first step in creating an environmentally friendly first generation of truly "green" RF electronics and modules. Not to mention that paper is one of the lowest cost materials produced. The inkjet printing technology, which is a much faster and cleaner method than the conventional wet etching techniques, which uses several "etchant"chemicals, can serve as low-cost mass-deployment technology for fabricating RFID tags to be produced in large reel to reel processes on paper. It can be easily verified that direct-write technologies such as inkjet printing, which consists of depositing nano-silver-particles, can be a very critical technology for the quick developing of next-generation flexible "cognitive" electronics due to its speed and ease of prototyping process. A "Global" UHF passive RFID antenna using the classic T-match approach has been presented as the benchmarking prototype for this novel technology featuring an excellent performance for a wide ("universal") frequency range. Last but not least, the authors have reported the " first ever 2D integrated-sensor with UHF RFID capabilities completely on paper" , that could potentially set the foundation for the truly convergent wireless sensor ad-hoc networks of the future.

References

1. Finkenzeller, K.: RFID Handbook, 2nd edn. Wiley, Chichester (2004)
2. Basat, S., Bhattacharya, S., Rida, A., Johnston, S., Yang, L., Tentzeris, M.M., Laskar, J.: Fabrication and Assembly of a Novel High-Efficience UHF RFID Tag on Flexible LCP Substrate. In: Proc. of the 56th IEEE-ECTC Symposium, pp.1352–1355 (May 2006)
3. UHF Gen-2 System Overview. Texas Instruments, HTTP (September 2005), http://rfidusa.com/superstore/pdf/UHF_System_Overview.pdf
4. Lessard, M., Nifterik, L., Masse, M., Penneau, J., Grob, R.: Thermal aging study of insulating papers used in power transformers. In: IEEE Annual Report of the Conference on Electrical Insulation and Dielectric Phenomena, vol. 2, pp. 854–859 (1996)
5. Yang, L., Rida, A., Vyas, R., Tentzeris, M.M.: RFID Tag and RF Structures on a Paper Substrate Using Inkjet-Printing Technology. Microwave Theory and Techniques, IEEE Transactions 55(12), Part 2, 2894–2901 (2007)
6. Rida, A., Yang, L., Vyas, R., Basat, S., Bhattacharya, S., Tentzeris, M.M.: Novel Manufacturing Processes for Ultra-Low-Cost Paper-Based RFID Tags With Enhanced Wireless Intelligence. In: Proc. Of the 57th IEEE-ECTC Symposium, Sparks, NV, pp. 773–776 (June 2007)
7. Pique, A., Chrisey, D.B.: Direct-write Technologies for Rapid Prototyping Applications. Academic Press, London (2002); International Standard Book Number: 0-12-174231
8. Carter, M., Colvin, J., Sears, J.: Characterization of conductive inks deposited with maskless mesoscale material deposition, TMS 2006, March 12-16, San. Antonio, Texas, USA (2006)

Smart Sensing and Sensor Data Collection on the Move for Modelling Intelligent Environments

Prem Prakash Jayaraman[1], Arkady Zaslavsky[1], and Jerker Delsing[2]

[1] Caulfield School of IT, Monash University, Melbourne, Victoria 3145
[2] Lulea University of Technology, Lulea, Sweden
{prem.jayaraman,arkady.zaslavsky}@infotech.monash.edu.au,
jerker.delsing@infotech.monash.edu.au

Abstract. With advent of pervasive computing and considerable acceptance of sensor networks, smart sensing techniques and data collection have been topics of interest. This paper presents a smart sensing and data collection technique from sensor networks using context aware high powered mobile objects within the environment. The paper proposes CAM-*R* a context aware robot that can move within smart environments sensing new sensor sources and collecting sensory originated data efficiently. Based on these sensed data sources, we propose an extension to context spaces model that builds a virtual model of the intelligent environment. This intelligent environment model built using extended context spaces can be used by number of context aware applications to efficiently query and retrieve data from the sensor network using CAM-*R* based data collection approach. We also present a prototype implementation of CAM-*R* built using off-the-shelf hardware and a context based cost function used to compute data collection decisions. We validate our system by implementing the virtual modelling of the intelligent environment based on simulated input obtained from CAM-*R* and sensors. We also evaluate CAM-*R* by simulating and comparing the energy spent by the sensor nodes during data collection process using our proposed approach and traditional fixed sink based approach.

1 Introduction

Technological advances in processing, sensing, communication and storage have rendered the development of tiny, powerful sensing devices with communication capabilities namely wireless sensor networks. These networked sensors open up a wide range of opportunity to build intelligent environments based on the physical world surrounding us. In future, these networked sensors play a pivotal role in interacting between this physical and virtual world. Wireless sensor network have application spread across number of fields hence finding themselves deployed in homes, highways, building, cities and other monitoring and controlling application [1, 3]. Sensor networks enable acquisition of data which previously were expensive, difficult or even impossible [2]. With the wide acceptance of sensor networks, it is likely that there would be an exponential increase in sensor network deployments [4]. The exponential increase in deployments will also result in exponential increase in the amount of data generated by

S. Balandin et al. (Eds.): NEW2AN 2008, LNCS 5174, pp. 306–317, 2008.
© Springer-Verlag Berlin Heidelberg 2008

these smart sensing devices. The inherent characteristics of sensor nodes depict them as revolutionary data sources for a number of applications. Hence arises, the challenge of efficiently collecting data from these sensor networks based data sources [2]. One of the primary sources of context information is sensors in the environment. For this context information to be available to context aware applications, the application needs to be aware of these sensor based information sources which may not be available before. Also there exists the challenge of efficiently collecting and delivering sensor data to these applications.

In this paper we propose the use of Context Aware Mobile Robot (CAM-R) to efficiently discover and collect sensor data. Our proposed approach of building a virtual environment is an extension to the Context Spaces theory [10, 11], incorporating dynamic discovery of context that aids while reasoning under uncertainty. This dynamic discovery of context is the process of creating / modifying the intelligent environmental model to incorporate newly discovered sensory sources within the environment.

The rest of the paper is organised as follows. Section 2 contains related works. Section 3 is divided into two sub sections one proposing and describing the architecture of CAM-R and the other proposing extension to context spaces used to model the intelligent environment. Section 4 elaborates on proposed context spaces extensions and CAM-R implementation and evaluation. Finally Section 5 concludes the paper.

2 Related Work

Lot of focus and work has gone into the area of data collection in sensor network. Direct Diffusion [5] is the general purpose data collection approach that disseminates queries in a sensor network using direct or multi hop communication. Our approach looks into using intelligent heterogenous mobile objects as data collectors for sensor data. Heterogenous here represents the class of high powered mobile devices that can act as data collectors. Since our approach looks at using a mobility layer, our literature focuses mostly on mobile based data collection techniques in sensor networks. We can classify these approaches as random mobility approach [9] predictable mobility approach [7] and controlled mobility approach [6, 8].

Shah et.al. and Sushant et al. [9] propose data mule, a three layer architecture comprising mobile elements, sensor nodes and data sinks. This approach uses random mobility and focuses on modeling a random walk based data collector and sensor data buffering issues. The types of sensors used in this approach are homogeneous (identical). Mule needs to have specialised receivers and transmitter to collect the data and need knowledge of sensor locations to collect data. The system does not look at context aware approaches to extend the sensor network lifetime and does not deal with inter mule communication which we believe can help these intelligent data collectors to collect sensor data more efficiently. Kansal et. al. [8] uses controlled mobility to collect data from sensor networks. They propose the use of specially designed robot that travels along a specified path colleting sensor data. The robot used in this implementation is designed specifically to collect sensor data in rough terrains. But we argue that in future pervasive environments, any high powered mobile object will be a part of the environment and hence does not require introducing new infrastructure (specially designed robots). Also Packbot is not context aware and hence cannot harvest on the benefits obtained from context aware decisions e.g. dynamically changing the wake/sleep time of the sensor node based on collected data over a period of time. The next section of our literature focuses on context modelling approaches.

Context models based on Logic and ontology have been presented in [14, 15]. Logic based models represent context in a formalized way but heavily depend on application specific rules. This makes them unsuitable for changing applications and environments. The Starthclyde Context Infrastructure [16] focuses on developing a scalable and adaptable infrastructure. The above discussed models are application specific and do not perform reasoning under uncertainty. Padovitz et. al. [10, 11] presents Context Spaces representing context using a spatial metaphor representing context in a multi-dimensional space. Each attribute defined within a situation's context will take a specific dimension to the context model. Padovitz et. al. [10, 11] has proposed context algebra used to perform complex operations over situations defined as situation spaces. Context spaces theory uses probabilistic reasoning approach to reasons a specific situation hence arriving at a probabilistic confidence value. This confidence values increases or decreases the occurrence of a specific situation. Context spaces works on known attributes that correspond to a specific situation.

The key contributions of this paper are 1) A proposed extension to context spaces model used to build a intelligent virtual environment comprising sensor sources discovered by smart sensing 2) The proposal of a context aware robot based sensor data collector (CAM-R) for smart sensing and efficiently collecting sensory data 3) The proposal of a novel approach of using mobile objects to build an intelligent environment over which context aware applications can perform better reasoning under uncertainty.

3   CAM-R and Dynamic Context Discover in Context Space

This section presents our proposed approach. The section is divided into two parts. The first section proposes the CAM-R architecture providing insights into the key components of the system framework. The second section describes our proposed extension to context spaces and how we use it to build a virtual model of the intelligent environment based on sensor inputs.

3.1   CAM-R System Architecture

CAM-R is our mobile robot platform based data collector that smartly senses sensor sources within the infrastructure. The context spaces extension proposed build an intelligent environment based on newly discovered context information obtained from CAM-R. These sensor inputs can be from a variety of heterogeneous sensors that are both mobile and static with different sensing capabilities. The system architecture of CAM-R is shown in Figure 1. CAM-R architecture is modular consisting of three key modules namely robot module that takes care of robots movement, data collection module that takes care of collecting sensor data and managing sensor nodes and the context and localization module that is responsible for obtaining location and context information required for robot navigation.

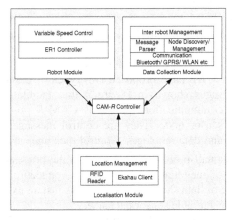

Fig. 1. CAM-R System framework

The Robot Module controls the robot hardware obtaining information from on board IR sensors that helps the robot to avoid obstacles. For our proof of concept implementation, we use an of-the-shelf built robot ER1 [13]. The ER1 has a controller interface that is used to control the robot movement. The robot control module also has a variable speed control algorithm implementation that facilitates adapting robot's movement speed based on context. We use two simple techniques namely "Wait and Go" and "Slow and Proceed". The objective of both the techniques is to facilitate sensor data collection using a single connection avoiding energy spent in re-connection and re-transmission. Based on available context information like robots itinerary, time to reach destination, sensor's wake/ sleep interval, last data collection time, etc., CAM-R uses either of the technique while collecting sensor data. Figure 2 presents the variable speed control algorithm implemented in CAM-R.

The activation schedule mentioned in the pseudo code is a sleep/wake itinerary that controls the sensor nodes sleep/wake interval. Section 4 elaborates the functions of the activation schedule. The data collection module is responsible for collecting sensor data, managing sensor locations, sensor nodes activation schedules and delivering the data to the sink. The sink in our proposal is a high level context server that uses context spaces extension to model the intelligent virtual environment. The inter robot management which is a part of data collection is used to manage multiple CAM-R in the environment. This module facilitates robot communication using simple message passing technique that allows robots to exchange sensor information (location, activation schedules etc). This allows robots to be in sync with sensor node location and does not require a central server / sink to synchronize them. The node discovery and management takes care of maintaining sensor node list. As our system looks at developing an intelligent model of the physical environment, we consider sensors whose location are pre-known and dynamically discovered sensors. CAM-R adds this

```
if (resources available for variable speed)
then
    obtain existing speed
    get existing itinerary
    estimate arrival time (to destination)
    start time Counter thread
            <exit operation and proceed to des-
            tination on Counter thread signal>
    while elapsed time < T (allowed time)
            <Counter thread – interrupt>
        get curr-location
        get node-list where node-list-loc =
        curr-location
    for each node-list
            get activation-schedule
            compute amount of data D = (T
            (this collect – last collect) *
            sampling rate)
            Compute "Wait and Go" or
            "Slow and Proceed"
            Collect data
    end for
    end while
end if
```

Fig. 2. Variable speed control algorithm

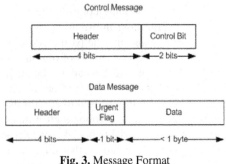

Fig. 3. Message Format

ability to discover sensors dynamically (smart sensing). The message parser component parses the data sent / received to and from the sensor node.

We have proposed two types of messages shown in Figure 3 that facilitate communication between CAM-*R* and sensor node. They are control messages and data messages. Control messages are small 6 bit messages used in the process of negotiating / initiating connection. The data message is the actual data. The size of data shown in the figure is just an example. The system is capable of handling considerable sensor data sizes.

The Localization module is responsible to provide CAM-*R* with location information. It involves obtaining existing location, path planning and navigating the robot to the destination using the shortest route. While planning to navigate from location A to B, the system also considers the list of active sensor nodes in path that can off-load data (computed using context and sensor node information). Hence the best route is not just chosen based on shortest path but also considers amount of data that can be collected while the path is traversed. Figure 4 gives the pseudo code of the context aware path planning algorithm implemented in CAM-*R*. Our proposed framework is under the assumption that CAM-*R* has a map of the physical environment.

```
Obtain itinerary
Compute time by shortest path arrival
If (Time (short path) < Time at Destination)
   //Re compute route by taking into consid-
   eration nodes in the surrounding and time
   to collect the data
      Node-list = list of nodes in the path
      For each node-list
         Obtain activation-schedule
         If (activeTime is between (start time,
         Time at Destination)
            Add node to collectnodelist
         End if
      End for
      For each collectnodelist
         Compute time at node to collect data
         (approximate)
         Plan path such that
         Time (navigate path) + Time (col-
         lect) < = Time at Destination
      End for
      Compute new itinerary with navigation
      map
End if
```

Fig. 4. Algorithm for Localized Path Planning

3.2 Context Spaces Extension to Build a Virtual Model of the Environment

The previous section described the CAM-*R* architecture which is used to discover and collect sensor data efficiently. In this section we will present the context spaces extension incorporated to build a virtual model of the environment. Figure 5 gives an overview of our proposed system. Context spaces that we are extending in this paper to build the virtual intelligent environment present an approach for representing context using geometrical metaphors, describing context and related contextual objects in multi-dimensional space. Context spaces works on the concept of using existing context information within the system to reason situations under uncertainty.

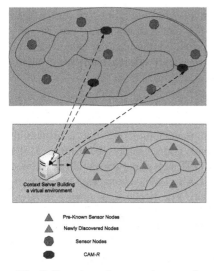

Pre-Known Sensor Nodes
Newly Discovered Nodes
Sensor Nodes
CAM-R

Fig. 5. Overview of proposed approach

We argue that, in most cases this may not be enough as additional context information would help increase the systems reasoning capabilities under uncertainty. The proposed extension built on context spaces theory adds new dimensions to the existing context model which is used to build an intelligent virtual environment. This virtual environment is not a part of context spaces and we propose this concept as our extension to context spaces in this paper. Using this newly built intelligent environmental model, the system can reason situations with more confidence. These additional discovered context attributes presented in this paper are primarily sensor data but not confined to that. The intelligent virtual model of the physical environment can be

extended / re-built based on discovered contextual objects within the infrastructure. This paper contributes the novel idea of building this intelligent environment built using sensor data sources discovered by CAM-*R* smart sensing approach. Figure 6 is an example of how our proposed context spaces extension approach builds a virtual model of the environment. The newly discovered contextual attributes adds new dimension to context spaces existing known attributes. In figure 6, C1, C2, C3 are predefined attributes existing in the virtual model while C4, C5, C6 are newly discovered attributes, hence re- building the virtual model dynamically.

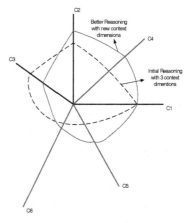

Fig. 6. Extended Context spaces modeled on the dynamically built Virtual Space

4 Implementation and Evaluation

In this section, we present the prototype implementation of our proposed system CAM-*R* which is built using off-the-shelf hardware and virtual intelligent environment modeling using proposed context spaces extension. Our proposal is oriented towards future pervasive environments and since context differs from one environment to another, our prototype system presented in this paper narrows to a specific scenario. This scenario specific context assumption does not restrict the framework to particular environments. The rest of the discussion presented are specific to a pervasive building environment where we have sensor sources at various locations and

CAM-*R* as a part of the pervasive environment involved in functions like surveillance or guide robot. The primary objective of CAM-*R* is not data collection but, as mentioned before, we look at harvesting existing mobility, communication and processing capacity within pervasive environments to achieve smart sensing facilitating intelligent environment modeling.

4.1 Mulle

Table 1. Mulle Sensor node's power consumption

Mode	Power
MCU 10.0 MHz, BT off	25.1 mW
MCU sleep, BT active	132.9 mW
MCU sleep, BT sniff (210 slots)	27.8 mW
MCU on, BT on, Tx/Rx	160mW

For the implementation we use a Bluetooth based sensor node namely the Mulle. Mulle [12] is a COST sensor network node developed at the EISLAB, Luleå University of Technology, Sweden. The Mulle is a generic wireless sensor node that uses a Bluetooth radio for communication. It works on a 3.3V powered by a lithium or lithium ion battery ranging from 120 mAh to 2200 mAh. The Mulle can work with variety of sensing devices that can be connected to the expansion port. The Mulle that we have used has a temperature sensor on-board. Table 1 shows the power consumptions values of the Mulle.

4.2 Activation Schedule

The activation schedule [12] controls the Mulle's operational modes. The Mulle operates in three modes: Active, Passive and Time Synchronous. In Active mode, the Mulle listens for incoming connection and constantly advertises itself. In Passive mode, the Mulle goes into low power mode and does not listen nor accept any connections. In Time Synchronous mode, the Mulle moves from the pass passive to the active schedule at defined intervals. This defined interval is specified by the activation schedule. For our implementation we use the time synchronous operational mode of the Mulle with activation schedule specifying when it has to change modes.

This allows us to achieve great flexibility by changing the Mulle sleep/wake time based on context hence achieving better network lifetime. We also assume that the Mulle sensor node can vary its Bluetooth output power based on request from CAM-*R*. Assume that initial Mulle monitoring schedule is x times in y units of time, we then have a duty cycle window $w = x/y$. The average of monitored vales is given by avg. From this information, we compute the efficient window interval times (duty cycle) hence resulting in lesser energy spent by reducing window intervals (duty cycle).

4.3 Robot, Localization Hardware and Software

The robot used for implementation is a commercially available off-the-shelf build robot ER1 developed by Evolution Robotics [13]. ER1 is provided with a set of programmable API's that provides an interface to control the motors, web-cam and the IR-sensors equipped on the robot. ER1 is equipped with 3 infra-red based sensors and a webcam used for obstacles avoidance and object recognition. The other deployed part of our prototype is the localization hardware and software used for CAM-*R* navigation. We use Ekahau positioning engine and RFID tags for robot navigation and localization. We assume the existence of RFID tags within future pervasive environment providing location information. The Ekahau positioning engine used in our prototype is software based system that provides location information using triangulation. It triangulates location information of the robot by measuring signal strengths between access point and WiFi card equipped on the robot. Ekahau with our test setup has an error of up to 1 meter which can make a significant impact in the localization process. Hence we use the idea of RFID tags deployed at various locations providing location information (X, Y coordinates) that gets mapped to Ekahau's location coordinates to minimize error during localization process.

4.4 Implementation Scenario and Assumptions

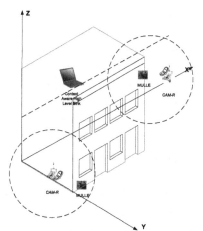

Fig. 7. Building Scenario

As mentioned, to represent context with some valid assumptions and model an intelligent environment based on sensor sources, we chose a building scenario represented in Figure7 where the CAM-*R* in a futuristic pervasive environment is a part of the environment. The building has wireless sensor nodes at specific location (e.g. rooms, corridors, etc) monitoring various parameters. The system also maps mobile sensor nodes that appear and disappear within the physical environment to the intelligent virtual environment. An example application of our proposed approach is a query processing request within a wireless sensor network. Traditional methods use direct diffusion [5] to disseminate the query and use the same to obtain the result. In such an approach, the sink disseminates the request and sensors nodes forward the request until the point of interest is reached. This requires nodes to be active when a broadcast happens and results in most nodes spending energy to process and forward the request. With our novel approach, using the intelligent environmental model the system can identify context sources and use CAM-*R* to efficiently collect data from the sensor source.

4.5 Prototype Implementation

The previous section laid a foundation on the various hardware, software and implementation scenario of our proposed system. For the prototype system, the Mulle is preloaded with the activation schedule which can also be updated by CAM-*R* dynamically. The sensor nodes are named EISMULLE1 to n. We use Bluetooth serial port profile for data exchange and the Mulle uses a maximum speed of 56.7 kbps. The reason for using the serial port profile is due to the simplicity of setting up and using the RFCOMM between CAM-*R* and the Mulle. The RFID tags implemented provide

X, Y coordinates which are compared with X, Y coordinates obtained from Ekahau positioning engine. We assume that the environment map is pre-loaded into the system and is available to CAM-*R*. CAM-*R* performs path planning based on existing location and the location of the sensor node from which data needs to be collected. The paper does not focus on robot path planning as it is research topic by itself and is out of scope of the paper. The CAM-*R* module implementations are done using Java. The system implemented works in two stages. The initial stage is the pre-collection stage where CAM-*R* senses nodes in the environment that are not already part of the model. In stage two based on newly discovered nodes and their characteristics, the systems builds a virtual model of the environment using proposed context spaces extension. We consider wireless sensor nodes that know their locations and nodes that are unaware of their location. For nodes that are not aware of their locations, we assume that CAM-*R* is equipped with range based techniques to identify sensor node location approximately. The negotiation between CAM-*R* and Mulle uses the message formats proposed in section 3. The process of building the intelligent environment is done by the context server (running extended context spaces). The communication available on the ER1 depends on the connectivity available on the laptop that controls ER1. We use IBM T42 laptop as the controller which is equipped with Bluetooth and 802.11 WLAN connectivity. The data received from the Mulle are time stamped and are stored on CAM-*R* and the context server. Both CAM-*R* and the context server have the context information to modify the Mulle's activation schedule based on collected data.

4.6   Simulation and Evaluation

The context server that builds the intelligent environment based on sensor sources input received from CAM-*R* has been developed using Visual Studio 2005. We evaluate the system by presenting implementation screenshots of the context server that builds a virtual model of the environment using context sources discovered by CAM-*R*. To simulate an environment model with more sensor nodes, the discovery process of CAM-*R* and sensors in the environment are simulated. The system uses this simulated input to construct the intelligent environment. Figure 8 illustrates the result of the simulation. As you can see, as new nodes are discovered, the context spaces extension pr.posed adds the new context attribute as a new dimension and updates their corresponding confidence and weights [10,11]. To evaluate the proposed CAM-*R* efficient data collection, we simulate data collection using traditional sink based approach and our proposed approach. Our evaluation parameter is the energy spent by the Mulle sensor to offload data either to sink using direct communication or to CAM-*R* using our proposed technique.

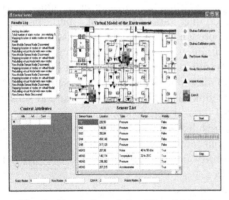

Fig. 8. Simulation of the Virtual Model Building using CAM-R

We evaluate the total energy spent by the Mulle using our context aware approach using a cumulative cost function illustrated below.

$$T_{Energy} = E_f(Send\,\mathrm{Re}\,cv(X_{bits}),d) + E_{con} * N_R + E_{cpu}$$
$$+ C_f(Activation_S, Power_{mode}) * w_i$$

Here C_f is a context function that adapts the sensor nodes activation schedule and power modes based on context. This function is associated with a weight factor which increases or decreases the significance of the context function while computing the cost under any given situation. In the cost function, we do not consider the cost of moving CAM-R as an additional cost as we are looking at a pervasive environment with such high power mobile objects. The energy function takes into consideration, signal strength, bit error rate, attenuation to compute distance from the sensor node. These attributes are derived from system context depending on situation and environment. The context function alters both the activation schedule and also the Bluetooth's power mode. We assume that the Mulle can change its Bluetooth power mode from Class A to Class B which allows the radio to run from 100mW to 2.5 mW at the same time reducing its communication range. Each simulation run simulates a 20 to 30 minutes window (data cycle) after which the sensor wakes and changes state to listen or communicate state. The traditional approach is simulated as a fixed sink approach with direct communication over Bluetooth with no modification in wake/ sleep schedule or Bluetooth radio power. The simulation using CAM-R approach presents results based on activation schedule and Bluetooth power adaptation based on context. The simulation result is obtained from three temperature sensor node from which CAM-R collects data. We do not take into consideration the energy consumed by the on board sensors since sensors like GPS require more power than simple temperature based sensors. Figure 9 illustrates the activation schedule adaptation used in the simulation and Figure 10 presents the energy consumption graph. The result of simulation is encouraging showing the amount of energy

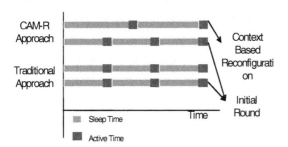

Fig. 9. Timeline chart of Modified Sleep/Wake Schedule

that can be saved on each of the node using our proposed CAM-R approach. As we can see that with initial simulation run, the power consumed are almost identical until our context aware approach alters the activation schedule based on context. This results in significant energy saving by allowing the sensor node to stay in sleep mode for duration longer than the traditional approach. From this energy consumptions results, we also conclude that our proposed extension to context spaces to model an intelligent environment with CAM-R results in greater efficient query processing within sensor network than traditional approach both in terms of time and cost.

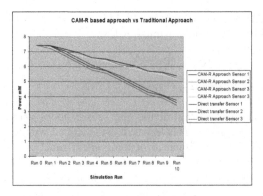

Fig. 10. Power outputs obtained from Simulation using Traditional and CAM-R approach

5 Conclusion

The paper has proposed the novel idea of using mobile objects to model an intelligent environment. We propose CAM-*R* a context aware intelligent robot based data collector that is uses smart sensing approach to identify sensor sources and efficiently collect data from these sources. The paper has also proposed and incorporated an extension to context spaces theory used to build the intelligent environment with sensory data sources obtained from CAM-*R*. The concept of building a virtual model of the environment is useful for a number of context aware applications whose reasoning processes can be vastly improved by the addition of newly discovered sensor/ context data sources. Our approach extends context spaces theory by building/updating existing multi dimensional context model adding newly discovered context attributes. The virtual model of the intelligent environment built on static, mobile and dynamically existing sensors node sources is a solution for future pervasive environments where identifying and discovering context is a major challenge. This model can help applications identify both type of context and its location by querying the intelligent environmental model and hence increasing their reasoning capabilities.

We have also presented a prototype implementation of CAM-*R* built using off-the-shelf hardware. The Mulle sensor node used for the implementation enables us to extend this concept to any high powered mobile object due to the widely accepted Bluetooth radio medium. CAM-*R* presented in this paper is one such example of a high powered mobile object. The paper has presented the implementation of the context server built on top of context spaces implemented in using Vb.Net. The simulation outputs have presented the intelligent environment model built from simulated inputs of sensory sources. The paper has also proposed a context based cost function used to evaluate the CAM-*R* approach against traditional fixed sink based approach. The results show great benefits in terms of extended network and sensor node lifetime achieved by using context for data collection. Our future work in this area includes the development of an intelligent environment model sharing algorithm that can be used by mobile objects to exchange context information sources extending our fixed scenario approach to a self learning approach to adapt to new environments.

References

[1] Gharavi, H., Kumar, S.P.: Special issue on sensor networks and applications. Proceedings of the IEEE 91(8), 1151–1153 (2003)

[2] Chu, D., Deshpande, A., Hellerstein, J.M., Hong, W.: Approximate Data Collection in Sensor Networks using Probabilistic Models. In: Proceedings of the 22nd international Conference on Data Engineering ICDE, April 3-7, 2006, IEEE Computer Society, Washington (2006)

[3] Xu, N.: A Survey on Sensor Network Applications, `http://courses.cs.tamu.edu/rabi/cpsc617/resources/sensor%20nw-survey.pdf`

[4] Jurdak, R., Lopes, C., Baldi, P.: A Framework for Modeling Sensor Networks. In: Workshop on Building Software for Pervasive Computing OOPSLA 2004 (2004)

[5] Intanagonwiwat, C., Govindan, R., Estrin, D.: Directed diffusion: A scalable and robust communication paradigm for sensor networks. In: Proceedings of ACM MOBICOM, Boston, MA (August 2000)

[6] David, D.J., Somasundara, A., Mani, B.S.: Multiple Controlled Mobile Elements (Data Mules) for Data Collection in Sensor Networks. In: International Conference on Distributed Computing in Sensor Systems (2005)

[7] Chakrabarti, A., Sabharwal, A., Aazhang, B.: Using Predictable Observer Mobility for Power Efficient Design of Sensor Networks. In: Information Processing in Sensor Networks (2003)

[8] Kansal, A., Somasundara, A.A., David, D.J., Deborah, E., Mani, B.S.: Controllably Mobile Infrastructure for Low Energy Embedded Networks. Mobile Computing, IEEE Transactions 5, 958–973 (2006)

[9] Sushant, J., Rahul, C.S., Waylon, B., Gaetano, B., Sumit, R.: Exploiting mobility for energy efficient data collection in wireless sensor networks. Mob. Netw. Appl. 11, 327–339 (2006)

[10] Padovitz, A., Loke, S.W., Zaslavsky, A.: Towards a Theory of Context Spaces. In: Das, C., Kumar, M. (eds.) Proceedings of the Second IEEE Annual Conference on Pervasive Computing and Communications Workshop, Orlando, Florida, March 14-17, pp. 38–42. IEEE Computer Society, USA (2004)

[11] Padovitz, A., Loke, S.W., Zaslavsky, A., Burg, B.: Verification of Uncertain Context Based on a Theory of Context Spaces. International Journal of Pervasive Computing and Communications (JPCC), Troubador Publishing (accepted for publication, 2006)

[12] Johansson, J., Völker, M., Östmark, A., Lindgren, P., Delsing, J.: Mulle: A minimal sensor networking device - implementation and manufacturing challenges. In: IMAPS Nordic 2004, pp. 265–271 (2004)

[13] Evolution Robotics, ERSP-NC 3.0, `http://www.evolution.com/education/erspnc30/hardware.masn`

[14] Ranganathan, A., Campbell, R.H.: A Middleware for Context-Aware Agents in Ubiquitous Computing Environments. In: ACM/IFIP/USENIX International Middleware Conference, Rio de Janeiro, Brazil, pp. 16–20 (June 2003)

[15] Chen, H., Finin, T., Anupam, J.: An Intelligent Broker for Context-Aware Systems, In: Proc. of Fifth International Conference on Ubiquitous Computing (Ubicomp 2003) (October 2003)

[16] Glassey, R., Stevenson, G., Richmond, M., Nixon, P., Terzis, S., Wang, F., Ferguson, I.: Towards a Middleware for Generalised Context Management. In: First International Workshop for Middleware for Pervasive and Ad Hoc Computing. In: Middleware 2003 Companion, pp. 45–52, Rio De Janeiro, Brazil (June 2003)

Ubi-Board: A Smart Information Diffusion System

Michel Banâtre, Mathieu Becus, and Paul Couderc

INRIA Rennes / IRISA
http://www.irisa.fr/aces

Abstract. In this paper, we present the design and implementation of Ubi-Board, a context sensitive and multimodal information diffusion system. Typically, Ubi-Board consists of smart displays with the ability to sense the profile of nearby users, to adapt the information content so as it is understandable and accessible to the people, and to dynamically distribute the information to the appropriate devices, including users mobile devices when necessary.

1 Introduction

Providing pertinent and accessible information to people in a complex or unknown environment, such as airports, has always been a challenge. In particular, people may speak different languages, have a disability, and will be interested by different information depending on the context. The current strategy to address these issues when designing on-site information systems is usually to provide *variants*, such as versions of different languages, or versions for different profiles (children, adult...). These variants are either presented to the public all at once (for example on a single board or a set of multiple boards), or in alternance, with a time-division scheme on a screen. Time-division is the only possible way for some media, such as public audio messages.

The common problem with this method is that the number of variants can quickly become unpractical: consider for example a museum with support for 4 languages and specific content for children; this means eight possible variants per description. As a general rule, fewer variants means reducing the accessibility for part of the public, while more variants tends to flood pertinent information for a given target, hence reducing the relevance for everyone.

Explicit selection of the information by the visitor, using a touch panel for example, offers a solution but also has limitation. First, it assumes that the service machine will be noticed, and that the selection interface is accessible to the user, which may not be always the case. For example, a blind visitor will likely not be aware of the availability of the information. Also, some people can be intimidated by using an electronic system, or simply a system which they are not familiar with. Second, this approach is limited to provide information requested by visitors, not unsolicited information, such as a notification message of a canceled flight for airport passengers for example.

S. Balandin et al. (Eds.): NEW2AN 2008, LNCS 5174, pp. 318–329, 2008.

A more advanced solution for this problem was introduced from ubiquitous computing. The general idea is to have a sensing infrastructure on the site in order to detect visitors, identify them, and provide them with pertinent information. However, implementing this idea into an effective and practical service is not so simple. To begin with, the very idea of monitoring the environment to identify people raises a strong privacy concern. Next, assuming that people profiles are known, providing information in an open space with many people requires a strategy in order to avoid information flooding while providing convenient presentation. The purpose of this paper is to present our experience with the design and experimentation of such a system: Ubi-Board. The rest of the paper is organized as follows: the second section presents the Ubi-Board service and a typical scenario, the third section examines the system architecture and the implementation issues, the fourth section presents a performance evaluation, and the fifth section discuss related work.

2 The Ubi-Board Service

Ubi-Board is a context-sensitive information diffusion system with multi-modal capability. The key feature of the system is to address both groups and individuals, and to distribute the information provisioning process on both collective and personal devices depending on the distribution of the people in the space and the suitable media.

2.1 Motivation

While personal communicating devices such as cellphones offers an ubiquitous target for information provisioning, their convenience and ergonomy are limited, especially in situations of mobility. The screen is small and the quantity of information that can be presented is limited. Reading a mobile screen distract the visitor from the physical environment, unlike boards and panels arranged for head up reading

These limitations suggest that the mobile terminal should not be considered as the ideal device to presents on-site information, especially when large flat panels are available in the surrounding for example.

A conflicting constraint is that collective presentation devices (such as public speakers and large wall panels) used for delivering information targeted to a specific profile is perceived as "noise" for other profiles.

The Ubi-Board principle is to determine people's profiles in a given spatial area. A provisioning strategy is then decided to target the people in an efficient way. By efficient, we mean allocating the collective resources to larger groups and using personal/mobile devices for smaller groups or individuals. A simple strategy is the majority rule: when a majority threshold is reached (such as 60%), the collective device is allocated to the dominant profile, while variants for minorities are forwarded to other devices, typically the personal mobile phones of the people.

2.2   A Typical Scenario

Peter is an English visitor to Finland. After his arrival, he look for a way to reach its hotel. An LCD panel displays information about transportation systems and costs. As he approaches the panel, the language switches to English and the costs are converted to English Pound currency.

As he reads, a group of four Japanese tourists join him in front of the panel. Their presence makes the panel switch to the Japanese language and Yen currency, becoming unreadable for Peter, except for a small English flag blinking near a mobile phone icon on the bottom right corner of the screen. Peter looks at his mobile phone screen, which shows the previous information in English. Peter can keep for reference the bus number line and direction which is going to his hotel, as he moves away from from the panel.

Later on Carole arrives, a French woman with a visual disability. She's wearing a Bluetooth headset allowing her to receive audio messages on sites enabled with enhanced-accessibility devices. The panel detects her, and offers the transportation information in spoken French directly to her headset.

2.3   Profiling Policy

Ubi-Board, as a context sensitive service, requires some information about nearby people in order to select the appropriate information and presentation medium. Of course, determining profiles by monitoring the environment raises strong privacy concerns. Profiles are usually considered in conjunction with the notion of identity: a global identifier is associated to with given user, as well as with a profile, which is a set of information characterizing the user. Collecting global identifiers from the environment is a privacy concern as it enables user tracking, when associated with logging.

Profiles used in Ubi-Board are anonymous, in the sense that Ubi-Board does not use identity information: it only uses anonymous attributes, such as "speaks Japanese" or "has a visual disability". The notion of profile in Ubi-Board is not associated with an individual, but also span groups of people, as the service tries to address both collective and individual needs.

However, such a system still introduces some potential risk for privacy. As we will see in the next section, performance optimization for the Bluetooth technology may favor to associate profiles with Bluetooth MAC level identifiers, which are global. Even though Ubi-Board only uses anonymous attributes, it would still be possible for other systems to masquerade as Ubi-Board information points in order to collect the public anonymous profiles and for example associate them with global identifiers (MAC level identifiers). But these type of "digital observation" is possible with virtually any digital wireless communication.

3   System Architecture and Implementation Issues

The Ubi-Board service consists of a set of autonomous Ubi-Board information points (UBIP), optionally managed by a remote administrative station. Each UBIP includes four functional parts, as shown in figure 1.

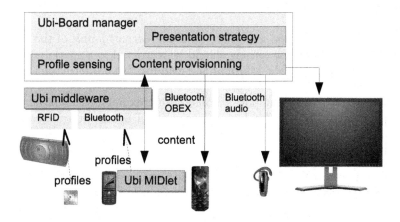

Fig. 1. Ubi-Board architecture

1. "Ubi" middleware
2. Profile sensing
3. Presentation strategy
4. Information feeding/provisioning

3.1 Ubi Middleware

This module is the runtime part of the framework upon which Ubi-Service is built. It is an evolution of the SPREAD [3] system, a programming model for pervasive computing based on the interactions of physical objects and LINDA-like tuple spaces [4]. It is beyond the scope of this paper to describe this framework, however, it is interesting to note that the framework virtualizes the lower level technology used, such as Bluetooth and RFID, into a single and simple view for the application designer. The latter is effectively exposed to an unified tuple space, and profile sensing is simply expressed as reading tuple patterns.

The Bluetooth interface of the Ubi-middleware is interesting as it uses an optimized discovery strategy, and a scalable architecture for handling Bluetooth connections. Instead of using a single Bluetooth chip for both discovering and servicing devices, the Ubi middleware can use a dedicated Bluetooth unit for performing the inquiry scan (discovery) and an arbitrary number of Bluetooth units for handling connections. This architecture has several advantages. First, it allows connections to be handled while being engaged in an inquiry scan. Second, the availability of other Bluetooth units enable to launch SDP (service discovery protocol) connection on devices as soon as they are discovered by the inquiry scan procedure, optimizing the discovery. Third, several service units allows to scale to the number of simultaneous connections required. It is especially useful for supporting Bluetooth audio devices, as discussed later in section 3.4.

3.2   Profile Sensing

The profile sensing function must determine the *current profile* of the potential users. For this purpose, the sensing should be restricted to a spatial area where people are able to access the information (read for visual information, or listen for audio information).

Three type of sensing is supported: Bluetooth proximity, RFID proximity and RFID perimetric. These technologies were selected because they are currently widely available. However, the principles discussed here could be used for similar communication technologies.

Bluetooth proximity. Most current mobile phones are enabled with Bluetooth short distance radios. Provided that the Bluetooth interface is configured as "discoverable", the UBIP is able to discover these devices. The discovery of the devices is not sufficient for profiling, as it only provides a low level contact address. Determining the user's profile by Bluetooth can follow two alternative approaches.

The first one consists of having a profile database, with each profile associated with the Bluetooth address. This approach is a problem for privacy, as explained in section 2.3. Moreover, a UBIP needs to look up the database on each device discovery, which is expensive in terms of communication. However, this approach can present a significant performance advantage as we will show later. In the case of a closed environment where people have to register (so that their profile is known in this environment), this approach can be practical.

The second approach for determining the profile does not associate profile data with an identifier: it collects profile attributes in the surroundings that the users decide to make public. To this end, a small application has to be installed on the user's device to register profile's attributes and make them visible to the Ubi-Board service. The user may of course decide to change the attributes of his profile, or change their visibility status. In this mode, the UBIP only collects the visible attributes in order to determine the global profile of the nearby population. While more privacy friendly, with Bluetooth this approach involves a penalty on performance: after discovering a device Bluetooth address, a UBIP has to open a Bluetooth connection with the device to ask for the profile, increasing discovery latency.

Given the range of Bluetooth radio and its variability (around 10 meters), profile sensing with Bluetooth can only be considered for UBIP targeting numerous people, such as a whole room, large outdoor display panels, or speakers for public messages for example. The best results for Bluetooth sensing are obtained using a majority based presentation strategy, where the content is only influenced when a non ambiguous majority is detected; even when a few devices that should not be integrated in the profile sensing are seen (such as a device in an upper or lower floor), the majority rule provides some resistance.

Proximity sensing. A much more accurate way of sensing is to read a user's profile in a determined spatial area is by using RFID technology. UHF readers can use directive antenna, providing effective reading between a few centimeters up to several meters depending on power and the antenna used. The longest range reader we used (Deister UDL500) is able to sense profile in front of a screen up to 8 meters.

Several antenna can be used to determine direction of movement of people passing near a UBIP, which is useful for navigation information, such as in airports.

RFID sensing assumes that the user is carrying an RFID tag. In the same way as the Bluetooth sensing method explained previously, the tag is not used as an identifier. Instead, its memory is used to directly store profile's attributes. The tag may be part of a physical object related to the profile it represents: for example, a person's cane may embeds a tag describing the visual disability.

In addition to the profile description, the tag is also used for individual addressing of information by the UBIP. For example, in the scenario described in section 2.2, when the public panel switches its content to Japanese, the English content is sent to Peter's phone. This function requires the UBIP to know the address of Peter's phone when using perimetric sensing (as Bluetooth discovery is not used). For this purpose, we include the Bluetooth address of the device in the tag. Of course, other types of device addresses can be used, such as a phone number allowing SMS/MMS sending. Bluetooth presents the advantage of being free of charge, and more privacy respectful. As a side advantage, device discovery using RFID sensing does not suffer from the high latency of Bluetooth inquiry scans.

Bluetooth address discovery from RFID sensing is also useful for devices that are typically not in discoverable mode, such as Bluetooth headsets. In the airport scenario, the discovery of Carole's headset is performed using this method of logically associating a Bluetooth address with a tag.

Perimetric sensing. In perimetric sensing, access to the area where a UBIP is working are determined typically by a door or a unique passage. RFID antenna are installed at these points in order to detect people entering or leaving the area.

Perimetric sensing is a good alternative to Bluetooth proximity sensing for indoor applications, as entrance and exit are determined points and it avoids detections Bluetooth enabled devices from other rooms or floors.

3.3   Presentation Strategy

Once the profiles from nearby users have been discovered, the Ubi-Board manager sends them to the presentation strategy module. Its function is to select the appropriate content for the current profiles. In its simplest form, this selection is made from a map which associates profile attributes with compatibles variants. While Ubi-Board does not provide by itself any automatic variant generator for generic content, the presentation strategy could optionally be fed by a remote service providing such a function.

From this first process, we obtain a set of variants suitable for the different users. The presentation strategy has to decide how to distribute these contents on the available medium, which may include both public devices controlled by the UBIP as well as personal devices. This decision is delegated to a user-defined policy function. We essentially experimented majority based policies, where the public devices are used in priority for people in major profiles, while people in minor profiles are addressed using their personal devices or public individual devices when available. This is logical from the motivation explained in section 2.1, as public devices target all surrounding people.

A content for the major profile does not necessarily requires the exclusivity of the public device: adaptations for minorities can (for example), be added as subtitles on the main device, or can be allocated a small part on a screen. These decisions are heavily dependent on the application and the content designer; this is why the strategy is designed to support user defined policies, providing them a view of the current profiles.

3.4   Information Provisioning

After deciding which content to deliver for each device, we have to effectively send them. For devices that are directly connected to a UBIP, such as a screen, this is not a problem. However, integrating user terminals such as cell phones into the global Ubi-Board information system is more complex, as it involves distributed cooperation over a dynamic set of devices.

Data provisioning to user terminals can be done using three alternatives: OBEX push, dedicated client, or Bluetooth audio channel. The three alternatives use Bluetooth for data transmission.

OBEX push. OBEX is a standard initially proposed in the context of the IRdA technology to support exchanges of digital objects, such as visit card (vCard), pictures etc. between mobile appliances. It is also widely available on Bluetooth devices, which allows generic data exchanges for the end user without requiring a particular application. Ubi-Board can take advantage of OBEX to send information on compliant devices, in particular text messages, pictures, video samples, and audio messages.

One limitation of OBEX push is the lack of interactivity: once the data is received by the target device, there is no opportunity for the users to interact with the UBIP from its device. To this end, a dedicated client must be installed on the user's device.

Dedicated client. A dedicated mobile client is interesting for supporting interactivity between the Ubi-Board service and the users. It is especially useful in context of crowded places, where digital kiosks (such as ticketing machines) can be busy with long waiting queues. Deporting service access from these kiosks to mobile devices can be very useful to provide better service.

We already explained in section 3.2 that determining profiles using the Bluetooth discovery requires a mobile application to expose the public attributes of the user's profile. The same MIDlet application also includes support for interaction with an UBIP, such as navigating through the available information. Ideally, the best would be to be able to point a mobile web browser to an embedded server on the UBIP, but current mobile devices usually prevent such type of interaction over Bluetooth and restrict HTTP access to GPRS/UMTS connections.

As example of interactions, we developed two mobile interfaces for Ubi-Board driven services: a menu ordering interface for restaurants, and a cash retrieval interface (so that the user can prepare safely his transaction away from the cash machine, and only approach it to validate the transaction and collect the cash).

Audio support. A UBIP may stream audio directly to a device using Bluetooth audio channels. However, this form of communication has the following limitation: a Bluetooth device has 7 available time slots, and synchronous audio channels (SCO) each require 2 time slots. Moreover, common audio devices such as headsets allocate two audio channels in order to support duplex communications, even when the microphone is not used. Hence, to support one headset, we need to allocate four time slots (and one additional time slot for control), a total of five time slots. This mean that supporting Bluetooth audio devices quickly saturates Bluetooth units on the UBIP side.

The Ubi middleware on which Ubi-Board is built supports an arbitrary number of Bluetooth units to service the Bluetooth connections. The Bluetooth audio resource requirements mean that for supporting n simultaneous audio devices, the system would need n Bluetooth units.

4 Evaluation

Several Ubi-Board systems were experimented in various real settings, in particular in two European Parliament sessions, with very positive feedback from the users. However we have not yet performed a real user validation, especially regarding the Human Computer interaction aspects, from which we could draw definitive conclusions. However, we performed many evaluations of sensing performance, which has a direct impact on profile accuracy as any "missed" user in sensing contributes to an error in profiling.

4.1 RFID Sensing

With RFID, sensing latency is a non-issue as detections remain below 1s even with multiple devices in the area. When using proximity mode instead of perimetric, UHF RFID reading can be screened by people. We generally consider this as a good property as this mode of sensing is typically used for accurate spatial proximity detection in front of a UBIP, where the collected profiles should reflect the people in front of the information point.

4.2   Profiling Using Bluetooth from Global Identifiers

In this mode, profiles are mapped to Bluetooth addresses in a database back-end. The latency of the profiling is related to performance of the Bluetooth inquiry scan procedure. Figure 2 show the distribution of the time needed to determine the profiles, assuming a negligible contribution for database look-up. We can see that 95% of the devices are detected in the first 5 seconds, and that the 50% are detected within the first second of the inquiry scan. This means that when numerous devices are present, majority profiles can be determined quickly as a representative sample of the population is collected almost immediately.

The inquiry scan procedure uses an asynchronous method to allow incremental detection, instead of the common inquiry scan procedure which requires waiting for the end of the inquiry scan before getting access to the results. This method was implemented with a slightly modified Linux Bluetooth stack (Bluez).

4.3   Profiling Using Bluetooth from Public Attributes

In this mode, profiles are determined from profile attributes exposed by the user through the Ubi Client application on their devices. Profiling involves scanning for devices, and opening a Bluetooth connection to each of them to ask for their profiles. With a standard Bluetooth architecture relying on a single Bluetooth unit, latency increases linearly with the number of devices after the initial delay of 10s required to complete the inquiry scan. From our experimental measures, we determined that the latency L is $L = I + n.C_t$ where $I = 10.24$ (inquiry scan length), n is the number of devices and $C_t = 1.4$, the mean connection time required to get profile attributes from a device. For example, collecting half of the profiles for 50 devices would take nearly 50s!

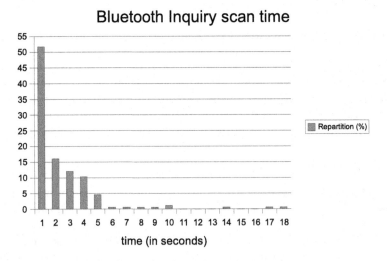

Fig. 2. Bluetooth inquiry scan time distribution

With the Bluetooth architecture used in the Ubi middleware, these connections to get the profiles are staged in parallel, interleaved with the inquiry scan procedure thanks to the asynchronous handling. Performances for 25 and 50 devices are shown respectively in figures 3 and 4 ; we used 1 to 3 Bluetooth connection handling units, in addition the the Bluetooth unit dedicated to Bluetooth discovery.

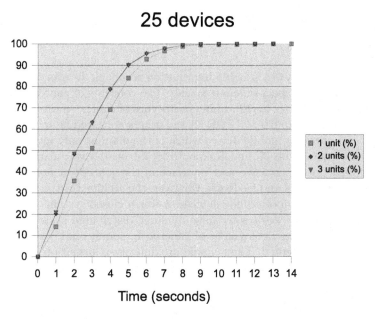

Fig. 3. Profile sensing performance (25 devices)

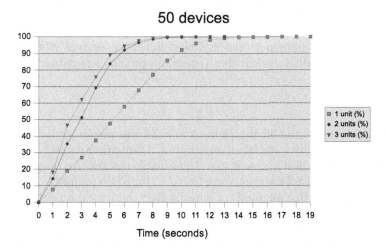

Fig. 4. Profile sensing performance (50 devices)

Here, we can see that half of the profiles are determined after less than 5 seconds for 50 devices, instead of 45s with a single unit architecture. It is also interesting to note that using more than one unit dedicated to connections handling provides a relatively modest improvement, and only with a large number of devices. In the 25 devices case, going from 1 to 2 service units only reduces the time required to collect a given proportion of profiles by at best one second; 3 units does not provide any improvement. This can be explained by the short duration connection of profile requests.

5   Related Work

Proposing smarter ways to provide information to people in everyday environments is an essential aspect of smart spaces. The concepts of context-awareness was introduced by Weiser in [12]. Many prototypes of context-aware information systems such as Parctab [10], GUIDE [2] and others appeared later. In these systems, the user carries a mobile device which is the main or only interface to the information system. Ishii [5] and Rekimoto [8] advocated very early the use of the environment as a pervasive user interface in the concept of *tangible bits*, to support implicit control and feedback. However, the architecture did not support the dynamic integration of ad hoc devices: it was a closed system. Spontaneous collaboration of mobile devices to make dynamic distributed systems were discussed in [11]. With the deployment of Bluetooth, commercial systems such as Kameleon [6] have emerged. They enable proximatity information broadcasting to nearby mobile phones. However, these systems focus on the mobile terminal and do not offer integration in existing on-site information systems such as public display panels, and multi-modality. DataTiles [9] is similar in some ways to Ubi-Board as it supports dynamic mapping of a display system to multiple tiled screens that can be re-arranged. However, personal devices such as mobile phones cannot be used as "opportunistic tiles" like in Ubi-Board. Many works related to distributed user interfaces address the issues of adaptation to heterogenous devices and interaction over a federation of devices. Examples include [1],[7]. However, we have not seen systems addressing content distribution to collective and individual devices that depend on profiles repartition.

6   Conclusion

Ubi-Board is a simple and flexible element to build information systems for smart spaces. It offers context sensitive selection of information as well as context-sensitive adaptation of its presentation to available devices. An original feature is its ability to direct information targeted to a group in a majority on collective devices while redirecting adapted variants for users in minor profiles to their personal devices. In order to support current Bluetooth enabled devices, we propose an optimized architecture for profiling and connection handling, and examined its performance. We also propose alternative sensing methods using RFID for accurate spatial detection and low latency sensing. Finally, we discussed the privacy issue raised by the profiling strategies. Future work will have to address the study of user experience with Ubi-Board systems, in order to design appropriate information distribution and provisioning strategies besides the simple majority rule we currently use.

References

1. Alimohideen, J.: Pavis - pervasive adaptive visualization and interaction service. In: CHI Workshop on Information Visualization and Interaction Techniques for Collaboration Across Multiple Displays, Montreal, Canada (2006)
2. Cheverst, K., Davies, N., Mitchell, K., Friday, A.: Experiences of developing and deploying a context-aware tourist guide: The guide project. In: MOBICOM 2000 (2000)
3. Couderc, P., Banatre, M.: Ambient computing applications: an experience with the spread approach. In: Proceedings of the 36th Annual Hawaii International Conference on System Sciences, Big Island, Hawaii (January 2003)
4. Gelernter, D.: Generative communication in Linda. ACM Transactions on Programming Languages and Systems 7(1), 80–112 (1985)
5. Ishii, H., Ullmer, B.: Tangible bits: Towards seamless interfaces between people, bits and atoms. In: CHI, pp. 234–241 (1997)
6. Kameleon, http://www.kameleon-media.com/
7. Luyten, K., Coninx, K.: Distributed user interface elements to support smart interaction spaces. In: Proceedings of the Seventh IEEE International Symposium on Multimedia (ISM 2005), Irvine, California, USA (2005)
8. Rekimoto, J.: Multiple-computer user interfaces: A cooperative environment consisting of multiple digital devices. In: Streitz, N.A., Konomi, S., Burkhardt, H.-J. (eds.) CoBuild 1998. LNCS, vol. 1370, Springer, Heidelberg (1998)
9. Rekimoto, J., Ullmer, B., Oba, H.: Datatiles: a modular platform for mixed physical and graphical interactions. In: CHI, pp. 269–276 (2001)
10. Schilit, B.N., Adams, N., Want, R.: Context-aware computing applications. In: Proceedings of the Workshop on Mobile Computing Systems and Applications, Santa Cruz, CA (December 1994)
11. Touzet, D., Menaud, J.-M., Banâtre, M., Couderc, P., Weis, F.: SIDE Surfer: a Spontaneous Information Discovery and Exchange System. In: Proceedings of the Second International Workshop on Ubiquitous Computing and Communications (WUCC 2001), Barcelona, Spain (September 2001)
12. Weiser, M.: Some Computer Science Issues in Ubiquitous Computing. Communication of the ACM 36(7), 75–83 (1993)

Author Index

Lecture Notes in Computer Science

Sublibrary 5: Computer Communication Networks and Telecommunications

Vol. 4396: J. García-Vidal, L. Cerdà-Alabern (Eds.), Wireless Systems and Mobility in Next Generation Internet. IX, 271 pages. 2007.

Vol. 4373: K.G. Langendoen, T. Voigt (Eds.), Wireless Sensor Networks. XIII, 358 pages. 2007.

Vol. 4357: L. Buttyán, V.D. Gligor, D. Westhoff (Eds.), Security and Privacy in Ad-Hoc and Sensor Networks. X, 193 pages. 2006.

Vol. 4347: J. López (Ed.), Critical Information Infrastructures Security. X, 286 pages. 2006.

Vol. 4325: J. Cao, I. Stojmenovic, X. Jia, S.K. Das (Eds.), Mobile Ad-hoc and Sensor Networks. XIX, 887 pages. 2006.

Vol. 4320: R. Gotzhein, R. Reed (Eds.), System Analysis and Modeling: Language Profiles. X, 229 pages. 2006.

Vol. 4311: K. Cho, P. Jacquet (Eds.), Technologies for Advanced Heterogeneous Networks II. XI, 253 pages. 2006.

Vol. 4272: P. Havinga, M. Lijding, N. Meratnia, M. Wegdam (Eds.), Smart Sensing and Context. XI, 267 pages. 2006.

Vol. 4269: R. State, S. van der Meer, D. O'Sullivan, T. Pfeifer (Eds.), Large Scale Management of Distributed Systems. XIII, 282 pages. 2006.

Vol. 4268: G. Parr, D. Malone, M. Ó Foghlú (Eds.), Autonomic Principles of IP Operations and Management. XIII, 237 pages. 2006.

Vol. 4267: A. Helmy, B. Jennings, L. Murphy, T. Pfeifer (Eds.), Autonomic Management of Mobile Multimedia Services. XIII, 257 pages. 2006.

Vol. 4240: S.E. Nikoletseas, J.D.P. Rolim (Eds.), Algorithmic Aspects of Wireless Sensor Networks. X, 217 pages. 2006.

Vol. 4238: Y.-T. Kim, M. Takano (Eds.), Management of Convergence Networks and Services. XVIII, 605 pages. 2006.

Vol. 4235: T. Erlebach (Ed.), Combinatorial and Algorithmic Aspects of Networking. VIII, 135 pages. 2006.

Vol. 4217: P. Cuenca, L. Orozco-Barbosa (Eds.), Personal Wireless Communications. XV, 532 pages. 2006.

Vol. 4195: D. Gaiti, G. Pujolle, E.S. Al-Shaer, K.L. Calvert, S. Dobson, G. Leduc, O. Martikainen (Eds.), Autonomic Networking. IX, 316 pages. 2006.

Vol. 4124: H. de Meer, J.P.G. Sterbenz (Eds.), Self-Organizing Systems. XIV, 261 pages. 2006.

Vol. 4104: T. Kunz, S.S. Ravi (Eds.), Ad-Hoc, Mobile, and Wireless Networks. XII, 474 pages. 2006.

Vol. 4074: M. Burmester, A. Yasinsac (Eds.), Secure Mobile Ad-hoc Networks and Sensors. X, 193 pages. 2006.

Vol. 4033: B. Stiller, P. Reichl, B. Tuffin (Eds.), Performability Has its Price. X, 103 pages. 2006.

Vol. 4026: P.B. Gibbons, T. Abdelzaher, J. Aspnes, R. Rao (Eds.), Distributed Computing in Sensor Systems. XIV, 566 pages. 2006.

Vol. 4003: Y. Koucheryavy, J. Harju, V.B. Iversen (Eds.), Next Generation Teletraffic and Wired/Wireless Advanced Networking. XVI, 582 pages. 2006.

Vol. 3996: A. Keller, J.-P. Martin-Flatin (Eds.), Self-Managed Networks, Systems, and Services. X, 185 pages. 2006.

Vol. 3976: F. Boavida, T. Plagemann, B. Stiller, C. Westphal, E. Monteiro (Eds.), NETWORKING 2006. Networking Technologies, Services, and Protocols; Performance of Computer and Communication Networks; Mobile and Wireless Communications Systems. XXVI, 1276 pages. 2006.

Vol. 3970: T. Braun, G. Carle, S. Fahmy, Y. Koucheryavy (Eds.), Wired/Wireless Internet Communications. XIV, 350 pages. 2006.

Vol. 3964: M.Ü. Uyar, A.Y. Duale, M.A. Fecko (Eds.), Testing of Communicating Systems. XI, 373 pages. 2006.

Vol. 3961: I. Chong, K. Kawahara (Eds.), Information Networking. XV, 998 pages. 2006.

Vol. 3912: G.J. Minden, K.L. Calvert, M. Solarski, M. Yamamoto (Eds.), Active Networks. VIII, 217 pages. 2007.

Vol. 3883: M. Cesana, L. Fratta (Eds.), Wireless Systems and Network Architectures in Next Generation Internet. IX, 281 pages. 2006.

Vol. 3868: K. Römer, H. Karl, F. Mattern (Eds.), Wireless Sensor Networks. XI, 342 pages. 2006.

Vol. 3854: I. Stavrakakis, M. Smirnov (Eds.), Autonomic Communication. XIII, 303 pages. 2006.

Vol. 3813: R. Molva, G. Tsudik, D. Westhoff (Eds.), Security and Privacy in Ad-hoc and Sensor Networks. VIII, 219 pages. 2005.

Vol. 3462: R. Boutaba, K.C. Almeroth, R. Puigjaner, S. Shen, J.P. Black (Eds.), NETWORKING 2005. XXX, 1483 pages. 2005.